Nightraiders

Heritage Books by Cdr. David D. Bruhn, USN (Retired)

Battle Stars for the "Cactus Navy":
America's Fishing Vessels and Yachts in World War II

Eyes of the Fleet:
The U.S. Navy's Seaplane Tenders and Patrol Aircraft in World War II

Ingram's Fourth Fleet:
U.S. and Royal Navy Operations Against German Runners,
Raiders, and Submarines in the South Atlantic in World War II

MacArthur and Halsey's "Pacific Island Hoppers":
The Forgotten Fleet of World War II

Home Waters:
Royal Navy, Royal Canadian Navy, and U.S. Navy
Mine Forces Battling U-Boats in World War I
Cdr. David D. Bruhn, USN (Retired) and Lt. Cdr. Rob Hoole, RN (Retired)

Nightraiders:
U.S. Navy, Royal Navy, Royal Australian Navy, and
Royal Netherlands Navy Mine Forces Battling the
Japanese in the Pacific in World War II
Cdr. David D. Bruhn, USN (Retired) and Lt. Cdr. Rob Hoole, RN (Retired)

We Are Sinking, Send Help!:
The U.S. Navy's Tugs and Salvage Ships in the African,
European, and Mediterranean Theaters in World War II

Wooden Ships and Iron Men:
The U.S. Navy's Ocean Minesweepers, 1953–1994

Wooden Ships and Iron Men:
The U.S. Navy's Coastal and Motor Minesweepers, 1941–1953

Wooden Ships and Iron Men:
The U.S. Navy's Coastal and Inshore Minesweepers,
and the Minecraft that Served in Vietnam, 1953–1976

Nightraiders

**U.S. Navy, Royal Navy, Royal Australian Navy,
and Royal Netherlands Navy Mine Forces
Battling the Japanese in the Pacific
in World War II**

**Cdr. David D. Bruhn, USN (Retired)
and
Lt. Cdr. Rob Hoole, RN (Retired)**

HERITAGE BOOKS
2018

HERITAGE BOOKS
AN IMPRINT OF HERITAGE BOOKS, INC.

Books, CDs, and more—Worldwide

For our listing of thousands of titles see our website
at
www.HeritageBooks.com

Published 2018 by
HERITAGE BOOKS, INC.
Publishing Division
5810 Ruatan Street
Berwyn Heights, Md. 20740

International Standard Book Number
Paperbound: 978-0-7884-5843-9

To all the sailors and RAF, RAAF, and USAAF airmen who engaged in mining and mine clearance in the Pacific in World War II, and to the associated bomb and mine disposal/render mine safe personnel, and Minemen who defused a variety of deadly enemy ordnance.

Also, a tip of the hat to two other veterans, the authors' fathers. Daniel Arent Bruhn served as a merchant seaman aboard the USC&GS Ship *Explorer* in the Aleutian Islands during World War II.

Leading Aircraftman Tyson Charles Austin Hoole was posted to India as a newly trained radio and radar technician in 1945 and the Japanese surrendered within three days of his arrival. However, the Royal Air Force kept him there for a further eighteen months just to make sure.

Contents

Foreword by Dr. Edward Marolda   xiii
Foreword by Rear Adm. Paul Ryan, USN (Retired)   xv
Foreword by Commodore Hector Donohue, AM, RAN (Retired)   xvii
Acknowledgements   xix
Preface   xxi
1. Mining the "Tokyo Express"   1
2. U.S. Navy Minelayers   11
3. Royal Navy Minelayers   29
4. Royal Australian and Royal Netherlands Navy Minelayers   45
5. Fall of Hong Kong and Singapore   53
6. Fall of the Netherlands East Indies   73
7. Attack on Pearl Harbor   81
8. Second Attack on Pearl Harbor/   101
    Mining of French Frigate Shoals
9. Build-up of Station Defenses/Forces at Midway Atoll   109
10. Mine Warfare in the Aleutians   115
11. Mine Division Two in the South Pacific   129
12. Minelayers Sent Back to the South Pacific   137
13. Consolidation of Southern Solomons   149
14. New Georgia Campaign   157
15. Occupation and Defense of Cape Torokina   165
16. Central Pacific Campaign   173
17. "Aussie Cat" Minelaying Operations   183
18. Capture of the Marianas and South Palau Islands   195
19. USN Auxiliary Minelayers Pressed into New Duties   209
20. Royal Navy Submarines/Dutch O-19 Lay Mines   217
    in Far East Waters
21. Liberation of the Philippines   225
22. Assault and Occupation of Iwo Jima   241
23. Assault and Occupation of Okinawa   249
24. Third Fleet Operations against Japan   277
25. War's End   285
Postscript   291
Appendices
  A. Japanese Ships Sunk by Allied Mines   305
  B. RAAF Personal Awards   313
Bibliography/Notes   315

Index   347
About the Authors   374

Photos and Illustrations

Preface-1: British Fairmile B motor launches   xxv
Preface-2: HNLMS *Abraham Crijnssen* covered with foliage   xxxi
1-1: "Mining the Tokyo Express" by Richard DeRosset   xxxvi
1-2: USS *Tracy* (DM-19), circa 1937-1940   2
2-1: USS *Terror* (CM-5) operating off the U.S. coast   14
2-2: USS *Salem* (CM-11) under way off Norfolk, Virginia   15
2-3: USS *Adams* (DM-27) off San Francisco, California   17
2-4: USS *Aaron Ward* (DM-34) barely afloat   18
2-5: USS *Picket* (ACM-8), circa 1945   20
2-6: USS *Perry* (DMS-17) being abandoned   21
3-1: HMS *Adventure* (M23) under way   31
3-2: HMS *Ariadne* (M65), circa 1943   33
3-3: HMS *Miner VI* in port   38
3-4: HMS Vernon's mining tender *Skylark*   39
3-5: HM Trawler *Vernon*, tender to HMS Vernon   39
3-6: Controlled minefield cabling embarked in boats   40
4-1: HMAS *Australia* under way   45
4-2: HMAS *Bungaree* (M29)   47
4-3: U.S. PBY-5A Catalina patrol bomber   48
5-1: HMS *Adventure* (M23) under way   54
5-2: British "S" class destroyer in the Dardanelles   55
5-3: British Mk 17L mine on a Mk 17 anchor   56
5-4: HMS *Redstart* (M62), circa 1939   58
5-5: Hong Kong Harbor, circa 1931-1933   60
5-6: SS *Kung Wo* at Shanghai in the 1920s   61
5-7: Okosuka, Japan, after the end of World War II   65
6-1: HNLMS *Abraham Crijnssen* under way   75
7-1: Japanese aircraft preparing to take off from a carrier   82
7-2: USS *Condor* (AMc-14) in 1941   83
7-3: View from Ten-Ten Dock, Pearl Harbor, 7 December 1941   88
7-4: Rear Adm. William R. Furlong, USN   89
7-5: Japanese Type A midget submarine sunk by USS *Monaghan*   93
7-6: USS *Gamble* (DD-123) and USS *Breese* (DD-122), circa 1919   95
7-7: Vice Adm. William F. Halsey, USN   95
7-8: Minelayers nested together in the Hawaiian Islands   96
8-1: Japanese Kawanishi H8K2 ("Emily") flying boat   102
8-2: French Frigate Shoals airfield on Tern Island   105

9-1: Midway Atoll, 24 November 1941   110
9-2: "Night Action off Tulagi" by Richard DeRosset   112
10-1: "Ice Floes, Kodiak" by Edward T. Grigware, 1943   120
10-2: "Adak Harbor" by William F. Draper, 1942   122
10-3: U.S. Mk 12 Mine   123
10-4: USS *Pruitt* leading assault craft toward landing beaches   125
10-5: Kiska Harbor, Aleutian Islands   128
11-1: Pago Pago Harbor, Tutuila Island, Samoa, 1899   129
11-2: Apia facing the Entrance of Harbor   131
11-3: Men pushing U.S. Mk 6 mines at Tongatabu, Tonga Island   135
12-1: Rear Adm. John S. McCain and Vice Adm. Aubrey Fitch   140
12-2: Destroyer USS *Tucker* under tow by USS *YP-346*   141
12-3: Destroyer minelayers USS *Gamble* under way   145
13-1: World War II poster of Adm. William F. Halsey, USN   150
13-2: Light cruiser USS *Honolulu* conducting bombardment   155
16-1: Adm. Raymon A. Spruance, USN   173
16-2: USS *Terror* loading mines at Yorktown, Virginia   176
16-3: Coastal transport USS *APc-108* under way   180
16-4: Navy PB4Y patrol plane taking off from Guadalcanal   182
17-1: RAAF Catalina with a mine under each wing   183
17-2: Crew of a PBY Catalina of No. 43 Squadron RAAF   186
17-3: RAAF armourers loading a mine onto a PBY   187
18-1: Destroyer minesweeper USS *Perry* abandoned and sinking   202
18-2: Minesweeper USS *Auk* off the Norfolk Navy Yard   205
18-3: Yard minesweeper USS *YMS-12* at anchor   207
19-1: Minelayers USS *Chimo* and USS *Obstructer* in company   209
19-2: Rear Adm. Arthur D. Struble, USN   212
19-3: USS *LCS(L)-50* under way off Portland, Oregon   215
20-1: Adm. Sir James Sommerville, RN, aboard USS *Saratoga*   219
20-2: HMS *Trespasser* under tow off Barrow, England   220
20-3: Submarines HMS *Porpoise* and HMS *Salmon*   222
20-4: Mk 16 moored contact, and M Mk 2 magnetic mines   223
20-5: Japanese Nakajima B6N2 Tenzan torpedo-bomber   224
21-1: A UDT team's LCPR towing a rubber boat   227
21-2: Minesweeper USS *Requisite* (AM-109) in a harbor   229
21-3: Kamikaze diving on light cruiser USS *Columbia* (CL-56)   234
21-4: Coastal minelayer USS *Monadnock* (CMc-4)   236
21-5: Lt. Col. Russel Volckmann, U.S. Army   238
21-6: Lt. Harold L. Billman, RANVR   239
21-7: Japanese Shinyo suicide motorboat at Lingayen Gulf   240
22-1: Landing craft staging for assault landings on Iwo Jima   242
22-2: Bombardment of Iwo Jima with YMS in foreground   243

22-3: "Nick" Kawasaki Ki-45 Toryu Army heavy fighter   244
22-4: Minelayer USS *Thomas E. Fraser* (DM-24) in the Atlantic   246
22-5: American flag flying over Mount Suribachi, Iwo Jima   247
23-1: Battleship USS *New Mexico* (BB-40) hit by a kamikaze   250
23-2: Minelayer USS *Dorsey* (DMS-1) under way   252
23-3: Painting "Kerama Retto" by C. W. Smith, 1945   254
23-4: Minesweeper USS *Skylark* (AM-63) under way   255
23-5: Escort carrier USS *Sangamon* (CVE-26) under attack   259
23-6: A Japanese "Zero" fighter about to crash into the sea   259
23-7: Motor minesweeper USS *YMS-103* under way   262
23-8: Destroyer USS *Emmons* (DD-457) at anchor   264
23-9: Minesweeper USS *Swallow* (AM-65) in the South Pacific   268
23-10: Kamikaze attack on shipping at Kerama Retto anchorage   269
23-11: Minelayer USS *Lindsey* (DM-32) at Kerama Retto   271
23-12: Adm. Bruce Fraser RAN, inspecting 21st MS Flotilla   272
23-13: HMAS *Ballarat* moors at Yokosuka Naval Dockyard   273
23-14: Japanese piloted-Baka bomb captured on Okinawa   274
24-1: DeRosset painting "Hidden Menace at Sin-Do Island"   280
25-1: Fleet Adm. Chester W. Nimitz arriving at Tokyo Bay   288
25-2: Nimitz signs the formal instrument of Japanese surrender   288
25-3: HMAS *Ballarat*, part of the 21st Minesweeping Flotilla   289
Postscript-1: Lt. Comdr. Leon V. Goldsworthy, RANVR   291
Postscript-2: Lt. Comdr. Ouvry showing King George VI a mine   293
Postscript-3: Vernon Enemy Mining Group, Lowestoft, 1943   294
Postscript-4: British mine recovery vessel *Esmeralda*   295
Postscript-5: Sydney Arnold and Maurice Batterham, RANVR   300
Postscript-6: A Japanese sea mine ashore in New Britain   300
Postscript-7: Lt. Charles George Croft, RANVR   301
Postscript-8: Fairey "Seafox" seaplane aboard HMS *Orion*   303
Postscript-9: POW Interrogation Center at Ofuna, Japan   304

Maps and Diagrams

Preface-1: British Mk 14 moored contact mine   xxvi
1-1: Guadalcanal area   3
1-2: Solomon Islands   4
5-1: Singapore, BCC   57
5-2: Hong Kong, BCC   59
6-1: Movements of Japanese forces into the Dutch East Indies   74
6-2: Route taken by HNLMS *Abraham Crijnssen* to escape Java   76
6-3: Western Australia   79
7-1: U.S. Navy ships at Pearl Harbor on 7 December 1941   91

8-1: Oahu Island, Hawaiian Islands   103
8-2: Outlying Hawaiian Islands   104
8-3: French Frigate Shoals Atoll   107
10-1: Aleutian Islands   117
10-2: Western side of Attu Island   127
11-1: Tutuila Island   130
11-2: New Hebrides Islands (today Vanuatu)   133
11-3: Fijian and Tonga Islands   134
12-1: Use of mooring lines to clear an oil lighter   138
12-2: Guadalcanal area of the Solomon Islands   143
13-1: Southern Solomon Islands   152
13-2: Kolombangara Island and Blackett Strait   153
14-1: New Guinea and Solomon Islands area   158
14-2: Georgia Islands group   159
15-1: Bougainville, Solomon Islands   166
16-1: Gilbert and Marshall Islands   174
16-2: Principal route of the Central Pacific forces toward Japan   175
18-1: Mariana Islands   197
18-2: South Palau Islands   200
18-3: Echelon Formation Minesweeping with Double-O sweep   201
18-4: U.S. Navy Double-O (Oropesa) moored minesweep   204
19-1: Eastern China in 1941   216
23-1: Minesweeping areas at Okinawa   251
24-1: Okinawa   278
25-1: Cover of a report by Vice Adm. John S. McCain, USN   286
Postscript-1: Allied invasion of Borneo in 1945   298

U.S. Mine Force Motto
(Pride in opening the seas for the Fleet)

Where the fleet goes, we've been

U.S. Minesweeping Ditty
(Serving as "maids of all duties")

Oh, the mines go bang and the winches clang,
And the guns they blaze away,
We rescue men and pick up stiffs
To pass the time of day.
We carry freight and stay out late,
No matter what the mess.
The tighter the spot when things are hot
Is the place for a YMS.

Fourteen Little Sweepers
**(difficulties of post-war minesweeping amid
Royal Australian Navy fleet reductions)**

Fourteen little sweepers went south in a row
And after leaving Hobart they started to go
They went up to Sydney to go north again
Four had a refit and then there were ten
On the way to Townsville, a tropical heaven
Three went to pay off and then there were seven
Up off Bougainville the boss got in a fix
Had to have a refit and then there were six
Six little sweepers in Rabaul once more
Two left for Moresby and then there were four
Four little sweepers working hell for leather
Staying in the tropics forever and ever
Two little sweepers in the tropical sun
Ararat's coming down and that leaves one
One little sweeper such a lot to do
Doesn't look like coming south until 1952

Foreword

In *Nightraiders: U.S. Navy, Royal Navy, Royal Australian Navy, and Royal Netherlands Navy Mine Forces Battling the Japanese in the Pacific in World War II*, Commander David D. Bruhn, USN (Ret.) has authored a comprehensive volume on the vital contribution to victory of Allied mine warriors and their offensive and defensive minelaying forces. Bruhn's focus is clearly on the efforts of U.S., British, Australian, and Dutch surface minelayers whose actions, he argues, have been "largely consigned to the dustbins of history." He demonstrates convincingly that these forces denied the enemy merchant ships safe ports and seaways and compelled them to operate farther offshore where they became prey to Allied aircraft and submarines. Throughout much of 1943 and 1944, U.S. surface ships laid close to 3,000 moored-contact and magnetic-ground mines in fields throughout the operational theaters. At the same time, Bruhn does not slight the work of the Allied air forces, including U.S. B-29 bombers that dropped by far the greatest number of mines in the war and sank the most enemy combatants and merchant ships, especially during the last six months of the war.

Nightraiders is a worthy follow-on to the author's *Wooden Ships and Iron Men* trilogy that covers mine warfare during the Korean War and the Vietnam War; and *Home Waters* which, like this book, he co-authored with Rob Hoole. *Home Waters* focuses on the German submarine offensive of WWI, which failed to bring Britain to her knees, in large part because of the efforts of the British, Canadian, and U.S. Navy mine warfare communities to deny the U-Boats unrestricted access to Allied home waters and greater devastation of shipping. Indeed, these works and *Nightraiders* provide anyone interested in the topic with a storehouse of critical information and analysis of mine warfare during the momentous 20th century.

Edward J. Marolda, Former Senior Historian of the Unites States Navy

Foreword

Winston Churchill once remarked, "The further backward you look, the further forward you can see."

When I think about World War II in the Pacific, I think about Pearl Harbor, unrestricted submarine warfare against Japan, the battle of Midway, Guadalcanal, the Solomon Islands campaign, the Marianas Turkey Shoot, Iwo Jima, Okinawa, Hiroshima and Nagasaki. I think about carrier aviation and amphibious warfare coming of age. Until reading this book I never gave much thought to mining and countermine warfare and their contributions to victory in the Pacific campaign.

Once again Commanders David Bruhn and Rob Hoole have done a masterful job giving us a detailed history of Allied mining and mine clearing. Over 40,000 mines were deployed for offensive and defensive purposes by Allied ships, submarines and aircraft during the course of the war in the Pacific. The Japanese reportedly deployed over 51,000 mines.[1] Several dozen converted and purpose-built minesweepers cleared over 12,000 mines. Explosive Ordnance Disposal (EOD) personnel rendered safe hundreds of mines to clear mine fields, and so that some of them could be exploited for intelligence purposes. This certainly was the era of "wooden ships and iron men."

As a submariner I was trained on the use of modern mine hunting sonar and how to traverse a minefield...not an experience I'd voluntarily undertake with a billion-dollar warship without a lot of soul searching! As the Commander of the U.S. Navy's Mine Warfare Command, I developed a deep respect for the complexity, challenges and dangers of minesweeping. For the men who crewed our World War II minesweepers, it took immense courage to drive through a minefield to locate and sweep enemy mines, much of the time within gunshot range of shore batteries and with the ever-present threat of attack from the air. They had rudimentary sweep gear, coarse sonar, and a vital mission to accomplish, whatever the risks. Twenty-six U.S.

---

[1] Arnold Lott, *Most Dangerous Sea* (U.S. Naval Institute, 1959), p. 263.

Navy mine warfare ships were lost due to enemy action during the Pacific campaign, clearly illustrating the hazard of mine sweeping in a combat zone. (Several others were also lost due to typhoons, collisions, foundering and other non-combat causes.)

I commend David and Rob for another volume of outstanding research and historical enlightenment. In the U.S. Navy today, mine warfare doesn't have a dedicated mine warfare flag officer to advocate for and lead the community, and doesn't have a recognized career path for officers or enlisted personnel. The U.S. Navy puts very few resources and very little effort into improving the capability and capacity of our mine force, while potential adversaries have on the order of 400,000 mines. As Santayana is often quoted, "Those who cannot remember the past are condemned to repeat it."[2] Let's not forget the importance of mine warfare in the high-tech age we live in.

Paul J. Ryan
Rear Admiral, U.S. Navy, ret.

---

[2] George Santayana, *The Life of Reason,* the Phases of Human Progress, Vol 1, Reason in Common Sense, 1905-1906.

Foreword

Major advances in mine warfare before and during World War II included the widespread use of influence ground mines that allowed greater flexibility for offensive mining. The impact of mines during World War II in the Pacific was significant and, in particular, saw the extensive use of aircraft-laid offensive ground mines. However, the majority of mines laid in the Pacific were defensive moored mines: more than 87,000 defensive mines compared to about 27,000 offensive mines.

Whilst most books on mining during World War II concentrate on the impact of offensive mining by aircraft and submarines, the many defensive minefields laid by surface forces has received little attention. Indeed, Gregory Hartman in his seminal book, *Weapons That Wait*, devotes only two paragraphs to the subject. He sums up by saying that 'U.S. defensive minefields seem to be better at sinking their own ships than those of the enemy, although they can perhaps be credited with keeping the enemy out'.

Night Raiders is a welcome addition to the literature on mine warfare as it covers the Pacific War at sea through the eyes of the mine warfare forces and, for the first time, describes in detail the many surface-laid fields. The book title effectively encompasses the context as minelaying ships were referred to as 'night raiders' due to the requirement to undertake operations at night in enemy waters to avoid detection.

The book summarises the total allied minelaying effort against the Japanese, covering USN, RN, USAAF, RAF, RAAF and Dutch minelaying. It provides a comprehensive listing of forces involved and, also, honours and awards. However, the main focus is the thorough description of surface-laid minefields. The U.S. Navy's surface minelaying force consisted of—at different times—seven minelayers (CM), 20 light minelayers (DM), three coastal minelayers (CMc), and 10 auxiliary minelayers (ACM). The total number of ships was 37, and much of the material in this book is devoted to U.S. Navy light minelayers.

A protective minefield has the capability to sink ships and submarines, and this capability makes its threat credible. However, the real function of a protective minefield is not the sinking of ships but the control of the movement of enemy assets. An ideal protective minefield deters the enemy from challenging the minefield, so does not sink any ships. Control does not arise from the number of mines and mine technology per se, but results from the enemy's perception of potential damage. It is human nature to overestimate the threat from weapons which cannot be seen. The mine can be considered to have a psychological warhead, and a USN report on this topic states that the real effect of a minefield derives from a more subtle influence—an exaggerated fear. Minefields work more on the mind than on ships.

Between 1942 and 1945, Queensland became an important strategic hub and a support base for the war effort in the South West Pacific. The ports in Brisbane, Townsville, and Cairns were rapidly expanded to become significant naval bases. There was a major USN submarine base at Brisbane, and by 1944 Brisbane had major headquarters establishments, including Allied Headquarters.

HMAS *Bungaree*, the Australian surface minelayer, laid 5,226 Mk 14 moored mines to provide a 500-mile safe passage inside the Great Barrier Reef north of Townsville, the major assembly port for the U.S. Army. In undertaking this minelaying task, during 1942-43, *Bungaree* made at least 38 passages along the east coast of Australia, where Japanese submarines had attacked 29 ships during this period. There was 0.05 kiloton of TNT in mines on the mine deck, 0.05 kiloton of TNT in mines in the four forward holds, equivalent to a 0.1 kiloton tactical nuclear weapon. The crew were always conscious that they were in a very large 'powder keg'. The end result was that the extensive defensive minefields ensured troop and support ships transiting north to New Guinea were never attacked by Japanese forces.

Night Raiders provides a valuable insight into the sea war in the Pacific and in addition will be the authoritative source on the mine warfare forces which participated. The early chapters summarise Royal Navy, Royal Australian Navy and Royal Netherlands Navy contributions before describing the full extent of the USN mine warfare forces' contribution to the war in the Pacific. This will be the source document for future researchers which, apart from being easy to read, provides extensive detail on the forces involved and awards won.

Hector Donohue AM
Commodore RAN Rtd

Acknowledgements

Brilliant maritime and aviation artist Richard DeRosset created the stunning painting, titled "Mining the Tokyo Express," which graces the cover. It depicts the USS *Tracy*, USS *Montgomery*—shielded from view by the *Tracy*—and USS *Preble*, laying mines off Guadalcanal the night of 1 February 1943, trying to prevent a force of nineteen Japanese destroyers from evacuating enemy troops from the island. This work honors the sailors that went in harm's way aboard slow, heavily-laden vessels carrying cargoes of high explosives into enemy waters.

Many thanks to the Australian Naval officers who provided much information about Royal Australian Navy and Royal Australian Air Force mine warfare in World War II. Commodore Hector Donohue, AM, RAN (Retired), a well-known expert on the subject, was kind enough to pen a foreword for the book. He began his career in the RAN in 1955 as a seaman officer and subsequently sub-specialized as a clearance diver and torpedo and anti-submarine officer. While assigned to exchange duty with the Royal Navy, Donohue served as squadron mine countermeasures officer and executive officer of the RN's first operational minehunter, HMS *Kirkliston*. His service in the RAN included command of the destroyer escort HMAS *Yarra* and the guided missile frigate HMAS *Darwin*. Ashore, Donohue held a number of senior positions in Defense policy and force development prior to retirement in mid-1991.

Those desiring to learn more about the rich history of Australian mine warfare will enjoy *Australian Minesweepers at War* by Mike Turner and Hector Donohue, published by Sea Power Centre – Australia in 2018. The poem titled "Fourteen Little Sweepers," found a few pages earlier, is from this book. Donohue, who had authored many articles and several books, generously shared his work with the authors.

Cdr. Greg Swinden, RAN (Retired) and Warrant Officer Mine Warfare Andy Perry, RAN, helped with research and lent their expertise. Swinden served at sea in HMA Ships *Jervis Bay*, *Swan*, *Melbourne* and *Kanimbla* (twice), and is currently Senior Naval Historical Officer at the Sea Power Centre – Australia. Perry is on active duty, currently assigned to the Commander Australian Mine Warfare and Clearance Diving Task Group (COMAUSMCDTG) based at Fleet Headquarters, HMAS Kuttabul in Sydney.

Turning to the American mine warfare community, Dr. Edward J. Marolda and Rear Adm. Paul J. Ryan, USN (Retired) were kind enough to also pen forewords, reflecting their unique perspectives and adding richness to the book. Dr. Marolda served as the Acting Director of Naval History and Senior Historian of the Navy. In 2017 the Naval Historical Foundation honored him with its Commodore Dudley W. Knox Naval History Lifetime Achievement Award. Marolda has written scores of books and articles. An expert on Vietnam and Mine Warfare, he taught courses on the Cold War in the Far East and the Vietnam War at Georgetown University in Washington, D.C.

Rear Admiral Ryan, a nuclear engineer and career submariner, commanded the attack submarine USS *Philadelphia* and the submarine tender USS *L. Y. Spear*. Ashore, he served in a variety of assignments while assigned to the Navy and Joint Staffs in the Pentagon, and on the staff of commander in chief, U.S. Atlantic Fleet. In February 2002 he assumed command of Mine Warfare Command in Corpus Christi, Texas, where he was responsible for the deployment of mine warfare forces in support of Operation Iraqi Freedom in the Persian Gulf. Ryan retired in December 2003.

Cdr. Ron Swart, USN (Retired) provided expertise on various mines and their use. After achieving the rank of Mineman Chief Petty Officer, he was commissioned to Ensign, (LDO Surface Ordnance). Subsequently, he served three tours as a Mobile Mine Unit commanding officer, two tours on the Mine Warfare Command staff, as chief staff officer to Commander, Mobile Mine Assembly Group, and finally as Commander, Mobile Mine Assembly Group, before retiring in 2003.

Dr. George Pollitt and Dr. Scott Truver also provided assistance. Pollitt is a former technical director for the U.S. Navy Mine Warfare Command, and currently serves as the mine warfare subject matter expert for the Johns Hopkins University Applied Physics Laboratory. Truver is director of Gryphon Technologies' TeamBlue National Security Programs group and is the co-author of the book, *Weapons that Wait: Mine Warfare in the U.S. Navy* (1991 edition).

Finally, I am grateful to my son Michael and to Lynn Marie Tosello. Michael assisted with the maps in the book, and Lynn lent her keen eye, intellect, writing skills, clarity, and passion to editing the manuscript.

Preface

The authors of this book do not claim that mines won the war against Japan. Mines were eminently successful. What they did do was provide the knock-out blow in the maritime blockade of Japan which had almost been completed by the submarines of the United States Navy. Even at the very end of the war our submarines were unable to penetrate the Inner Zone of Japan and were thus incapable of completing the blockade by themselves. Mines played a very important, though not a decisive part, in the war.

—Statement made in the foreword of *Mines Against Japan*, a book prepared by Ellis A. Johnson and David A. Katcher in classified form in 1947. The Naval Ordnance Laboratory White Oak, Silver Spring, Maryland, published it in 1973.[1]

There are few books devoted to the contributions to naval history of minelayers—ships, submarines, and aircraft—perhaps because the use of mines is considered ungentlemanly. Within the U.S. Navy, there has in the past, and there remains today among some senior leaders, disinterest or even disdain for mine warfare. Such attitudes date back to the American Civil War. Faced with the Confederate Navy's use of mines (then termed torpedoes), Adm. David Glasgow Farragut, USN, only grudgingly decided to employ these same weapons in retaliation. On 12 August 1864, he wrote to Secretary of the Navy Gideon Welles, "Torpedoes are not so agreeable when used on both sides, therefore I have resultantly brought myself to it. I have always deemed it unworthy of a chivalrous nation, but it does not do to give your enemy such a decided superiority over you." This perspective was not unique to Americans. During the Russo-Japanese war of 1904-05, British Admiral Sir Algernon Frederick Rous de Horsey, RN, declared that the laying of mines in the open sea was "the act of an enemy to the human race."[2]

Such attitudes prevailed at the onset of both world wars. Following Germany's attack on France through Belgium in 1914, and Britain's resultant entry into World War I, the Royal Navy's initial use of mines was defensive in nature. Fields were put down in the Dover Straits and in the approaches to British naval bases, ports and harbors in an attempt to protect the Fleet and merchant shipping from attack.

Despite these initial efforts, German submarines continued to sink vessels with near impunity, because only relatively small portions of the British Isles and the North Sea were then protected by mines. Forced to find other means to hold the submarines at bay, Britain began offensive mining in German home waters. This action also helped to blockade merchant shipping carrying desperately needed food, supplies, and war materials from entry into enemy ports.

This same progression from defensive to offensive mining took place in the Pacific in World War II. After Japan invaded China in 1937, she required resources available in Southeast Asia (colonized by the British, French, and Dutch) to sustain her war effort. In recognition of increased Japanese aggression, and to help prepare for the possibility of invasion, Britain and the Netherlands laid defensive minefields at Hong Kong, Singapore, and the Netherlands East Indies including Borneo. America followed suit in the Philippine Islands, and Australia in waters off that country and New Zealand.

During the three months following the attack on Pearl Harbor on 7 December 1941, Japanese forces captured Wake Island, blitzed the Philippines, and seized Guam, Hong Kong, Thailand, North Borneo, and Singapore. Finally, the loss of the Netherlands East Indies on 9 March 1942 knocked the British and Dutch out of the war in the Pacific. The U.S. lost Wake Island and Guam; and would later lose the Philippines on 8 May. The U.S. Pacific Fleet worked to prevent further expansion by Japanese forces, while Australia prepared for the possibility of an invasion.

Over the course of the war, a total of 21,000 U.S. mines were used against Japan, of which about 3,000 were laid by the Navy. Some were employed defensively. As Halsey and MacArthur's forces advanced up the Solomon Islands and New Guinea, and through the Bismarck Archipelago, and Spruance's forces through the Central Pacific toward Japan, the Allies laid mines to protect ports and harbors used by their naval forces from entry by Japanese submarines.[3]

In 1942, few aircraft or bases were available for aerial offensive minelaying, and accordingly, U.S. submarines were pressed into service. Seventh Fleet submarines based in Australia placed minefields along Japanese shipping routes in the Gulf of Thailand and the Gulf of Tonkin. Submarines from Pearl Harbor conducted similar operations off the coast of Japan and China. Because submarines were so profitably engaged in hunting ships with torpedoes, their mining effort was never large. But their intermittent minelaying throughout the war (carried out on 33 patrols) contributed a total of 658 mines in 36 minefields.[4]

With the exception of one operation in August 1942, early 1943 marked the start of offensive mining by surface minelayers. Between February 1943 and May 1944, U.S. Navy destroyer minelayers (DMs) and infantry landing craft (LCIs) laid 2,817 moored-contact mines and 12 magnetic-ground mines in seventeen fields in the South and Southwest Pacific.[5]

Also in early 1943, an aerial minelaying offensive began which would continue throughout the war, increasing in intensity and transitioning from a series of tactical operations to a strategic campaign.[6]

USN, USAAF, AND RAAF MINING AGAINST JAPAN

In early 1945, when U.S. Army Air Force B29 bombers became available, they sowed thousands of mines in Japanese home waters in an effort to force an end to the war. This effort to induce Japan to capitulate was codenamed Operation STARVATION. Between March 1945 and the end of the war, B29s mined every important port in the Japanese homeland, her Inner Zone. Japan's outer defense perimeter, or Outer Zone, roughly formed a 3,000-mile radius arc.

Quantities of mines cited in the following tables do not include British, Dutch, and Australian defensive mining, nor RN and RAF offensive mining in the Pacific. Also, a few more mines were dropped than those annotated "in target areas," which apparently reflect pilots not always including in their count, ones that fell astray.

U.S. Army Air Force Minelaying in Japan's Inner Zone

Aircraft Sorties	Total Mines	Enemy Vessels Sunk/Damaged	Notes
1,529	12,135*	294/376	Damaged ships include 137 that were knocked out of the war[7]

* Mines laid by aircraft in target areas; number expended unknown

USN, USAAF, and RAAF Minelaying in Japan's Outer Zone

Minelayer	Total Mines	Enemy Vessels Sunk/Damaged	Notes
Submarines	658	27/27	Records from only 21 of 36 minefields
Ships	2,829	8/3	In support of tactical operations, and often in collaboration with aircraft
Aircraft	9,254*	186/154	Includes RAAF planting 2,512 U.S. mines in SWPA, and USAAF British mines in CBI
Totals	12,741**	221/184[8]	

USAAF: U.S. Army Air Force   CBI: China-Burma-India Theater
RAAF: Royal Australian Air Force   SWPA: Southwest Pacific Area
* Mines laid by aircraft in target areas; 9,829 mines were expended
** Includes 1,791 British mines and 106 dummy mines

A total of 1,075 Japanese ships were sunk or damaged by the 24,876 mines laid by USN ships, submarines, and aircraft, USAAF bombers, and Royal Australian Air Force PBY-5 Catalina flying boats over the course of the war. Included in this total were at least 109 combat vessels, and about one-quarter (by tonnage) of the prewar strength of the Japanese merchant marine. A comparison of the percentage of mines laid by aircraft, ships, and submarines and percentages of resultant enemy ship casualties (sunk or damaged) follows:

- Aircraft: 86% mines   94% ship casualties
- Ships: 11% mines   1% ship casualties
- Submarines: 3% mines   5% ship casualties[9]

Even a cursory review of this data might well prompt the question, why in the world did the U.S. Navy employ surface minelayers, given the high risk to them, and apparent low yield of Japanese ship casualties? The answer is, ships could carry large quantities of moored mines into enemy waters, thus creating a complementary danger to that of ground mines laid by aircraft or submarines. However, they were at great risk when operating independently because, being neither as fast as planes nor as stealthy as subs, they could be easily detected and attacked.[10]

Aircraft became the minelayer of choice because they could rapidly sow mines in enemy waters and reseed those areas without fear of being endangered by ones they had previously laid. However, plane losses were much greater than those of ships or submarines conducting mining, owing to the high number of sorties they flew in hostile areas.[11]

Submarines could lay mines with great secrecy, and at distances beyond the range of aircraft operating from particular bases. However, they carried relatively few mines, were at risk when operating in shallow waters to deposit them, and thereafter had to avoid those areas. Moreover, commanders were averse to leaving on war patrols without a full load of torpedoes, having left some of their precious "fish" behind to free up space for mines.[12]

EMPLOYMENT AND CHARACTERISTICS OF MINES

Sea mines are categorized by means of positioning and means of detonation. The body of a moored mine is suspended in the water column, tethered to an anchor on the sea floor. Bottom mines (sometimes called ground mines) rest on the floor of a body of water. Moored mines, intended for use against ships or submarines operating near the surface, must be close enough to cause their targets harm. Being buoyant, moored mines typically have smaller explosive charges than do bottom mines. Bottom mines are primarily used against surface targets. Pressures at depths greater than a few hundred feet inhibit pressure and acoustic sensors, and the explosive force is suppressed.[13]

Photo Preface-1

British Fairmile B motor launches, laden with moored mines and ground mines.
From the collection of Rob Hoole

A third type of mine is the drifting mine. Unanchored mines not affixed to or embedded on the bottom must sterilize themselves (become non-functional or self-destruct) within an hour after deployment (Hague Convention VIII of 1907 guidance). Because these type mines are indiscriminate in what they might damage or destroy, a Mine Danger Area must be clearly established and its boundaries widely disseminated in a Notice to Mariners (to prevent neutrals and innocents from entering it). Some countries ignore the requirement for the ordnance to disable itself in an effort to gain advantage over, or parity with their adversaries. Drifting mines may also result from moored mines having broken adrift due to strong winds, waves or currents.[14]

Mines may either be controlled or independent and may be used offensively or defensively. Those which are controlled (command activated) are usually used in defensive fields to protect friendly harbors from enemy ships or submarines and are manually detonated from shore. Offensive mines are employed against opposing forces, targeting shipping routes, submarine operating areas, and harbors and ports utilized by them. Such ordnance operates autonomously. Once in the water, mines await unsuspecting target vessels to make contact or, for some types, pass close enough to trigger them.

Diagram Preface-1

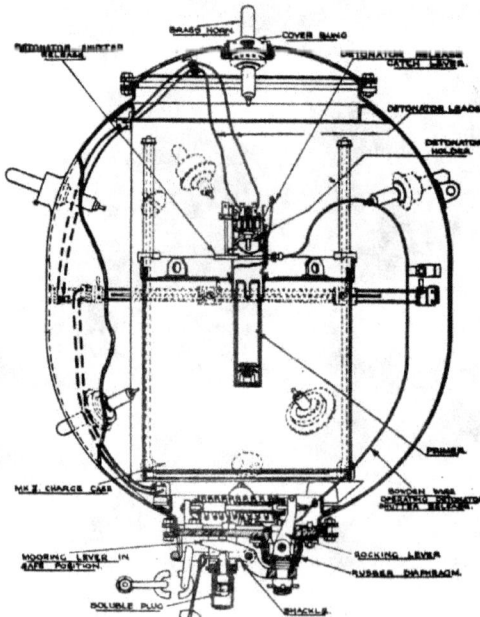

British Mk 14 moored contact mine used in World War II.
Collection of Rob Hoole

Contact mines are designed to explode when a ship strikes one of several projections on the outside of the mine body. The earliest mines, such as the British Mk 14 (illustrated), utilized protruding horns for this purpose. Each horn contained the components of an electrical battery: a carbon plate, a zinc plate, and (inside a sealed glass tube) a bichromate solution. When a ship bumped against the mine and bent a horn, the glass tube inside fractured allowing the liquid to come into contact with the two plates. This completed the battery, producing the voltage necessary to detonate the mine.[15]

More sophisticated "influence mines" do not require physical contact. They are activated by one or more predetermined criteria. These included: (1) the disturbance the steel hull of a passing ship or submarine makes to the earth's magnetic field, (2) the sound its propeller cavitation and/or hull resonance (created by operating machinery) produces, (3) the change in water pressure caused by passage of the vessel, or (4) some combination of these influences.

BRITISH, DUTCH, AND AUSTRALIAN MINEFIELDS

Almost all of the mines collectively sown by British, Dutch, and Australian forces in the Pacific were put down in defensive fields. These included fields laid off Hong Kong, Singapore, and the Netherlands East Indies as part of unsuccessful defenses against Japanese invasion; and in fields off Australia and New Zealand to help guard against possible invasions of these countries—which did not materialize.

BRITISH OFFENSIVE MINING

In 1944, No. 159 and 355 Squadrons, Royal Air Force, and two flotillas of Royal Navy submarines began offensive mining in the South-East Asia Command (SEAC) theater; created by the Allies in August 1943, following the fall of Singapore and the Dutch East Indies. The SEAC encompassed the land areas of India, Burma, Ceylon (Sri Lanka), Malaya, northern islands of Sumatra, and Siam (Thailand), the Indian Ocean and the Arabian Sea. The RAF was generally allocated targets in Burma and Siam, while the submarines were left to mine the Malacca Straits, at that time outside air range.[16]

The Royal Navy and Royal Air Force laid a total of 263,850 mines in all theaters of war, sixty percent for defensive purposes. In the Far East, a handful of World War I vintage British destroyer-minelayers laid 6,273 mines in the approaches to Hong Kong and Singapore, doing a fine job. Unfortunately, Japanese invasion forces arrived over land, not by sea. In 1944-45, as part of Allied offensive mining, RN submarines sowed 501 mines in the Malacca Straits. The only offensive mining by a ship in the Far East took place in December 1941, when HMS *Express* laid eighteen mines in the southern approach channel to Penang.

British Minefield Types/Numbers of Mines

Type Minefield	Far East	Elsewhere	#Mines
Defensive	6,273 ship-laid	153,719	159,992
Deep Trap		26,546	26,546
Offensive	519 ship & sub-laid	76,793	77,312
Total number of mines	6,792	257,058	263,850

Ships/Submarines/Aircraft	#Mines
HMS *Adventure,* and auxiliary minelayers (converted merchantmen)	149,720
fast minelayers	18,450
coastal minelayers, local defense destroyers, and mining tenders	22,448
destroyers	7,143
coastal forces craft	7,014
submarines	3,429
aircraft	55,646
Total numbers of mines	263,850[17]

A portion of the 55,646 mines planted by British aircraft (RN and RAF) in the war were 3,235 of 4,374 mines laid by Allied aircraft within the SE Asia Command between February 1943 and July 1945. Use of "Deep Trap" minefields cited in the first table (unrelated to the subject matter of this book) was initiated in August 1944 to endanger U-boats which had penetrated anti-submarine defenses to the west of the British Isles and had started to take a toll on cross-channel shipping.[18]

AUSTRALIAN MINELAYING

A single surface minelayer HMAS *Bungaree* was responsible for all the defensive fields sown in Australian and New Zealand home waters, and off Allied bases at (1) Noumea, New Caledonia, (2) Port Moresby (on the Papuan Peninsula of New Guinea), and (3) in the Torres Strait between Australia and New Guinea. Her first operation was at Port Moresby on 14 August 1941, and her last mining of the war at the same location, 29-31 December 1943. *Bungaree*, a converted 357-foot cargo vessel, laid a total of 9,289 mines in five areas.

- 5,226 Mk 14 along the Great Barrier Reef
- 1,468 Mk 14 and 593 Mk 17 at Noumea
- 910 Mk 14 at Port Moresby
- 680 Mk 14 off New Zealand
- 412 Mk 14 in the Torres Strait[19]

The heavy cruiser HMAS *Australia* (D84) carried out the only offensive mining by a ship—eighteen A Mk 1 magnetic ground mines laid at anchorages in the French Kerguelen Islands, South Atlantic, in

November 1941. This operation was conducted in an effort to prevent continued use of the remote area by German raiders for resupply. The term "raider" referred to converted freighters armed with hidden guns and torpedoes ("wolves in sheep's clothing") preying on Allied shipping in the Pacific and Indian Oceans.[20]

Royal Australian Air Force PBY-5 Catalinas performed all other offensive mining, a truly herculean effort involving 1,130 sorties (by RAAF No. 11, 20, 42 and 43 Squadrons) to plant a total of 2,512 mines between April 1943 and March 1945. The Australian "Cats" prowled the night skies, carrying U.S. ground mines from Mine Assembly Depot No. 1 at Darwin. The mining was over a wide area from Kavieng, New Ireland to Wenchow, China, with most of the effort devoted to four general areas:

- Celebes   519 mines
- Borneo   489 mines
- China   436 mines
- Java   425 mines
- All other   646 mines[21]

The fabric-clad Catalina patrol planes were slow, cumbersome and poorly armed; requiring them to rely upon stealth by arriving over their targets at night in the dark and at a low altitude. These planes carried two mines each (one under each wing) into almost every Japanese-held harbor of importance. By laying their minuscule loads with great precision, they forced enemy shipping into deeper waters to become the prey of USN submarines and easier targets for USAAF bombers.[22]

Remaining undetected was paramount to success, and mostly the Japanese didn't know the planes were there or how they got there. By flying low and slow (sometimes at less than 200 feet), the PBYs' pilots sought to avoid discovery by enemy radar and fighters. Nevertheless, a considerable number of aircraft were lost (32 CATS and 330 crew) on the dangerous long-range missions—many in excess of twenty hours.[23]

COMPARISON OF RAAF, RN, USAAF MINELAYING

Although numbers of mines laid by the RAAF was small in comparison to that of the Royal and U.S. Army Air Forces, the number of ships sunk or damaged, sea traffic disrupted, harbors closed and shipping delayed was high in proportion to the effort expended. The Royal Air Force, operating from bases in India and Ceylon, made fewer sorties than did the RAAF, but their B24 Liberators carried much larger loads—3,235 mines in 663 sorties. Royal Air Force mining began on 10 January 1944 in Burma at Moulmein (present day Mawlamyine) on the Irawaddy River and ended on 10 July 1945 at Bangkok.[24]

DUTCH DEFENSIVE MINELAYING

Prior to the capture of the Netherlands East Indies by Japanese forces on 9 March 1942, Royal Netherlands Navy mine warfare ships (and other Allied naval units) were divided between bases on Java Island, Tandjong Priok (the port of Batavia, today Jakarta) in the west and Soerabaja (today Surabaya) in the east. In addition to measures taken before the war, the Dutch mined Soerabaja and Tandjong Priok with four fields each in December 1941. Other fields were laid in Madoera Strait, separating Java and Madoera, and elsewhere in the Netherlands East Indies: (1) the entrances to the Pelambang River and off Tuban Island, Sumatra, (2) Ambon Island, and (3) Tarakan Island, off northeastern Borneo.[25]

Early 1942 found British, Dutch, and American forces fighting to hold the Malay Barrier, the term for a string of big islands stretching from the Malay Peninsula to New Guinea, most of them belonging to the Netherlands. If Japan gained possession of these rich territories, teaming with oil, rubber, and other strategic materials, she would be practically self-sufficient.   Once in control of Molucca ("Spice Islands") and the Sunda, Lombok and other straits between the islands, Japanese forces might pour into the Indian Ocean and threaten British India and Australia. An ABDA (American-British-Dutch-Australian) combined command set up on 15 January 1942 to try to prevent this eventuality lasted only six weeks. The short-lived ABDA force included the small U.S. Asiatic Fleet, based at Soerabaja after fleeing the Philippines.[26]

Rapid advances down through the Netherlands East Indies by two powerful Japanese attack groups (the Eastern, under Vice Adm. Ibo Takahashi, and the Western, under Vice Adm. Jisaburo Ozawa, comprised of heavy cruisers and destroyers escorting army transports), culminated in capture of the Dutch possessions. Following Japanese victory over the ABDA Fleet in the Battle of the Java Sea on 27 February, invasion forces began landing on Java the following day. As ensuing naval actions (the Battle of Sunda Strait and Second Battle of the Java Sea, 28 February-1 March) resulted in additional Allied losses, remnants of the ABDA Naval Force tried to escape to Australia. Many of the fleeing ships were sunk by the enemy in route, others made it.[27]

The commanding officers of some ships—ones damaged, with insufficient fuel to make the voyage, or which were judged to have little chance of surviving an encounter with enemy forces out on the open sea, due to their light armament—scuttled their vessels to prevent capture by the enemy. Their crews then had to hope to catch a ride on some other departing ship, or face death or capture on the island. In similar circumstances at Guam, Hong Kong, Singapore, and in the Philippines, American and British crews had, or would face this same unpalatable dilemma.

Of the eight Dutch Navy minelayers, only NLMS *Willem van der Zaan* escaped the Netherlands East Indies.

Date	Minelayer	Disposition
11 Jan 42	*Prins Van Oranje*	Sunk by the Japanese destroyer *Yamakaze* and Patrolboat *38* while proceeding to Soerabaja after laying mines near Tarakan, Borneo
15 Feb 42	*Pro Patria*	Scuttled in the mouth of the Musi River near Palembang, Sumatra
1 Mar 42	*Willem van der Zaan*	Left Java area on 1 March, arrived unscathed at Colombo, Ceylon, on the 9th and joined British Royal Navy's East Indies Station
2 Mar 42	*Rigel*	Scuttled at Batavia
6 Mar 42	*Serdang* (Lt. C. A. E. Rhee)	Scuttled near Soerabaja
7 Mar 42	*Gouden Leeuw*	Scuttled at Soerabaja
8 Mar 42	*Krakatau*	Scuttled at Madura, an island off the northeastern coast of Java[28]

Photo Preface-2

Dutch minesweeper HNLMS *Abraham Crijnssen* covered with tropical foliage and camouflaged as a jungle island while escaping from Soerabaja, Java, in March 1942. Naval History and Heritage Command photograph #NH 87905

Some ships decided to take their chances at sea, and gamble on avoiding death by Japanese naval gunfire or aircraft bomb during the 850-nautical mile trek to Australia.   On the afternoon of 3 March, three 80-ton auxiliary minesweepers of the 4th Minesweeper Division stood out of Soerabaja.   Despite seemingly impossible odds, the 74-foot *Merbaboe* (HMV10), *Rindjani* (HMV11), and *Smeroe* (HMV12) reached Broome, on the northwest coast of Australia, on 10 March. During a week of chugging along at 10 knots, propelled by a single Caterpillar diesel engine, members of each 14-man crew had manned the two 30-caliber Lewis machine guns fitted in the diminutive minesweepers.[29]

The Dutch minesweeper HNLMS *Abraham Crijnssen*—at 186 feet in length, a behemoth in comparison—was the last ship to successfully escape from Java.   She accomplished this by sailing close to the coast at night and hiding, camouflaged as an island with trees and foliage, by day.   She eventually reached Fremantle, some 900 nautical miles down the coast from Broome, on 20 March 1942.[30]

DUTCH OFFENSIVE MINELAYING

The Dutch submarine HNMS *O-19* was one of those to escape the Netherlands, and subsequently reached Ceylon to continue operations under British control.   She was refitted in the UK from February 1943 to May of the following year, and arrived at Fremantle, Australia, in September 1944.   Capable of carrying forty Vickers T III mines, she carried out two mining missions.   In her maiden operation, she laid mines in the main channel between Babi Island and the Java coast, about thirty miles west of Batavia, on 3 January 1945.[31]

On 9 January, during a patrol off the south coast of Borneo, *O-19* sank the converted Japanese gunboat *Shinko Maru No. 1*, but sustained severe damage in an ensuing counterattack from the escort.   Following temporary repairs at Darwin, the submarine was able to make it back to Fremantle.   She carried out her second and final mining mission on 13 April, laying forty mines in the northern entrance to the Banka Strait, separating Sumatra from Bangka Island in the Java Sea.   While en route, *O-19* sank the tanker *Hosei Maru* with gunfire on the 10th.   It is possible that the tanker *Yuno Maru* and store ship *Hyaski Maru* were both sunk in her minefield.[32]

However, *O-19*'s luck had nearly run out.   After making a torpedo attack on a *Nachi*-class cruiser, she was depth-charged and forced to bottom, suffering damage, but once again made it back to Fremantle. The submarine was docked, but due to poor overall condition, she was deemed unfit for further war patrols.   In July, while running stores to

Subic Bay, Philippines, for use by the 8th Flotilla, *O-19* grounded. After determined efforts by the submarine USS *Cod* (SS-224) to pull her free, *O-19* was declared a total loss and destroyed with explosives two days later. The *Cod* is today a museum ship in Cleveland, Ohio.[33]

STAND OUT TO SEA ABOARD NAVY MINELAYERS

It is not possible in a reasonably-sized book, to provide amplifying information about all the events and actions that made up the Allied mining campaign against Japan in the Pacific. As set down in this preface, massive mining by U.S. Army bombers in Japan's Inner Zone, in the last few months of the war, accounted for nearly one-half of all the mines laid by aircraft, ships and submarines up to that point. Moreover, this period accounted for well over half of all Japanese vessels sunk or damaged by mines during the war. Up until March 1945, the bulk of offensive mining was also carried out by aircraft, followed by ships, and submarines, in that order. While enemy losses due to surface minelayers lagged behind even those of submarines, mining by Allied ships denied ports, harbors, and sea passages to enemy forces, and forced Japanese shipping further offshore, where it could be attacked by submarines and aircraft. This is the story of the surface minelayers which have been largely consigned to the dustbins of history.

Chapter 1 takes readers to Guadalcanal, where three old "flush deck" U.S. Navy destroyers, of World War I vintage, were sent in late January 1943 (with a full load of American Mk 6 moored, contact mines) to help prevent the "Tokyo Express" from evacuating Japanese troops from the island. Forces of enemy warships regularly raced down "the Slot" (New Georgia Sound) between the Solomon Islands to attack any ships caught out at night off Guadalcanal. The brilliant cover art by Richard DeRosset depicts the *Tracy* (DM-19), *Montgomery* (DM-17), and *Preble* (DM-15) laying mines at 20 knots, to hurry the operation, as Japanese destroyers hurl toward them at 35 knots—a combined closing speed of 55 knots. Off the ships' starboard bows, PT boats engage in combat with the leading ships of the force, as it draws ever nearer.

Following this riveting action, Chapters 2, 3, and 4 offer summary information about U.S. Navy, Royal Navy, Royal Australian Navy and Royal Netherlands Navy minelayers, before progressing to Chapter 5, the beginning of chronological operations. Some readers may choose to move on to the heart of the book, and return to the earlier chapters as necessary, which also provide an account of the minelaying by HMS

Australia in the Kerguelen Islands, and a listing of the battle stars and Battle Honours garnered by Allied minelayers.

With this overview in their wake, readers may now stand out to sea (vicariously) seven decades ago, with sailors of the Allied Mine Forces in the Pacific Theater.   One or more of the following definitions may prove useful while progressing through the book:

NAUTICAL TERMS AND USE OF THE BOOK'S INDEX

- Abaft: Toward the stern, relative to some object ("abaft the deckhouse").
- Armada: A large fleet of warships, though the term may be used symbolically to signify any large moving group of vessels.
- Athwartships: Across the ship from side to side.
- Atoll: A ring-shaped coral reef or a string of closely spaced small coral islands, enclosing or nearly enclosing a shallow lagoon. The largest island of an atoll often has the same name as the atoll, just as the largest island of an island chain often has the same name as the chain.
- Caliber: The bore-to-barrel-length ratio of a naval gun, obtained by dividing the length of the barrel (from breech to muzzle) by the barrel diameter to give a dimensionless quantity.   For example, a 3-inch/50-caliber gun has a barrel length of 150 inches.
- Scuttle: To cause a vessel to sink by opening the seacocks or making holes in the bottom of its hull.
- D-Day: The unnamed day on which a particular operation commences or is to commence.
- Dead in the water: Not moving (used only when a vessel is afloat and neither tied up nor anchored).
- Deckhouse: An enclosed structure built on the ship's upper or main deck, usually the navigating station though the term can refer to any simple superstructure on deck.
- Fathom: A unit of measurement equal to six feet, used to measure water depth.
- General Quarters: Battle Stations.
- Gun(s) opened: To begin firing a gun or guns.
- Land: To put ashore. Disembark.
- Lighter: Flat-bottomed barge.
- Master: The commander of a non-military ship.

- Rating: The rating of a sailor is a combination of rate (pay grade, as indicated by the number of chevrons he or she wears) and rating (occupational specialty, as indicated by the symbol just above the chevrons).
- Roadstead: A sheltered offshore anchorage area for ships.
- Stand (past tense stood): Of a ship or its captain, to steer, sail, or steam, usually used in conjunction with a specified direction or destination, e.g., "stand into port."
- Stoker: An engineering rating responsible for feeding coal into the firebox of a boiler providing steam to the propulsion turbine. "Stoker" survives as an unofficial term for a marine engineering mechanic in the Royal Navy to this day.
- Vessel: Any craft (from largest ship to smallest boat) that is capable of floating and moving on the water.

Former sailors picking up a book such as this one often desire to ascertain whether or not it contains any references to a ship(s) in which they served. In acknowledgement of this fact, an extensive index is included. To reduce its size, multiple ships listed on the same page or pages in the text are combined into a single entry. Entries for American ships are located under their associated ship type headings. For example, the minelayer *Aroostook* can be found under Ships and Craft, as well as the sub-categories: United States, Navy, mine warfare, and minelayers. A reader searching for a particular foreign ship should review all entries under the heading for that country.

The names of HMS *Redstart* and HMS *Thracian* casualties, identified on pages 67-68, are not listed in the index unless there are additional references to these individuals in the book. The same is true for U.S. Mine Forces involved in Third Fleet operations against Japan (pages 279-282), Appendix A: Japanese Ships Sunk by Allied Mines (305-312), and Appendix B: RAAF Awards (313-314).

Photo 1-1

"Mining the Tokyo Express" by Richard DeRosset depicts the USS *Tracy* (DM-19), USS *Montgomery* (DM-17)—shielded from view by the *Tracy*—and USS *Preble* (DM-15), laying mines off Guadalcanal the night of 1 February 1943, trying to prevent an approaching force of nineteen Japanese destroyers from evacuating enemy troops from the island.

1

Mining the "Tokyo Express"

Yes, the [Japanese] troops brought from Guadalcanal went up to Bougainville Island from all the Solomon Islands. Remaining infantry troops gathered at Buin [a town on Bougainville Island].

January 1943 by destroyer. They evacuated Guadalcanal with 20 destroyers. Munda, Rendova and Vella LaVella troops were also evacuated by destroyer. The closer islands were evacuated by submarine and very small boats.

—Answers by Comdr. Okumiya Masatake, IJN, to the questions:
"Did the ground troops remain on Bougainville?" and "How did
they travel between the islands?" posed by Capt. C. Shands,
USN, during an interview at Tokyo on 12 October 1945.[1]

In mid-afternoon on 29 January 1943, three U.S. Navy minelayers— *Tracy* (DM-19), *Montgomery* (DM-17), and *Preble*—(DM-15) sailed from Noumea, the capital city of New Caledonia (a French colony) bound for Guadalcanal in the Solomon Islands, about 750 nautical miles to the north-northwest. The commanding officer of *Tracy* was in charge of the three ships, designated Task Group 66.2. Sealed orders opened after getting under way—originating from commander, South Pacific Force and South Pacific Area, Adm. William F. Halsey, USN—directed the ships to mine the mouth of the Tenambo River. This action was to be in support of the Guadalcanal Campaign. Begun on 7 August 1942, fighting between Allied and Japanese forces on and around the island of Guadalcanal included to date: three major land battles; seven large naval battles (five nighttime surface actions and two carrier battles); continual (almost daily) aerial battles; and most recently, the Naval Battle of Guadalcanal in early November 1942.[2]

Aboard each of the ships (converted World War I vintage "four-piper, flush deck" destroyers) was a full load of American Mk 6 moored, contact mines—each containing 300 lbs. of TNT high explosive. While at Noumea, *Montgomery* had taken aboard eighty-five at the mine assembly depot to fill out her allowance. Canvas covered the deck-loads of mines to shield them from view of any Japanese patrol planes the ships might encounter, which might otherwise disclose the mission and possible destination of the elderly destroyers.[3]

The commissioning dates of the units of Task Group 66.2, as well as the dates on which they were later reclassified as light minelayers (DM), and the identities of their commanding officers during the forthcoming operation, are provided in the table:

Task Group 66.2
(ex-*Clemson*- or *Wickes*-class Destroyers: 314 feet/1,215 tons/35 kts)

Ship	Commissioned/ Reclassified	Commanding Officer
Montgomery (DM-17) ex DD-121	26 July 1918 17 Jan 1931	Lt. Comdr. John Andrews Jr., USN
Preble (DM-20) ex DD-345	19 Mar 1920 30 June 1937	Lt. Comdr. Frederick Samuel Steinke, USN
Tracy (DM-19) ex DD-214	9 Mar 1920 30 Jun 1937	Lt. Comdr. John Leon Collis, USN

Photo 1-2

Destroyer minelayer USS *Tracy* (DM-19), circa 1937-1940.
Naval History and Heritage Command photograph #NH 89199

OVERVIEW OF THE GUADALCANAL CAMPAIGN

The Guadalcanal Campaign was spurred by the Japanese occupation of Tulagi, a small island nestled in a bay at Florida Island opposite Guadalcanal in the Solomons, on 3 May 1942. In response to the

Japanese establishing a toehold in the strategically located island chain, U.S. naval forces began building a base on nearby Espiritu Santo Island on the 28th. Japan was in need of an air base in the Solomons, from which its land-based bombers could provide cover for the advance of Japanese land forces to Port Moresby, the capital of Papua and site of an Allied base. The thousands of troops stationed there were the Allies' last line of defense before Australia. Having found Tulagi to be fit for only a seaplane base, Japanese forces landed "across The Slot" (New Georgia Sound) at Renandi, on the much larger island of Guadalcanal, on 5 July and began rapid construction of Lunga Point Airfield, from which the Empire's planes could menace the shipping lanes to Australia.[4]

Map 1-1

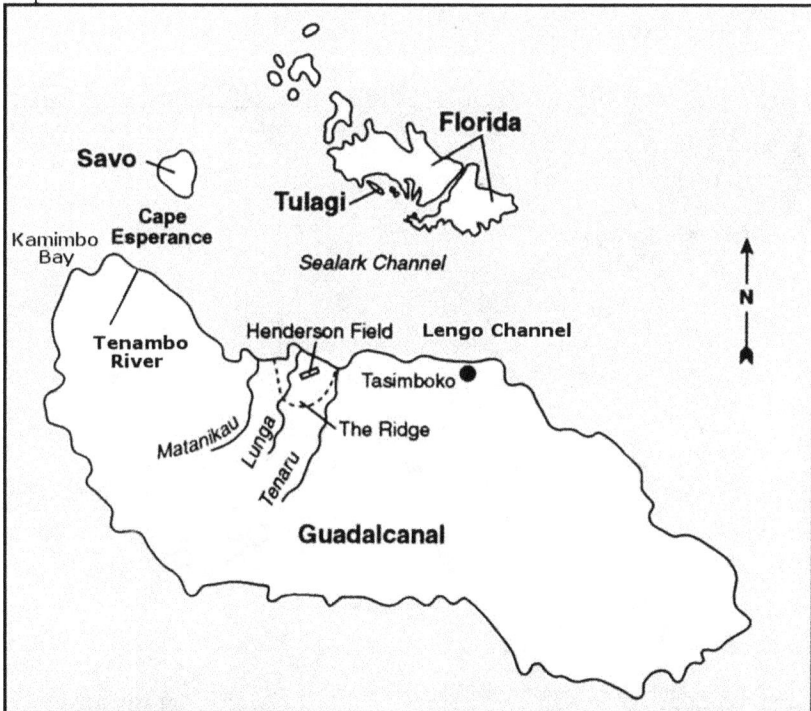

USS *Tracy*, USS *Montgomery*, and USS *Preble* laid mines off the Tanambo River mouth on the night of 1 February 1943, to try to restrict movements of the "Tokyo Express." This action coincided with the arrival of nineteen enemy destroyers sent to begin the evacuation, under the cloak of darkness, of 11,000 Japanese troops from Guadalcanal. www.nps.gov/history/history/online_books/npswapa/extContent/usmc/pcn-190-003130-00/sec6.htm

In the escalating contest for control of the Solomons, 11,000 members of the 1st Marine Division landed at Guadalcanal on 7 August and captured the airstrip at Lunga Point the following day, as well as the principal Japanese encampment at Kukum located on the west side of Lunga Point. That same afternoon, Marines discharged at Tulagi took the Japanese-held island after fierce fighting, as well as the smaller islands of Gavutu and Tanambogo. The captured airstrip on Guadalcanal was renamed Henderson Field, and its occupation and employment by allied forces temporarily halted Japanese expansion in the South Pacific. The significance of U.S. control of the island, from which the allies could expand their presence in the South Pacific, while thwarting the Japanese thrust, was not lost on the enemy and Guadalcanal became a pivotal piece of island real estate, one that both sides wanted to control and to which they were willing to commit large numbers of forces.[5]

Map 1 2

Guadalcanal in the Solomon Islands was a bitterly contested piece of real estate.
http://www.ibiblio.org/hyperwar/USN/ACTC/img/actc-35.jpg

The first U.S. Army reinforcements, the 164th Infantry, landed on 13 October to augment the Marines, while the Japanese built up their forces via the "Tokyo Express," destroyers that raced down The Slot between the Northern and Southern Solomons at night with supplies and troops. By day, aircraft from Henderson Field controlled the skies, which allowed U.S. Navy transports and small vessels to operate in the area with some degree of safety. At night, however, control of these waters shifted as Japanese warships, safe from attack from the sky, appeared out of the darkness, to assault allied vessels caught outside the protected harbor of the fortified island of Tulagi.[6]

The Japanese made several attempts between August and November 1942 to retake Henderson Field. The naval battles, land battles, and smaller skirmishes and raids that comprised the Guadalcanal Campaign culminated in the Naval Battle of Guadalcanal fought from 12-15 November, in which the last Japanese attempt to land enough troops to retake Henderson Field was defeated. The inability of the Japanese to capture Henderson Field doomed their effort on Guadalcanal, and they evacuated their remaining forces by 7 February 1943, conceding the island to the Allies.[7]

JAPANESE TROOP EVACUATION (OPERATION KE)

By the end of December 1942, the Japanese Imperial High Command had reluctantly accepted recommendations for the evacuation of Japanese forces on Guadalcanal and had informed Emperor Hirohito. In preparation for Operation Ke (Ke-gō Sakusen), the Japanese navy landed a fresh battalion of troops on Guadalcanal on 14 January to act as a rearguard. The Yano Battalion, commanded by Major Keiji Yano, consisted of 750 infantrymen and a battery of mountain guns crewed by another 100 men. While Japanese ships and aircraft staged at Rabaul and Bougainville to support the evacuation, the Japanese 17th Army, the bulk of the enemy's ground forces on Guadalcanal, withdrew to the island's western shore while staging counterattacks and fighting rearguard skirmishes. U.S. Army forces, which had relieved the 1st Marine Division, mistook the Japanese movements for offensive preparations and did not immediately pursue the retreating enemy.[8]

MINELAYERS ARRIVE OFF GUADALCANAL

Lt. Comdr. John Collis, USN (commanding officer of the minelayer *Tracy* and commander, Mine Division 1) had received information that the enemy intended to launch an all-out attack on Guadalcanal. The "Tokyo Express," a group of high-speed destroyers accompanied by CL (light cruisers) or DL (destroyer leaders), had been reinforcing Japanese ground troops at the northwest end of the island. In support of these operations, enemy aircraft had been making increasing attacks against U.S. supply lines in the vicinity of the southeast Solomons, and enemy subs were concentrated along these same routes. Intelligence also indicated that once the ships arrived off Guadalcanal, opposition from shore batteries could be expected in the area to be mined.[9]

Timing would be critical to the success of the three minelayers in running the gauntlet of enemy submarines, air and surface craft from Noumea to Guadalcanal. Collis planned to arrive at the limiting arc of Japanese air reconnaissance on the morning of 1 February, and then made a high-speed run at 20 knots through the aircraft and submarine danger areas. By this action, the ships would hopefully reach friendly air coverage by the time the enemy could organize an air attack against the minelayers. (Japanese aircraft spotted the minelayers in transit, but no attack developed.) Collis further planned to make the passage through Lunga Roads in darkness and lay the field at 2100.[10]

"TOKYO EXPRESS" COMING DOWN THE SLOT

As Task Group 66.2—*Tracy, Montgomery,* and *Preble*—approached Guadalcanal on 1 February 1943, sightings were made aboard *Montgomery* of San Cristobal Island at 0925, and of Guadalcanal, thirty-five miles distant, at 1215. At 1308, Maramasike Island came into view. In the early afternoon, Collis received information that fourteen Japanese CL (light cruisers) or DL (destroyer leaders), and eight DD (destroyers) were en route to Guadalcanal. A plot of the enemy's current speed and position revealed that this powerful force would arrive at about the same time as the minelayers. In order to avoid almost certain annihilation should this occur, and in anticipation of the receipt of orders to expedite laying the field, Collis directed the task group to increase speed to maximum (26 knots).[11]

Japanese Evacuation Force, 1-2 February 1943	
Cape Esperance Unit	**Kamimbo Bay Unit**
Screen	**Screen**
1st Unit: *Maikaze, Kawakaze, Suzukaze*	*Satsuki, Nagatsuki*
2nd Unit: *Shirayuki, Fumizuki*	
Transport Unit	**Transport Unit**
DesDiv 10: *Kazegumo, Makigumo,*	DesDiv 16: *Tokitsukaze, Yukikaze*
Yugumo, Akigumo	DesDiv 8: *Oshio, Arashio*[12]
DesDiv 17: *Tanikaze, Urakaze,*	
Hamakaze, Isokaze	

Unbeknownst to Collis, or anyone else for that matter, the enemy force en route to Guadalcanal was not preparing to conduct offensive operations. Instead, the ships were instead to begin evacuation of Japanese troops from the northwest coast of Guadalcanal at Cape Esperance and Kamimbo Bay. Rear Adm. Shintaro Hashimoto, embarked in the destroyer *Shirayuki,* commanded the Cape Esperance and Kamimbo Bay units, which were comprised of screening destroyers and others (those of Destroyer Division 8, 10, 16, and 17) functioning as transports.

Hashimoto had departed the Shortlands (a group of small islands located farther up the Solomons near Bougainville) at 1130 that morning with twenty-one destroyers for the first evacuation run. Two waves of Allied planes from Henderson Field (TBFs, SBDs, F4Fs, and USAAF P38s, P39s, and P40s) had attacked his naval force in the late afternoon near Vangunu Island, northwest of Guadalcanal. Aircraft scored a near miss on *Makinami*, Hashimoto's flagship, heavily damaging it. Hashimoto transferred to *Shirayuki* and detached *Fumizuki* to tow *Makinami* back to base—leaving nineteen destroyers.[13]

Eleven U.S. Navy motor torpedo (PT) boats awaited Hashimoto's destroyers between Guadalcanal and Savo Island and, beginning at 2245, engaged them in a series of running battles over the next three hours. The boats were units of Motor Torpedo Boat Flotilla One, based at Tulagi Island, twenty miles from Guadalcanal across the New Georgia Sound (termed "the Slot"). The destroyers, with help from "R" Area aircraft from the Eighth Fleet, sank three of the PT boats. *PT-111* (Lt. John H. Claggett, USN) was sunk by gunfire from *Kawakaze* at 2254, with two crewmen killed. *PT-37* (Ens. James J. Kelly, USNR) was sunk by destroyer gunfire sometime later, with nine of her ten-man crew killed. *PT-123* (Ens. Ralph L. Richards, USNR) was bombed and sunk by a reconnaissance floatplane, killing four.[14]

FINAL APPROACH TO GUADALCANAL

In early evening, a sighting of six unidentified planes at 1711 sent the minelayers' crews to battle stations until the planes were identified as friendly. Drawing ever closer to Guadalcanal, the destroyers USS *Wilson* (DD-408), USS *Anderson* (DD-411), and four merchant ships were sighted at 1819. Shortly after Savo Island came into view, other ships— the destroyer minesweeper USS *Southard* (DMS-10) and one then two other merchant ships—were sighted. The merchantmen and their naval escorts were retiring from the area, heading away from imminent combat.[15]

The weather in the immediate vicinity was clear, but at a radius of about six miles, the area was ringed with cumulus thunderheads, violent lightning, and thunderstorms. The night was very dark except when broken by flashes of lightning. At 2000, off the northwest coast of Guadalcanal, the minelayers began receiving fire from enemy small caliber shore batteries. The ships withheld their fire to avoid detection by Japanese destroyers coming down the Slot, which were much in evidence eleven minutes later. Collis, in his after-action report, described the first engagements between U.S. and Japanese forces witnessed by the minelayers:

> At 2011 there was searchlight illumination and gunfire beyond Savo Island on our starboard bow about six miles distant. In a few minutes, a PT boat afire on the water [likely *PT-123*] in a glare of searchlights indicated an engagement was taking place between the Tokyo Express and PT boats.[16]

MINEFIELD LAID OFF TENAMBO RIVER MOUTH

Mining commenced at 2021, with underwater acoustics the primary means of communication, and TBS (talk between ships line-of-sight radio) as emergency and secondary. The operation was carried out in accordance with doctrine until 2031. Ships' speed was then increased to 20 knots, and time between mine drops decreased to 9 seconds. (Heavy gunfire coming from Cape Esperance, an engagement between units of the Tokyo Express and Allied aircraft off *Tracy*'s port bow, and anti-aircraft fire coming up from Henderson Field on Guadalcanal, precipitated this action.) The work ceased at 2035, with the three ships having laid three parallel rows of 85 mines each. The inboard row was on the 100-fathom curve (600-foot water depth), with the other rows to seaward on a line 305°T from Doma Reef, exactly as planned.[17]

At completion of the mining, shipboard radar indicated that the southernmost section of the Tokyo Express was a mere 12,000 yards away, bearing 290°T from *Tracy*—closing at a combined speed of 55 knots. Unless the minelayers immediately reversed course, and headed away at top speed, the destroyers would be upon them in nine minutes. However, Collis did not have even this amount of time. The destroyers only needed to sight *Tracy*, *Preble*, or *Montgomery*, which were already within range of their guns, to send one or more of the ships to the bottom, joining scores of other vessels previously sunk in the New Georgia Sound during the hard-fought Guadalcanal Campaign.[18]

At 2038, Collis ordered Task Group 66.2 to retire eastward toward Lengo Channel (bordering the northeast coast of Guadalcanal) at maximum speed. Seven minutes later, word was received that Japanese planes were in the vicinity, and anti-aircraft fire was again observed at Henderson Field. Believing that the enemy destroyers were not aware of their whereabouts, Collis ordered the minelayers to reduce speed to 20 knots and then to 15, in order to minimize the possibility that the ships' wakes might disclose their presence to enemy aircraft. At 0100, Collis received orders for the ships to return to Espiritu Santo, in the New Hebrides, instead of to Noumea, New Caledonia.[19]

AFTERMATH

It is a very difficult task for the army to withdraw under existing circumstances. However, the orders of the Area Army, based upon orders of the Emperor, must be carried out at any cost. I cannot guarantee it can be completely carried out.

—General Harukichi Hyakutake, IJA, 16 January 1943.[20]

With insufficient Allied aircraft and ships available at Guadalcanal to oppose Operation Ke, the Tokyo Express was able, in three consecutive nighttime excursions, to evacuate the Japanese troops on Guadalcanal to Bougainville, farther up the Solomons. On the night of 1-2 February, while maneuvering to avoid a PT-boat attack, *Makigumo* struck one of the mines sown by Task Group 66.2. The resultant explosion left the ship severely damaged, with three killed, two missing, and seven injured. *Yugumo* took off 237 survivors, including Comdr. Fujita Isamu, the ship's captain, and scuttled the destroyer with a torpedo, three miles south-southwest of Savo Island. A second destroyer also struck a mine but remained operational. Despite valiant efforts by the overmatched PT boats, the enemy managed to extricate 4,935 troops that night.[21]

Japanese destroyers returned on the night of 4-5 February to remove another 3,921 troops, and the final 1,796 men (including most of the rearguard) were evacuated in darkness on 7-8 February 1943. Organized resistance on Guadalcanal ended the following day. Over the course of the Guadalcanal Campaign, U.S. forces suffered 7,100 dead and nearly 8,000 wounded. Japanese losses were at least 19,200 dead and most likely a proportionate number of wounded.[22]

PRAISE FOR THE MINELAYING OPERATION

Bold execution of a sound plan was responsible for the success of this mission. Faced with the knowledge that a large enemy force was undoubtedly en route to the same destination, the Commander Mine Division ONE increased speed to arrive first on the scene, then proceeded to mine the area unobserved, while the enemy rapidly closed. Detection meant certain destruction.

—Adm. William F. Halsey, commander, South Pacific Force.[23]

This splendidly conducted operation was carried out by old ships, inadequate in speed and gun power, which had previously proved their worth in convoy duties and in transporting troops and supplies, especially in the early difficult days of Guadalcanal.

—Adm. Chester W. Nimitz, commander, U.S. Pacific Fleet.[24]

AWARDS FOR VALOR AND COMBAT SERVICE

Legion of Merit Medal

Asiatic–Pacific Campaign Medal
(with two battle stars)

Lt. Comdr. John L. Collis, USN, was awarded the Legion of Merit for the actions of the three minelayers he commanded. His medal citation reads in part, "Undeterred by the knowledge that a large Japanese naval force was approaching the exact position for his mine-laying operation, Commander Collis, as Commander of a mine division, proceeded to his destination ahead of the enemy. Throughout a tense period in which the hostile units closed rapidly, he accomplished his mission."[25]

Collis had been aboard the Mine Force flagship USS *Oglala* (CM-4) on 7 December 1941, when she was sunk during the attack on Pearl Harbor. He would remain in the Navy after the war, and ultimately achieve flag rank. Rear Adm. John L. Collis died on 12 December 1999, at the age of ninety-three, in Fairfield, California.[26]

All the officers and men aboard the *Montgomery*, *Preble*, and *Tracy* were authorized to affix a battle star to their Asiatic-Pacific Campaign ribbon for "Capture and Defense of Guadalcanal." One or more battle stars adorning this service ribbon denoted participation in one or more of the forty-eight Navy-Marine Corps official campaigns of the Pacific Theater during the war. Recipients were later authorized a full medal, in addition to their service ribbon, in 1947.

2

U.S. Navy Minelayers

For extraordinary heroism in action against enemy Japanese forces as Support Ship on Radar Picket Station and in the Transport Screen during the Okinawa Campaign from March 24 to June 11, 1945. One of the first ships to enter Kerama Retto seven days prior to the invasion, the U.S.S. HARRY F. BAUER *operated in waters protected by mines and numerous enemy suicide craft and provided fire support for our minesweeper groups against hostile attacks by air, surface, submarine and shore fire. Constantly vigilant and ready for battle, she furnished cover for our anti-submarine screen, served as an antiaircraft buffer for our Naval Forces off the Okinawa beachhead and, with her own gunfire, downed thirteen Japanese planes and assisted in the destruction of three others. A natural and frequent target for heavy Japanese aerial attack while occupying advanced and isolated stations, she defeated all efforts of enemy kamikaze and dive-bombing planes to destroy her. On April 2, she rendered invaluable service by fighting fires and conducting salvage operations on a seriously damaged attack transport. Although herself damaged by a Japanese suicide plane which crashed near her on June 6, she remained on station and escorted another stricken vessel back to port. A seaworthy, fighting ship, complemented by skilled and courageous officers and men, the* HARRY F. BAUER *achieved a notable record of gallantry in combat, attesting the teamwork of her entire company and enhancing the finest traditions of the United States Naval Service.*

—Presidential Unit Citation awarded by Secretary of the Navy John L. Sullivan to the officers and men of the light minelayer USS *Harry F. Bauer* (DM-26) for extraordinary heroism during the Okinawa Campaign.

In World War I, the U.S. Navy sent a squadron of minelayers to Europe to create, with the Royal Navy, a mine barrage stretching from the Orkney Islands to Bergen, Norway—in an effort to bottle German U-boats up in the North Sea and thereby prevent them from attacking Allied shipping in the Atlantic. The American squadron consisted of ten ships. Two of them, *San Francisco* and *Baltimore*, were old cruisers converted for use as minelayers. The others were merchant vessels rapidly acquired by the Navy for conversion and use in a new role.

Aroostook and *Shawmut* were fast passenger liners of the Eastern Steamship Corporation; *Canonicus, Housatonic, Roanoke*, and *Canandaigua*, Southern Pacific freight steamers; and *Quinnebaug* and *Saranac* coastal passenger ships, which provided ocean service between New York and the steamship and rail lines of Chesapeake Bay. The latter two vessels belonged to the Old Dominion Steamship Company, which in the early 1920s became a subsidiary of the Eastern Steamship Lines.[1]

Of these ten minelayers, at the time of the Japanese attack on Pearl Harbor on 7 December 1941, only the former *Shawmut*—renamed *Oglala* (CM-4) on 1 January 1928—remained in service. *Oglala* was the flagship of commander, Minecraft Battle Force, Rear Adm. William R. Furlong, USN. Like the *Oglala*, all the surface minelayers in World War II, with the exception of *Terror* (CM-5), were converted vessels originally constructed for other purposes. *Terror* was then and remains today the only U.S. Navy purpose-built minelayer to serve.[2]

Ships employed as minelayers were referred to as "night raiders," due to the necessity in enemy waters, to carry out operations at night to avoid detection and likely destruction. The advent of radar and other search devices rendered surface minelayers even more vulnerable to air, surface and submarine attack—necessitating use of other means to perform their mission. Allied aircraft configured to lay mines could rapidly arrive in designated areas, sow the waters below, and retire. Submarines were stealthy and could, with relative safety, deposit small loads of "weapons that wait."

During the war, the U.S. Navy's surface minelaying force consisted of at different times, 7 minelayers (CM), 20 light minelayers (DM), 3 coastal minelayers (CMc), and 10 auxiliary minelayers (ACM). The total number of ships was 37, not 40 as one might surmise, because coastal minelayers *Monadnock* and *Miantonomah* were reclassified as minelayers, and *Monadnock* a second time as an auxiliary minelayer.

MINELAYERS (CM)

- *Oglala* (CM-4), 386-foot ex-passenger/cargo ship, sunk during Japanese attack on Pearl Harbor, salvaged and reclassified an internal combustion engine repair ship on 21 May 1943
- *Terror* (CM-5), 455 feet, USN only purpose-built minelayer
- *Keokuk* (CM-8), 353-foot ex-ferry, former net layer AN-5, designated a minelayer 18 May 1942, reclassified net cargo ship (AKN-4) in November 1943

- *Monadnock* (CM-9), 292-foot ex-cargo ship, former coastal minelayer (CMc-4), reclassified a minelayer 1 May 1942, reclassified an auxiliary minelayer (ACM-10) 10 July 1945
- *Miantonomah* (CM-10), 292-foot ex-cargo ship, former coastal minelayer (CMc-5), reclassified a minelayer 15 May 1942, struck a mine and sank 25 September 1944 at Le Havre, France
- *Salem* (CM-11), 350-foot ex-ferry, renamed *Shawmut* 15 August 1945 to permit a new cruiser to be named *Salem* (CA-139)
- *Weehawken* (CM-12), 350-foot ex-ferry, ran aground off Okinawa in October 1945 during Typhoon Louise and was damaged beyond repair

With the exception of *Terror*, the navy's deep-water minelayers were all converted civilian vessels; one ex-passenger/cargo ship, two ex-cargo ships, and three former ferries. *Oglala*, built in 1907 as a passenger/cargo ship, served as the fleet's principal minelayer into the early 1940s; despite many deficiencies associated with her civilian origin and advanced age. In recognition of these shortfalls, the Chief of Naval Operations wrote to the Secretary of the Navy on 11 September 1936 that the navy's only minelayer then in service was not in a satisfactory condition to continue in commission much longer without a large expenditure of funds and then would not be entirely satisfactory. Noting that there was a lack of vessels on the Navy list suitable for conversion as minelayers, he requested that SecNav direct the General Board of the U.S. Navy (an advisory body) to recommend the military characteristics for design of a new minelayer.[3]

Following its consideration of input from commander in chief, United States Fleet, and those of mine warfare commands and experts, the General Board called for a ship not to exceed 7,000 tons standard displacement, carrying 600 assembled mines on at least four under-cover tracks, plus not less than 300 unassembled mines stowed in holds with assembly space for these. The board further suggested: a speed of at least 18 knots sustained, with an endurance of at least 10,000 miles at 15 knots; and an armament of four 5-inch/38-caliber dual-purpose guns. SecNav approved this recommendation on 17 November 1937.[4]

Terror was laid down at the Philadelphia Navy Yard on 3 September 1940 and commissioned on 15 July 1942. After participating in laying a defensive minefield off the port at Casablanca (as part of the support and reinforcement of Operation TORCH, the Allied invasion of French North Africa), she returned to America's East Coast. The minelayer received four 40mm quad mounts at the Navy Yard in Norfolk, Virginia, in April 1943. These anti-aircraft weapons replaced her original two 1.1-inch quad mounts and augmented her 20mm gun battery (which was also increased and rearranged at the same time, resulting in fourteen 20mm guns), greatly improving her self-defense capabilities.[5]

Terror was sent to the Pacific in October 1943, and used to transport mines, conduct defensive mining, and assist in cargo and personnel movements. She was fitted out as the flagship for commander, Minecraft, Pacific Fleet, in December 1944, and during the Iwo Jima and Okinawa invasions in February-May 1945, served as flagship and support ship for all the mine craft in the operations. At Okinawa on 1 May 1945, *Terror* suffered 171 casualties (41 dead, 7 missing, and 123 wounded) while sustaining severe damage from a kamikaze attack. Following repairs in San Francisco, she returned to duty in the Pacific until relieved in December 1945 by the amphibious command ship *Panamint* (AGC-13) as flagship for commander, Minecraft, Pacific Fleet.[6]

Photo 2-1

Minelayer USS *Terror* (CM-5) operating off the U.S. coast, 1942-1943.
National Archives photograph #80-G-411681

Formed on 25 May 1942, Mine Division 50 consisted of *Monadnock*, *Keokuk*, and *Miantonomah*, with *Weehawken* and *Salem* to join upon commissioning. *Keokuk* (CM-8) served as a minelayer between 18 May

1942 and November 1943. During this period, she mined waters off the Atlantic Coast and, as the war in Europe intensified, off Casablanca and later Gela, Sicily, prior to the landings there. During the latter operations, *Keokuk* was attacked on 11 July 1943 by six enemy planes, which anti-aircraft fire drove off. In addition to *Terror* and *Keokuk*, *Monadnock*, *Miantonomah*, *Salem*, and *Weehawken* also participated in minelaying at Casablanca—all of the Navy's CMs, save *Oglala* at Pearl Harbor.[7]

Monadnock was reassigned to the Pacific Fleet in late autumn 1943. She reached Pearl Harbor on 7 January 1944 and reported for duty to commander, Minecraft, Pacific Fleet. *Miantonomah* remained a unit of the Atlantic Fleet and participated in the landings at Normandy in June 1944. On 21 September 1944, she delivered port clearance materiels (in support of salvage and clearing operations) to Le Havre in northwestern France on the English Channel. The port city had been liberated by sea and land less than two weeks before. After exiting the harbor on the 25th, *Miantonomah* struck an enemy mine at 1415 and sank about twenty minutes later with the loss of some fifty-eight officers and men.[8]

Photo 2-2

Minelayer USS *Salem* (CM-11) under way off Norfolk, Virginia, on 29 April 1944. Navy photograph #80-G-229558, now in the collections of the National Archives

Salem laid 202 mines off Casablanca on 27 and 28 December, and helped fight off an air raid there on 31 December 1942. On 20 January

1943, she sailed from Casablanca and arrived at Norfolk on 9 February. She left the United States again on 13 June and arrived at Oran, Algeria, on 5 July 1943. *Salem* got under way the next day as part of the Sicily invasion force. On 11 July, she laid 390 mines off Gela, on the southern coast of Sicily, in company with *Weehawken* and *Keokuk*. Returning to Norfolk in autumn, she carried out local operations along the Atlantic coast until 11 May 1944, when she stood out of Hampton Roads for duty with Service Squadron 6 in the Pacific.[9]

Weehawken participated with other minelayers in sowing a defensive minefield off the harbor at Casablanca in late December 1942. Seven months later, *Weehawken* joined *Keokuk* and *Salem* in mid-July 1943 in laying defensive minefields around the invasion beaches at Gela, on the southern coast of Sicily. During the operation, *Weehawken*'s group underwent a series of heavy attacks by the Luftwaffe on 11 July, but she came through unscathed save for some fragments from a stick of bombs which exploded just off her starboard bow.[10]

On 20 April 1944, Mine Division 50—*Miantonomah*, *Salem*, and *Weehawken*—was dissolved and *Weehawken* was assigned to the Pacific Fleet to transport cargo, mines, and equipment to Pacific bases. After loading mines and cargo from 7 to 9 May, *Weehawken* cleared Hampton Roads, Virginia, on the 11th, in company with *Salem*, bound for the Pacific. The two minelayers reached Pearl Harbor on 14 June for duty with Service Squadron 6.[11]

LIGHT MINELAYERS (DM)

The Navy's twenty light minelayers were converted destroyers intended for use in mining enemy areas, usually at night, or enemy shipping lanes. Late in the war, some of the light minelayers were pressed into picket duty at Okinawa and suffered severe damage from kamikaze attacks. Four of the light minelayers (DM-15 through 18) were ex-*Wickes*-class destroyers, and another four (DM-19 through 22) ex-*Clemson*-class destroyers. During conversion of these eight World War I vintage, 314-foot, "four-piper, flush-deck" destroyers, their torpedo tubes and most anti-submarine warfare equipment was removed, and mine rails added.[12]

The remaining twelve light minelayers (DM-23 through 34) were newly constructed ships. After being laid down as *Allen M. Sumner*-class destroyers, they had been converted at the builders' yards for their new role. Spanning 376 feet in length, they were both larger and more heavily armed than the ex-*Wickes*- and *Clemson*-class destroyers.[13]

- *Gamble* (DM-15), decommissioned 1 June 1945 due to battle damage at Apra, Guam, towed outside the harbor and sunk 16 July 1945

- *Ramsay* (DM-16)
- *Montgomery* (DM-17), damaged beyond repair by a mine in Ngulu Lagoon, in the Caroline Islands, on 17 October 1944, decommissioned 23 April 1945
- *Breese* (DM-18)
- *Tracy* (DM-19)
- *Preble* (DM-20)
- *Sicard* (DM-21)
- *Pruitt* (DM-22)
- *Robert H. Smith* (DM-23)
- *Thomas E. Fraser* (DM-24)
- *Shannon* (DM-25)
- *Harry F. Bauer* (DM-26)
- *Adams* (DM-27)
- *Tolman* (DM-28)
- *Henry A. Wiley* (DM-29)
- *Shea* (DM-30)
- *J. William Ditter* (DM-31)
- *Lindsey* (DM-32)
- *Gwin* (DM-33)
- *Aaron Ward* (DM-34), severely damaged off Okinawa on 3 May 1945, decommissioned on 28 September 1945

Photo 2-3

Light minelayer USS *Adams* (DM-27) off San Francisco, California, 2 May 1945. Naval History and Heritage Command photograph #NH 77371

Much of the material in this book is devoted to U.S. Navy light minelayers. Accordingly, summary information about their operations is not provided here. Readers interested in an overview of their contributions to the war, may review the last few pages of this chapter, which provide a summary of the 95 battle stars earned by the minelaying force. Light minelayers accounted for 76 of the battle stars. Most significantly, three DMs received the Presidential Unit Citation—the highest unit award for heroism, and the equivalent of a Navy Cross awarded to an individual—and six DMs the Navy Unit Commendation. The Navy Unit Commendation is the equivalent of a Silver Star when awarded for heroism, or the Legion of Merit when awarded for meritorious service.

Photo 2-4

Light minelayer USS *Aaron Ward* (DM-34) severely damaged, and barely afloat in the Kerama Retto anchorage, Okinawa, 5 May 1945.
Naval History and Heritage Command Photograph #NH 62571

COASTAL MINELAYERS (CMc)

The Navy's coastal minelayers (CMc) were intended to sow mines along friendly or enemy shores, for defensive or offensive purposes, respectively. The Navy had minimal coastal minelaying forces in World War II, because the Army had primary responsibility for defensive coastal minelaying and enemy shores were more accessible by submarine or aircraft for offensive minelaying purposes. The 230-foot *Wassuc* was an ex-coastal passenger ship, and the 292-foot *Monadnock* and *Miantonomah*, former cargo ships.

- *Wassuc* (CMc-3)
- *Monadnock* (CMc-4), reclassified a minelayer (CM-9) 1 May 1942, and an auxiliary minelayer on 10 July 1945
- *Miantonomah* (CMc-5), reclassified a minelayer (CM-10) 15 May 1942, struck a mine and sank 25 September 1944 at Le Havre, France[14]

AUXILIARY MINELAYERS (ACM)

The U.S. Army's Coast Artillery Corps was responsible until 1949 for maintaining minefields to protect American harbors and coastlines from hostile ships. However, by 1944, it was clear that there was no longer any serious threat to American harbors and the Army began releasing its 188-foot twin-screw, steel-hulled, steam-propelled mine planters for other uses. The Navy received three of the 1,300-ton ships in 1944 and five more in 1945.[15]

The Army offered three of the mine planters to the Navy on 15 March 1944 but withdrew the offer five days later after it learned of a possible Navy requirement to rehabilitate the controlled minefield in the Manila-Subic Bay area. Having an urgent and immediate need for the three vessels to round out its combat minesweeping and disposal units in the Mediterranean, the Navy withdrew the Manila requirement, upon which the Army transferred the ships. *Chimo* was sent to England to support the Normandy invasion, and *Planter* and *Barricade* to the Med to support the landings in southern France, all after very brief conversions at Norfolk, Virginia.[16]

- *Chimo* (ACM-1)
- *Planter* (ACM-2)
- *Barricade* (ACM-3)
- *Buttress* (ACM-4), 184-foot ex-*PCE-878*
- *Barbican* (ACM-5)
- *Bastion* (ACM-6)
- *Obstructor* (ACM-7)
- *Picket* (ACM-8)
- *Trapper* (ACM-9)
- *Monadnock* (ACM-10), 292-foot former minelayer (CM-9)

On 30 December 1944, the chief of Naval Operations approved the acquisition of five more army mine planters for conversion to minesweep gear and repair ships. Each of the ACMs served as a tender and flagship for a squadron of 136-foot yard minesweepers (YMSs) and carried stocks of YMS minesweeping gear and diesel engine and other spare parts. The ACMs also provided for the YMSs medical and dental care, radio maintenance and repair work, handling of pay and pay accounts, radio guard services, furnishing of fresh water, cleaning of evaporators, furnishing of ice cream, and movie entertainment. ACM 1-3 returned to America from Europe in March, January, and June 1945, respectively. They underwent conversions similar to ACM 5-9 at Charleston, South Carolina, and then joined their sisters in the Pacific Fleet to also serve in flagship/tender roles.[17]

Photo 2-5

Auxiliary minelayer USS *Picket* (ACM-8), circa 1945.
Naval History and Heritage Command photograph #NH 79740

The remaining two auxiliary minelayers had a different pedigree than army mine planter. *Buttress* (ACM-4) was a former escort patrol craft, *PCE-878*. Following conversion to a drill-mine laying and recovery ship (from 27 March to 26 April 1944 at Mare Island Naval Shipyard), she was assigned to Service Squadron 6. *Buttress* thereafter served at advanced bases in the central and western Pacific through the end of the war. *Monadnock* (ACM-10) was the former minelayer CM-9, and former coastal minelayer CMc-4. Following her conversion at San Francisco, at war's end, to auxiliary minelayer, she transported troops to Eniwetok, Guam, and Okinawa en route to Japan for occupational duty. She remained in the Far East for several months, until 9 March 1946, when she began the return voyage to San Francisco.[18]

U.S. NAVY MINELAYERS LOST IN WORLD WAR II

Three minelayers were lost to combat action during the war. In the European Theater, *Miantonomah* (CM-10) was sunk by a mine off Le Havre, France, on 25 September 1944. In the Pacific, *Montgomery* (DM-17) was damaged by a mine off Palau on 17 October 1944 and was scrapped; *Gamble* (DM-15) was damaged by aircraft bombs off Iwo Jima on 18 February 1945 and was scuttled off Saipan on 16 July 1945.

Photo 2-6

High-speed minesweeper USS *Perry* (DMS-17) being abandoned after striking a mine off Angaur, 13 September 1944. Light minelayer USS *Preble* (DM-20) is standing by. Naval History and Heritage Command photograph #NH 92987

MINELAYING AIRCRAFT (21,389 TOTAL MINES)

Types of USN, USAAF, and RAAF aircraft employed for offensive mining against Japan (table does not include mines laid by RAF aircraft).

Aircraft	#Mines in Target (% of Total Mines)	Aircraft	#Mines in Target (% of Total Mines)
B24 USAAF	4,981 (23%)	PB4Y-1 USN	101 (0.005%)
B25 USAAF	101 (0.005%)	PB4Y-2 USN	186 (0.009%)
B29 USAAF	13,122 (61%)	PV-1 USN	18 (0.0008%)
PBY-5 RAAF	2,548 (12%)	TBF USN	320 (0.015%)
PB2Y-3 USN	12 (0.0006%)[19]		

MINELAYER UNIT AWARDS AND BATTLE STARS

Presidential Unit Citation

Okinawa

Ship	Award Period	Commanding Officer
Harry F. Bauer (DM-26)	24 Mar-11 Jun 45	Comdr. Richard Claggett Williams Jr., USN
Harry A. Wiley (DM-29)	23 Mar-24 Jun 45	Comdr. Paul Henrik Bjarnason, USN

Vicinity of Okinawa

Aaron Ward (DM-34)	3 May 45	Comdr. William Henry Sanders Jr., USN

Navy Unit Commendation

Angaur, Peleliu, and Kossol Passage

Montgomery (DM-17)	12-15 Sep 44	Lt. Dan Thomas Drain, USNR

Okinawa

Adams (DM-27)	24 Mar-1 Apr 45	Comdr. Henry Jacques Armstrong Jr., USN
Gwin (DM-33)	24 Mar-24 Jun 45	Comdr. Frederick Samuel Steinke, USN
Shea (DM-30)	24 Mar-4 May 45	Comdr. Charles Cochran Kirkpatrick, USN
Robert H. Smith (DM-23)	24 Mar-24 Jun 45	Comdr. Henry Farrow, USN / Comdr. Wilber Haines Cheney Jr., USN
J. William Ditter (DM-31)	24 Mar-6 Jun 45	Comdr. Robert Roy Sampson, USN

Battle Stars by Ship Type: DM (75), CM (14), ACM (5)

Asiatic-Pacific Campaign Medal
with no battle stars

European-African-Middle Eastern
Campaign Medal with 2 battle stars

European, African, and Mediterranean Theaters

North African Occupation: Algeria-Morocco Landings

Ship	Period	Commanding Officer (s)
Miantonomah (CM-10)	8-11 Nov 42	Lt. Comdr. Raymond Dorsey Edwards, USN
Monadnock (CM-9)	8-11 Nov 42	Lt. Comdr. Frederick O. Goldsmith, USNR

Sicilian Occupation

Keokuk (CM-8)	9-15 Jul 43	Comdr. Leo Brennan, USNR
Salem (CM-11)	6-15 Jul 43	Comdr. Henry Goodman Williams, USN
Weehawken (CM-12)	9-15 Jul 43	Lt. Comdr. Robert Edwin Mills, USNR

Invasion of Normandy

Chino (ACM-1)	6-25 Jun 44	Lt. Comdr. John Winston Gross, USNR
Miantonomah (CM-10)	6-25 Jun 44	Comdr. Austin Edward Rowe, USN

Invasion of Southern France

Barricade (ACM-3)	15 Aug-25 Sep 44	Lt. Charles Percy Haber, USN
Planter (ACM-2)	15 Aug-25 Sep 44	Lt. Theodore Thomas Scudder Jr., USNR; Lt. Richard Albert Knapp, USN

Pacific Theater

Pearl Harbor

Ship	Period	Commanding Officer
Oglala (CM-4)	7 Dec 41	Comdr. Edmond Pryor Speight, USN
Breese (DM-18)	7 Dec 41	Lt. Comdr. William Jenkins Longfellow, USN; Lt. Comdr. Herald Franklin Stout, USN
Gamble (DM-15)	7 Dec 41	Lt. Comdr. Donald Allen Crandell, USN
Montgomery (DM-17)	7 Dec 41	Lt. Comdr. Richard Allen Guthrie, USN
Preble (DM-20)	7 Dec 41	Lt. Comdr. Harry Darlington Johnston, USN
Pruitt (DM-22)	7 Dec 41	Lt. Comdr. Edwin Warren Herron, USN
Ramsay (DM-16)	7 Dec 41	Lt. Comdr. Gelzer Loyall Sims, USN
Sicard (DM-21)	7 Dec 41	Lt. Comdr. William Christian Schultz, USN
Tracy (DM-19)	7 Dec 41	Lt. Comdr. George Richardson Phelan, USN

Anti-submarine Operations

Ship	Period	Commanding Officer
Gamble (DM-15)	28 Aug 42	Lt. Comdr. Stephen Noel Tackney, USN

Guadalcanal-Tulagi Landings

Ship	Period	Commanding Officer
Tracy (DM-19)	7-9 Aug 42	Comdr. John Leon Collis, USN

Capture and Defense of Guadalcanal

Ship	Period	Commanding Officer
Gamble (DM-15)	30 Aug 42	Lt. Comdr. Stephen Noel Tackney, USN
Montgomery (DM-17)	31 Jan- 2 Feb 43	Lt. Comdr. John Andrews Jr., USN
Preble (DM-20)	31 Jan- 2 Feb 43	Lt. Comdr. Frederick Samuel Steinke, USN
Tracy (DM-19)	31 Jan- 2 Feb 43	Comdr. John Leon Collis, USN

Consolidation of Southern Solomons

Ship	Period	Commanding Officer
Breese (DM-18)	6-13 May 43	Lt. Comdr. Alexander Bacon Coxe Jr., USN
Gamble (DM-15)	20 Mar- 13 May 43	Lt. Warren Wilson Armstrong, USN
Preble (DM-20)	6-13 May 43	Lt. Comdr. Frederick Samuel Steinke, USN

Aleutians Operation: Attu Occupation

Ship	Period	Commanding Officer
Pruitt (DM-22)	11-29 May 43	Lt. Comdr. Richard Claggett Williams Jr., USN
Ramsay (DM-16)	13 May- 2 Jun 43	Lt. Comdr. Charles Helmick Crichton, USN

New Georgia Group Operation:
New Georgia-Rendova-Vangunu Occupation

Breese (DM-18)	29 Jun- 25 Aug 43	Lt. Comdr. Alexander Bacon Coxe Jr., USN
Gamble (DM-15)	29-30 Jun 43	Lt. Warren Wilson Armstrong, USN
Montgomery (DM-17)	24-25 Aug 43	Lt. Comdr. Dwight Lyman Moody, USN
Preble (DM-20)	29 Jun- 25 Aug 43	Lt. Comdr. Frederick Samuel Steinke, USN

Treasury-Bougainville Operation:
Occupation and Defense of Cape Torokina

Breese (DM-18)	1-8 Nov 43	Lt. Comdr. Alexander Bacon Coxe Jr., USN
Gamble (DM-15)	1-8 Nov 43	Lt. Warren Wilson Armstrong, USN
Pruitt (DM-22)	7-8 Nov 43	Lt. Comdr. Richard Claggett Williams Jr., USN
Sicard (DM-21)	1-8 Nov 43	Lt. Comdr. John Vavasour Noel Jr., USN
Tracy (DM-19)	7-8 Nov 43	Comdr. William Julius Richter, USN

Gilbert Islands

Terror (CM-5)	13-28 Nov 43	Comdr. Howard Wesley Fitch, USN

Marshall Islands Operation:
Occupation of Kwajalein and Majuro Atolls

Preble (DM-20)	29 Jan- 8 Feb 44	Lt. Comdr. Frederick Samuel Steinke, USN
Ramsay (DM-16)	29 Jan- 8 Feb 44	Lt. Comdr. Robert Henderson Holmes, USN

Marianas Operation: Capture and Occupation of Guam

Terror (CM-5)	4 Aug 44	Comdr. Horace William Blakeslee, USN

Western Caroline Islands:
Capture and Occupation of south Palau Islands

Montgomery (DM-17)	6 Sep- 14 Oct 44	Lt. Dan Thomas Drain, USNR
Preble (DM-20)	6 Sep- 14 Oct 44	Lt. Edward Francis Baldridge, USN

Leyte Operation: Leyte Landings

Breese (DM-18)	12-20 Oct 44	Lt. David Barney Cohen, USN
Preble (DM-20)	12-20 Oct 44	Lt. Edward Francis Baldridge, USN

Luzon Operation: Lingayen Gulf Landing

Monadnock (CM-9)	5-18 Jan 45	Lt. Comdr. Frederick O. Goldsmith, USNR
Breese (DM-18)	4-18 Jan 45	Lt. George W. McKnight, USNR
Preble (DM-20)	4-18 Jan 45	Lt. Edward Francis Baldridge, USN

Iwo Jima Operation: Bombardment of Iwo Jima

Gwin (DM-33)	24 Jan 45	Comdr. Frederick Samuel Steinke, USN

Iwo Jima Operation: Assault and Occupation of Iwo Jima

Terror (CM-5)	16-19 Feb 45	Comdr. Horace William Blakeslee, USN
Breese (DM-18)	16 Feb- 7 Mar 45	Lt. George W. McKnight, USNR
Gamble (DM-15)	16-19 Feb 45	Lt. Comdr. Donald Noble Clay, USN
Henry A. Wiley (DM-29)	16 Feb- 9 Mar 45	Comdr. Robert Emmett Gadrow, USN
Henry F. Bauer (DM-26)	19 Feb- 6 Mar 45	Comdr. Richard Claggett Williams Jr., USN
Lindsey (DM-32)	16-19 Feb 45	Comdr. Thomas Edward Chambers, USN
Robert H. Smith (DM-23)	19 Feb- 9 Mar 45	Comdr. Henry Farrow, USN
Shannon (DM-25)	19 Feb- 16 Mar 45	Comdr. Edward Lee Foster, USN
Thomas E. Fraser (DM-24)	19 Feb- 8 Mar 45	Comdr. Ronald Joseph Woodman, USN
Tracy (DM-19)	16 Feb- 7 Mar 45	Lt. Comdr. Richard Edward Carpenter, USNR

Okinawa Operation: Assault and Occupation of Okinawa

Monadnock (CM-9)	10 Apr- 27 May 45	Lt. Comdr. Frederick O. Goldsmith, USNR; Lt. Comdr. John E. Cole, USNR
Salem (CM-11)	26 Mar- 4 Apr 45	Lt. Comdr. George Carlton King, USN
Terror (CM-5)	25 Mar- 8 May 45	Comdr. Horace William Blakeslee, USN
Weehawken (CM-12)	10 Apr- 30 Jun 45	Comdr. Walter Patrick Wrenn Jr., USNR
Aaron Ward (DM-34)	25 Mar- 11 Jun 45	Comdr. William Henry Sanders Jr., USN
Adams (DM-27)	1-7 Apr 45	Comdr. Henry Jacques Armstrong Jr., USN
Breese (DM-18)	25 Mar- 30 Jun 45	Lt. George W. McKnight, USNR
Gwin (DM-33)	1 Apr- 30 Jun 45	Comdr. Frederick Samuel Steinke, USN
Henry A. Wiley (DM-29)	25 Mar- 30 Jun 45	Comdr. Paul Henrik Bjarnason, USN
Henry F. Bauer (DM-26)	25 Mar- 11 Jun 45	Comdr. Richard Claggett Williams Jr., USN
J. William Ditter (DM-31)	30 Jun 45	Comdr. Robert Roy Sampson, USN

Lindsey (DM-32)	16-19 Feb 45	Comdr. Thomas Edward Chambers, USN
Robert H. Smith (DM-23)	25 Mar-30 Jun 45	Comdr. Henry Farrow, USN; Comdr. Wilber Haines Cheney Jr., USN
Shannon (DM-25)	25 Mar-30 Jun 45	Comdr. Edward Lee Foster, USN; Comdr. William Thomas Ingram II, USN
Shea (DM-30)	25 Mar-16 May 45	Comdr. Charles Cochran Kirkpatrick, USN
Thomas E. Fraser (DM-24)	25 Mar-30 Jun 45	Comdr. Ronald Joseph Woodman, USN
Tolman (DM-28)	1 Apr-28 Jun 45	Comdr. Clifford Arthur Johnson, USN
Tracy (DM-19)	25 Mar-28 Jun 45	Lt. Comdr. Richard Edward Carpenter, USNR

3rd Fleet Operations against Japan

Breese (DM-18)	5-31 Jul 45	Lt. George W. McKnight, USNR; Lt. David James Pikkaart, USNR
Gwin (DM-33)	5-31 Jul 45	Comdr. Frederick Samuel Steinke, USN
Henry A. Wiley (DM-29)	5-31 Jul 45	Comdr. Paul Henrik Bjarnason, USN
Henry F. Bauer (DM-26)	10-31 Jul 45	Comdr. Richard Claggett Williams Jr., USN
Robert H. Smith (DM-23)	5-31 Jul 45	Comdr. Wilber Haines Cheney Jr., USN
Shannon (DM-25)	5-31 Jul 45	Comdr. William Thomas Ingram II, USN
Thomas E. Fraser (DM-24)	5-31 Jul 45	Comdr. Ronald Joseph Woodman, USN; Comdr. Brooke Atkins, USN

Minesweeping Operations Pacific: Nagasaki (Kyushu-Korea Area)

Tracy (DM-19)	10-16 Sep 45	Lt. Comdr. Richard E. Carpenter, USNR

Minesweeping Operations Pacific: "Skagway" (East China Sea-Ryukyus Area)

Chino (ACM-1)	7-9 Nov 45	Lt. Comdr. John Winston Gross, USNR
Breese (DM-18)	14-24 Aug 45	Lt. David James Pikkaart, USNR
Gwin (DM-33)	27 Oct 45	Comdr. Elmer Cecil Long, USN
Henry A. Wiley (DM-29)	14-24 Aug 45	Comdr. Paul Henrik Bjarnason, USN
Henry F. Bauer (DM-26)	20 Aug-9 Nov 45	Comdr. Richard Claggett Williams Jr., USN
Robert H. Smith (DM-23)	14-24 Aug 45	Comdr. Wilber Haines Cheney Jr., USN
Shannon (DM-25)	14-24 Aug 45	Comdr. William Thomas Ingram II, USN

Minesweeping Operations Pacific: Kagoshima (Kyushu-Korea Area)

Adams (DM-27)	3-7 Sep 45	Comdr. Donald Noble Clay, USN

Special Operations: Minesweeping Operations Pacific

Obstructor (ACM-7)	10-18 Dec 45	Lt. Sammie Smith, USN
Adams (DM-27) no BS	14-16 Dec 45	Comdr. Gerald Louis Christie, USN
Henry A. Wiley (DM-29) no BS	2-11 Nov 45	Comdr. Paul Henrik Bjarnason, USN

3

Royal Navy Minelayers

When Britain declared war on Germany on 3 September 1939, the Royal Navy was deficient in minelayers. If three small, specialized indicator loop minelayers are set aside—*Linnet*, *Redstart*, and *Ringdove*, which could carry only twelve mines each—*Plover* was the single minelayer operational. In fact, that day she laid mines off Bass Rock near the Firth of Forth. The other minelayer, *Adventure* (the Royal Navy's first purpose-built minelayer), was in reserve, awaiting re-commissioning. A number of E- and I-class destroyers built in the 1930s with minelaying capability, were nearing conversion to this role at the time of the outbreak of hostilities. However, whereas *Adventure* could carry 280 mines and *Plover* 100, the converted destroyers were limited to sixty. There were also six S-class destroyers, commissioned 1918-1920, that could carry forty mines. This situation would later improve with the commissioning between April 1941 and February 1944, of six fast (40 knot) minelaying cruisers of the *Abdiel*-class, each capable of carrying up to 160 mines; and of four O-class destroyers in 1942, which could carry sixty apiece.[1]

To address the immediate shortfall, the Royal Navy turned to the Merchant navy, requisitioning, first, the liner SS *Teviotbank* (built 1938). Commissioned HMS *Teviot Bank*, after conversion to carry 280 mines, she could muster only 12 knots. Two subsequent acquisitions requiring minimal modification were the Southern Railway's railway ferries SS *Hampton Ferry* and *Shepperton Ferry*, built in 1934. Their boilers exhausted through funnels athwartships, leaving completely open their main decks with four rail tracks and, aside from a low gate, sterns as well. Able to carry 270 mines apiece, the former railway ferries quickly joined *Adventure* and *Plover* at Dover. The four ships began a massive six-day minelaying operation in the Dover Strait on 11 September 1939, effectively closing the English Channel to German U-boats heading for the Western Approaches.[2]

The factors that made the railway ferries attractive applied equally to the newly built Stranraer-Larne car ferry *Princess Victoria*. Moreover, being diesel-driven there was no requirement for frequent boiler cleaning. Tragically, her service was short-lived. On 19 May 1940, the former stern-loading ferry struck a mine off the entrance to the Humber River and sank quickly.[3]

The Royal Navy also acquired the Prince Line cargo liner MV *Southern Prince*, Blue Funnel Line sister cargo-passenger ships MV *Agamemnon* and MV *Menestheus*, and merchant vessels *Port Napier* and *Port Quebec* for conversion to auxiliary minelayers. The latter two ships were still under construction when requisitioned.[4]

The SS *Kung Wo* and SS *Mao Yeung* were taken up in the Far East to assist the Admiralty S-class destroyers *Scout, Stronghold, Tenedos, Thanet*, and *Thracian* with minelaying associated with the defense of Hong Kong and Singapore. Summary information about the Royal Navy's surface ship minelayers and submarine-minelayers follows. These totals do not include the considerable numbers of smaller craft of the Coastal Forces that engaged in laying mines in hostile waters.

RN MINELAYERS (50 SHIPS/22 SUBMARINES)

- Cruiser-minelayer *Adventure* (purpose-built)
- Eleven merchant ships converted to auxiliary minelayers
- Six *Abdiel*-class cruiser-minelayers
- Two E-class destroyer-minelayers
- Four I-class destroyer-minelayers
- Four O-class destroyer-minelayers
- Six Admiralty S-class destroyer-minelayers
- Coastal minelayer leader *Plover*
- Two *Corncrake*-class coastal minelayers
- Three *Linnet*-class indicator loop minelayers
- Eight *M 1*-class coastal minelayers
- Two mining tenders, *Vernon* (later-*Vesuvius*) and *Nightingale*
- Six *Porpoise*-class minelaying submarines
- Thirteen T-class minelaying submarines
- Three S-class minelaying submarines

CRUISER-MINELAYER HMS *ADVENTURE*

The abbreviation "Comm." in the ensuing tables refers to the date a ship was commissioned; and "No. Mines" to the number of mines it could carry, and the total number laid during World War II (if known).

Adventure-class Cruiser-Minelayer
(520 feet, 6740 tons, 28 knots, 395 ship's complement)

Ship	Comm.	No. Mines	Disposition
Adventure (M23)	2 Oct 26	280 mines/ 12,401 mines	Damaged by mines, became a repair ship in 1944, sold 10 Jul 1947, broken up at Briton Ferry[5]

Photo 3-1

British Minelayer HMS *Adventure* (M23); location and date unknown.
National Archives photograph #80-G-165939

CONVERTED MERCHANT VESSELS

The information in parentheses in the "Former Ship Type" column refers to ship propulsion: MV (motor vessel) or SS (steamship), and date the ship was built, as opposed to when the RN commissioned it a minelayer.

Auxiliary Minelayers

Ship	Former Ship Type	Comm.	No. Mines	Disposition
Agamemnon (M10)	Blue Funnel Liner (MV/1929): 460 feet, 7593 tons, 16 kts	Oct 40	530/ 24,216	Became an amenities ship in 1944 for service with the Pacific Fleet, returned to owner on 26 Apr 1947
Hampton (M19)	Train-Ferry (SS/1934): 347 feet, 2839 tons, 16.5 kts	5 Sep 39	270/ 5,190	To the Ministry of War Transport (MoWT) in 1940

Kung Wo	River Passage Ship (SS/1921): 350 feet, 4636 tons, 10 kts	1941	248/ 224	Bombed by Japanese aircraft northwest of Pompong Island (near Lingga) and scuttled on 14 Feb 1942
Mao Yeung	Hong Kong River boat (SS): 371 tons	1939, 1941	100/ 476	Returned to owners in late Oct 1939, requisitioned again in Jan/Feb 1941
Menestheus (M93)	Blue Funnel Liner (MV/1929): 460 feet, 7494 tons, 16 kts	22 Jun 40	410/ 22,866	Damaged by bombing, amenities ship in 1944 for service with the Pacific Fleet
Port Napier (M32)	Mercantile (SS/1940): 503 feet, 9600 tons, 16 kts	12 Jun 40	550/ 6,331	Dragged anchor in gale, ran aground and set on fire at Kyle of Lochalsh on 27 Nov 1940
Port Quebec (M59)	Mercantile (SS/1940): 451 feet, 5936 tons, 14.5 kts	Jun 40	548/ 33,494	Repair ship *Deer Sound* (F99) in 1944, returned to owner on 20 Dec 1947
Princess Victoria (M03)	Car Ferry (SS/1939): 310 feet, 3167 tons, 16 kts	2 Nov 39	244/ 2,755	Sunk by mine off the mouth of the Humber River on 19 May 1940
Shepperton (M83)	Train-Ferry (SS/1935): 347 feet, 2839 tons, 16.5 kts	4 Sep 39	270/ 2,742	To the Ministry of War Transport (MoWT) in 1940
Southern Prince (M47)	Cargo Liner (MV/1929): 496 feet, 10917 tons, 17 kts	15 Jun 40	560/ 23,762	Damaged on 26 Aug 1941 by torpedo from *U-652*, accommodation ship in 1945
Teviot Bank (M04)	Liner (SS/1938): 424 feet, 5087 tons, 12 kts	Dec 39	280/ 15,865	Returned to owners on 29 Mar 1946[6]

FAST MINELAYERS

The *Abdiel*-class cruiser-minelayers were very fast at 40 knots, could do the Gibraltar to Malta run overnight, and were used to supply fuel and food to the island. They were built in two groups with different armament. *Abdiel*, *Latona*, *Manxman*, and *Welshman* were fitted with six 4-inch (3x2) anti-aircraft guns, four 2pdr (1x4) AA and eight 0.5-inch (2x4) AA guns. *Apollo* and *Ariadne* had four 4-inch (2x2) AA guns, four 40mm (2x2) AA, twelve 20mm (6x2) AA guns, and 160 moored mines.[7]

Abdiel-class Cruiser-Minelayers
(418 feet, 2650 tons, 40 knots, 242 ship's complement)

Ship	Comm.	#Mines	Disposition
Abdiel (M39)	15 Apr 41	160/ 2,207	Sunk in Taranto Harbor, Italy, 10 Sep 1943, by mines laid by the German motor torpedo boats *S-54* and *S-61*
Latona (M76)	4 May 41	160/ 0	Sunk 25 Oct 1941 north of Ras Azzaz, Libya, by a German Stuka-dropped bomb
Manxman (M70)	20 Jun 41	160/ 3,111	Damaged by torpedo 1 Dec 1942, broken up at Newport in October 1972
Welshman (M84)	25 Aug 41	160/ 3,275	Sunk 1 Feb 1943 east-northeast of Tobruk, Libya, by two torpedoes from *U-617*
Apollo (M01)	12 Feb 44	160/ 8,361	Arrived Blyth Nov 1962 for breaking up
Ariadne (M65)	9 Oct 43	160/ 1,352	Broken up Jun 1965 at Dalmuir and Troon[8]

Photo 3-2

British cruiser-minelayer HMS *Ariadne* (M65), circa 1943.
Naval History and Heritage Command photograph #NH 81892

DESTROYER MINELAYERS

The E-class destroyers *Esk* and *Express* were designed to be prepared in twenty-four hours for the minelaying role, by removing their 'A' and 'Y' 4.7-inch guns and all torpedo tubes and installing mine rails along their upper deck. Complete armament consisted of four 4.7-inch (4x1) gun mounts, eight 0.5-inch (2x4) machine guns, eight torpedo tubes (2x4), and 60 moored mines. The ships were capable of 36 knots, propelled by geared turbines producing 36,000 hp driving twin propellers.[9]

Impulsive, Ivanhoe, Intrepid and *Icarus*—I-class destroyers—were likewise fitted for rapid conversion. Two of their 4.7-inch guns had to be landed as well as both banks of torpedo tubes. The I-class ships were slightly smaller than those of the E-class. They could make 36 knots and had similar armament: four 4.7-inch (4x1) guns, two 0.5-inch (2x4) machine guns, ten 21-inch torpedo tubes (2x5), and 60 moored mines.[10]

E-class Destroyer-Minelayers
(329 feet, 1,370 tons, 36 knots, 145 ship's complement)

Ship	Comm.	#Mines	Disposition
Express (H61)	31 Oct 34	60/ 1,210	Damaged by mine, transferred to the Royal Canadian Navy as HMCS *Gatineau* on 3 Jun 1943, sold Vancouver 1956 and hulk sunk as breakwater
Esk (H15)	26 Sep 34	60/ 1,110	Sunk by a mine on 31 Aug 1940 northwest of Texel island, off the Dutch coast[11]

I-class Destroyer-Minelayers
(323 feet, 1,370 tons, 36 knots, 138 ship's complement)

Ship	Comm.	#Mines	Disposition
Icarus (D03)	3 May 37	60/ 1,196	Sold for scrapping on 29 Oct 1946, broken up at Troon.
Impulsive (D11)	29 Jan 38	60/ 892	Sold for scrapping on 22 Jan 1946, broken up in Sunderland.
Ivanhoe (D16)	24 Aug 37	60/ 754	Mined and damaged on 1 Sep 1940 northeast of Texel Island, off the Dutch coast, sunk later that day by the destroyer HMS *Kelvin*
Intrepid (D10)	29 Jul 37	60/ 1,592	Sunk on 26 Sep 1943 by German Ju 88 bombers in Leros Harbor, Dodecanese[12]

The O-class destroyers employed as minelayers—*Obdurate, Obedient, Opportune,* and *Orwell*—were armed with four 4-inch (4x1) guns, four 2pdr (1x4) AA, six 20mm (2x2, 2x1), eight 21-inch torpedo tubes (2x4), and 60 moored mines. Geared turbines producing 40,000 hp coupled to two shafts provided 37 knots top speed.

O-class Destroyer-Minelayers
(345 feet; 1,550 tons, 37 knots, 175 ship's complement)

Ship	Comm.	#Mines	Disposition
Obdurate (G39)	3 Sep 42	60/	Damaged 25 Jan 44 by a torpedo fired by *U-360*, arrived Inverkeithing 30 Nov 1964 for breaking up
Obedient (G48)	30 Oct 42	60/ 80	Arrived Blyth 19 Oct 1962 for breaking up
Opportune (G80)	14 Aug 42	60/ 120	Arrived Milford Haven 25 Nov 1955 for breaking up

Orwell (G98)	17 Oct 42	60/ 80	Arrived Newport 28 Jun 1965 for breaking up[13]

Thanet, *Tenedos*, *Scout*, *Stronghold*, *Thracian*, and *Sturdy* (remnants of a class of 67 destroyers ordered for the Royal Navy in 1917) were dispatched to the Far East in 1939 to form local defense flotillas at Hong Kong and Singapore. When war broke out, *Sturdy* was on passage in the Mediterranean and was detained there. These Admiralty S-class destroyers were characterized by two funnels, tall bridge, long fo'c'sle, and heavily raked stem and sheer forward. Propelled by three Yarrow boilers and Brown-Curtis single-reduction geared turbines, they had a top speed of 31 knots. Armament consisted of three 4-inch (3x1) gun mounts, four 21-inch torpedo tubes (2x2), one 2pdr AA gun, and 40 moored mines. Converting the destroyers to minelayers involved removing and landing ashore the after 4-inch gun, torpedo tubes, and depth charge racks and control units to enable mine rails to be fitted on either side. Twenty mines were carried in each set of rails.[14]

Admiralty S-class Destroyers
(265 feet, 905 tons, 31 knots, 90 ship's complement)

Ship	Comm.	#Mines	Disposition
Scout (H51)	15 Jun 18	40/ 772	Arrived Briton Ferry 29 Mar 1946 for breaking up
Stronghold (H50)	2 Jul 19	40/ 1,174	Sunk 2 Mar 1942 by a Japanese task group, while fleeing from Tjilatjap to Australia
Tenedos (H04)	11 Jun 19	40/ 266	Sunk 5 Apr 1942 by bombs from Japanese carrier aircraft while under repair at Colombo, Ceylon (now Sri Lanka)
Thanet (H29)	30 Aug 19	40/ 240	Sunk 27 Jan 1942 by Japanese destroyers *Fubuki*, *Asagiri*, *Yugiri*, *Shirayuki* off the east coast of Malaya while attacking Japanese transports in company with HMAS *Vampire*
Thracian (D86)	21 Apr 20	40/ 480	Scuttled by crew at Hong Kong on 16 Dec 1941, salvaged by the Japanese and placed in service as IJN patrol vessel No.*101*[15]

A total of 6,069 mines were laid by these five destroyers and by the *Mao Yeung*, *Teviot Bank*, and *Kung Wo*, at Hong Kong and Singapore:

Scout (Sep 39: 772) = 772
Stronghold (Jan 41: 186; Mar 41: 418; Feb & Jul 41: 456; Nov & Dec 41: 114) = 1,174
Tenedos (Sep 39: 266) = 266
Thanet (Oct 39: 240) = 240
Thracian (Oct 39: 320; Jan 41: 40; Feb 41: 40; Oct 41: 40; Dec 41: 40) = 480
Mao Yeung (Oct 39: 278; Jan 41: 98; Feb 41: 100; Oct 41: 100) = 576
Teviot Bank (Dec 41: 253; Dec 41:1,158; Dec 41: 450) = 1,861
Kung Wo (Mar 41: 224; Sep & Oct 41: 220; Jan 42: 256) = 700[16]

COASTAL MINELAYERS

The steam-propelled *Plover* was designed to undertake mining trials, and to serve as the leader of other coastal minelayers. For antiair defense, she had one 12pdr 3-inch antiaircraft gun, one 20mm Oerlikon, and two .303-inch machine guns; and for her primary mission, 100 moored mines.

Plover-class Coastal Minelayer Leader
(195 feet; 805 tons; 14.75 knots; 69 ship's complement)

Ship	Comm.	#Mines	Disposition
Plover (M36)	27 Sep 37	100/ 15,327	Sold 26 Feb 1969 and broken up at Inverkeithing in April 1969[17]

The two *Corncrake*-class coastal minelayers were ex-*Fish*-class anti-submarine trawlers built by Cochrane & Sons Shipbuilders Ltd., Selby, United Kingdom. A single boiler sending steam to a single vertical triple-expansion reciprocating engine, propelled the ships at a pedestrian 11 knots. Their armament was equally modest: three 20mm Oerlikon gun mounts and two twin-barrel .303-inch machine guns.

Corncrake-class Coastal Minelayers
(162 feet, 830 tons, 11 knots, 35 ship's complement)

Ship	Comm.	#Mines	Disposition
Corncrake (M82) ex-*Mackerel*	7 Dec 42	12 mines	Foundered in a storm in the North Atlantic on 25 Jan 1943
Redshank (M31) ex-*Turbot*	10 Jan 43	12 mines	Arrived at Sunderland for scrapping on 9 May 1957[18]

INDICATOR LOOP MINELAYERS

Unlike mines sown at sea (which could be either offensive or defensive in nature), the locations of controlled minefields, normally situated in the approaches to harbors, were chosen so that they could be under observation. One or more mines were manually detonated when a target vessel was observed to be within their effective range. For this reason, the mines were planted in predetermined locations with electrical cables connecting them to a centralized firing location. The mines, cables and junction boxes required maintenance. Specialized vessels undertook the hazards of planting and maintaining the fields.

The three ships of the *Linnet*-class were the largest of a dozen indicator loop minelayers built for the Royal Navy immediately before and during World War II. The vessels were designed to lay controlled mines, used in coastal defenses, as well as anti-submarine indicator loops to detect the presence of enemy submarines. They were very similar to the mine planters operated by the U.S. Army during the same era.[19]

Linnet-class Indicator Loop Minelayers
(164 feet, 498 tons, 11 knots, 24 ship's complement)

Ship	Comm.	#Mines	Disposition
Linnet (M69)	18 Jun 38	12 mines	Arrived at Dunston to be scrapped on 11 May 1964
Redstart (M62)	1 Nov 38	12 mines	Scuttled at Hong Kong on 19 Dec 1941 to prevent capture by the Japanese
Ringdove (M77)	9 Dec 38	12 mines	Served as tender to HMS Vernon; sold to the Pakistani government in 1950 as a pilot vessel; sold in 1951[20]

Eight *M 1*-class (later *Miner*-class) coastal minelayers were ex-*Isles*-class trawlers converted for use in sowing controlled minefields. Two Ruston & Hornsby 6-cylinder diesels, producing 60 bhp, propelled the diminutive craft at a modest top speed of 10 knots but, displacing only 8 feet, the minelayers could ply shallow waters. For anti-aircraft defense, the craft were fitted with one 20mm Oerlikon gun mount and two .303-inch machine guns.

M 1-class Coastal Minelayers
(122 ½ feet; 346 tons; 10 knots; 32 ship's complement)

Ship	Comm.	#Mines	Disposition
M 1 (M19)	26 Oct 39	10 mines	Renamed *Miner I* in 1942, *Minstrel* 7 Sep 1962, sold in 1967
M 2 (M34)	19 Jan 40	10 mines	Renamed *Miner II* in 1942, expended as target on 18 March 1970
M 3 (M53)	16 Mar 40	10 mines	Renamed *Miner III* in 1942, sold Feb 1977, broken up at Sittingbourne
M 4 (M68)	12 Nov 40	10 mines	Renamed *Miner IV* in 1942, scrapped in May 1964
M 5 (M74)	26 Jun 41	10 mines	Renamed *Miner V* in 1942, cable layer *Britannic* 1960, expended as target 6 Jun 1970
M 6 (M94)	20 May 42	10 mines	Renamed *Miner VI* in 1942, sold as a merchant vessel 16 Aug 1988 at Malta
Miner VII (M88)	15 May 43	10 mines	Launched 29 Jan 1944 Dartmouth, *ETV. VII* 1959, trials vessel *Steady*. Sold Mar 1980 for breaking up at Portsmouth

Miner VIII (M98)	31 Mar 44	10 mines	Launched 24 Mar 1943 Dartmouth, *Mindful* 7 Sep 1962, sold 22 Feb 1965 and renamed *Rawdhan*. Scrapped in 1981[21]

Photo 3-3

Coastal minelayer HMS *Miner VI* in April 1949.
Collection of Rob Hoole

Royal Navy Controlled Loop Minelayers also plied their trade outside the British Isles. HMS *Atreus* (base ship) and HMT *Alsey* (laying vessel) laid loop-controlled minefields at New Zealand in October 1942 and Brisbane, Australia in May/June 1942. The 443-foot *Atreus* was completed at the Scotts Greenock, Yard No. 432 in March 1911 as a steam cargo vessel for the China Mutual Steam Navigation Company Ltd., Liverpool. The Royal Navy requisitioned her in November 1939 for use as a minelayer base ship but could also operate as a minelayer. She was returned to her owner in May 1946, and later delivered to Rosyth in 1949 for scraping. *Alsey* was an Admiralty requisitioned motor launch built in 1940 and returned in 1945.[22]

MINING TENDERS

Minelaying tenders, essentially Admiralty trawlers, included HMT *Nightingale* and HMS *Skylark*, which were both built at the Portsmouth Royal Dockyard, in 1931 and 1932 respectively. *Skylark* was renamed *Vernon* in 1938—after the first *Vernon* was paid off—and later, in 1941, *Vesuvius*. *Nightingale* spent her entire life in Portsmouth until sold out of service in 1958. These small vessels operated from HMS Vernon, a shore establishment in Portsmouth and home of the Royal Navy's Mining, Torpedo, and Electrical departments. They usually supported mining trials, but also performed some minelaying.[23]

Photo 3-4

HMS Vernon's mining tender *Skylark* being re-christened *Vernon* on 9 December 1938. Her name was changed to *Vesuvius* in April 1941 owing to difficulties with the postal arrangements. The tender was launched at Portsmouth Dockyard on 15 November 1932 and served until sold on 5 July 1957. She was broken up in February 1958 at Pollock Brown in Southampton.
Collection of Rob Hoole

Photo 3-5

HM Trawler *Vernon* (ex-*Strathcoe*)—tender to HMS Vernon for minelaying trials between 1924 and 1938—with controlled mines.
Collection of Rob Hoole

The diminutive 106-foot *Nightingale* displaced 255 tons, drew 6 ½ feet of water, and could make 10 knots. She was propelled by a single triple-expansion engine, producing 400 bhp, coupled to a single shaft; and had no armament, other than the twenty mines she was capable of carrying. Her crew size was equally austere, only twelve men. *Vesuvius'* characteristics were very similar.

Photo 3-6

Controlled minefield cabling embarked in boats at HMS Vernon
Collection of Rob Hoole

COASTAL FORCES CRAFT

Coastal Force minelayers—comprised of a variety of craft—laid 6,900 mines in World War II. Coastal Forces was a division of the Royal Navy first established during World War I, and then again later in World War II under the command of Rear Admiral Coastal Forces. The last sailors to wear the 'HM Coastal Forces' cap tally were the ships companies of HMS *Dittisham* and HMS *Flintham* on these inshore minesweepers being taken out of reserve in 1968. (A cap tally is a black nylon ribbon woven with gold text, fastened around a Royal Navy Sailor's cap and tied in a bow. The gold text denotes the ship or unit to which the sailor belongs).

Type Craft	Displ. (tons)	Speed (knots)	No. Mines
Fairmile B Motor Launches (ML)	75.5	17.7	9 moored, 8 ground
Motor Torpedo Boats (MTB) – various	37-40	31-38	4 ground
Fairmile D Motor Torpedo Boats (MTB)	102	29	10 moored
Motor gunboats (MGB)	28	39-42	4 ground[24]

MINELAYING SUBMARINES

Eighteen Royal Navy submarines collectively laid 557 mines in Far East waters during the period March 1944 through May 1945. These operations were largely aimed at forcing enemy shipping seaward, where it could be attacked by conventional means. Comprising this force were two *Porpoise*-class (*Porpoise* and *Rorqual*) submarines, thirteen T-class, and three considerably smaller S-class submarines. *Rorqual*—by far the most prolific RN submarine-minelayer during the war—deposited fifty mines off Phuket Island, Thailand, on 3 January 1945, and twelve off Papra Strait—separating the island from the Malay Peninsula—the next day. Her last work in the Far East came on 10 May 1945, when she laid a field of twelve mines off the Two Brothers Islands in the Java Sea; and a second, of forty-eight mines to the north of Batavia (the capital city of Java, today Jakarta).[25]

During operations by *Rorqual* on 3 January 1945, the Netherlands HNLMS 019 (Lt. Comdr. J. F. Drijfhout van Hooff, RNLN) laid forty Vickers T3 mines off Batavia. Summary information about Australian and Netherlands' minelayers is provided in the following chapter. It is important to note that the mine totals associated with submarines in the below tables is for all theaters, not just the Far East; and while only two of the six *Porpoise*-class submarines laid mines in the Far East, summary information about the other four is provided for continuity.[26]

Porpoise-class Minelaying Submarines
(289 feet, 1500/2157 tons, 15.75/8.75 knots, 59 ship's complement)

Name	Comm.	No. Mines	Disposition
Porpoise (N14)	11 Mar 33	62/430 Mk 16, 71 Mk 2	Sunk by Japanese aircraft in the Malacca Strait on 19 Jan 1945
Narwhal (N45)	28 Feb 36	62/400 Mk 16	Believed sunk on 23 Jul 1940 off Aberdeen, Scotland by bombs from German aircraft
Cachalot (N83)	15 Aug 38	62/250 Mk 16	Rammed and sunk north of Benghazi, Libya, by the Italian torpedo boat *Generale Achille Papa* on 30 Jul 1941
Seal (N37)	24 May 39	62/50 Mk 16	Captured by the Germans 5 May 1940 after being damaged by a mine the day before
Grampus (N56)	10 Mar 37	62/50 Mk 16	Likely sunk 16 Jun 1940 off Syracuse by Italian torpedo boats *Circe* and *Cleo*
Rorqual (N74)	10 Feb 37	62/1292 Mk 16, 36 Mk 2	Went into reserve on 28 Jul 1945, sold 19 Dec 1945 and arrived Newport 17 Mar 1946 for breaking up[27]

The thirteen T-class submarine minelayers collectively sowed 178 mines during the war—all in Far East waters between 14 March and 23 December 1944. With the exception of *Taurus* and *Thorough*, which each laid two loads of mines, the other submarines each made a single lay—twelve Mk 2 ground mines. These efforts began with mining in the Malacca Strait; *Trespasser* laid a field off Mati Bank on 14 March, and *Taurus* one off Arca Island on 19 March. *Taurus* deposited another twelve mines south of Penang on 18 April, completing her wartime mining. *Tally-Ho* laid twelve mines off Buja Shoal, Malacca Strait on 14 May; and *Tactician* the same number at Langkawi Sound, Malaya, two days later, concluding their mining efforts as well.

Similar efforts by the other submarines are summarized below:

Mines Laid by T-class Submarines in Far East Waters

Date	Submarine	No. Mines	Location
14 Mar 44	*Trespasser*	12 Mk 2	Outer Mati Bank, Malacca Strait
19 Mar 44	*Taurus*	12 Mk 2	Aroa Islands, Malacca Strait
18 Apr 44	*Taurus*	12 Mk 2	south of Penang
14 May 44	*Tally-Ho*	12 Mk 2	Bunja Shoal, Malacca Strait
16 May 44	*Tactician*	12 Mk 2	Langkawi Sound, Malaya
2 Jun 44	*Tantalus*	12 Mk 2	Dindings, Malaya
4 Jun 44	*Templar*	12 Mk 2	Dindings, Malaya
7 Jun 44	*Tantivy*	12 Mk 2	Sembilan Islands, Malacca Strait
24 Jun 44	*Truculent*	12 Mk 2	south of Klang Strait, Malaya
16 Sep 44	*Trenchant*	12 Mk 2	Sembilang Channel, Sumatra
24 Sep 44	*Tudor*	10 Mk 2	off Pulo Lantar, Malaya
30 Oct 44	*Tradewind*	12 Mk 2	Mergui Islands, Burma (now Myanmar)
19 Nov 44	*Thorough*	12 Mk 2	Outer Mati Bank, Sumatra
16 Dec 44	*Thule*	12 Mk 2	off Pulo Terutan, Malaya
23 Dec 44	*Thorough*	12 Mk 2	off Pulo Terutan, Malaya[28]

Greater information about the characteristics of these submarines, including their commissioning dates and final disposition, follows.

T-class Submarine Minelayers
(275 feet, 1090/1575 tons, 15.25/9 knots, 61 ship's complement)

Name	Comm.	No. Mines	Disposition
Tactician (P314)	29 Nov 42	12/12 Mk 2	Arrived Newport 6 Dec 1963 for breaking up
Tally-Ho (P317)	12 Apr 43	12/12 Mk 2	Arrived Briton Ferry 10 Feb 1967 for breaking up
Tantalus (P318)	2 Jun 43	12/12 Mk 2	Scrapped at Milford Haven in Nov 1950
Tantivy (P319)	25 Jul 43	12/12 Mk 2	Sunk as an anti-submarine target in the Cromarty Firth in 1951

Taurus (P339)	3 Nov 42	12/24 Mk 2	Loaned to the Royal Netherlands Navy on 4 June 1948 and commissioned into the Royal Netherlands Navy as *Dolfijn* the same day. Decommissioned and returned to the Royal Navy on 7 December 1953. Recommissioned into the Royal Navy as HMS *Taurus* on 8 December 1953. Scrapped at Dunston-on-Tyne in April 1960.
Templar (P316)	15 Feb 43	12/12 Mk 2	Sunk as a target in Loch Striven, Scotland in 1954. Salvaged on 4 December 1958. Arrived at Troon, Scotland on 17 Jul 1959 to be scrapped.
Thorough (P324)	1 Mar 44	12/24 Mk 12	Scrapped at Dunston on Tyne on 29 June 1962
Thule (P325)	13 May 44	12/12 Mk 12	Arrived Inverkeithing on 14 Sep 1962 for breaking up
Tradewind (P329)	16 Sep 43	12/12 Mk 12	Arrived Charlestown on 14 Dec 1955 for breaking up
Trenchant (P331)	31 Jan 44	12/12 Mk 12	Sold 1 July 1963 and arrived Faslane 23 Jul 1973 for breaking up
Trespasser (P312)	11 Sep 42	12/12 Mk 2	Arrived Gateshead on 26 Sep 1961 for breaking up
Truculent (P315)	31 Dec 42	12/12 Mk 2	Sunk in the Thames Estuary on 12 Jan 1950 after a collision with the Swedish tanker ship *Divina.* Salvaged on 14 Mar 1950 and sold 8 May 1950 for breaking up at Grays, Essex.
Tudor (P326)	19 Dec 43	12/10 Mk 12	Sold 1 Jul 1963 and arrived Faslane 23 Jul 1963 for breaking up[29]

The three small S-class submarines laid a combined thirty-two mines in Far East waters during the war. HMS *Surf* planted eight Mk 2 mines off Pulo Terutan, Malaya, on 13 May 1944, and another eight there on 14 June 1944. *Sea Rover* sowed eight Mk 2 off the Sembilan Islands, Malacca Strait, on 18 May 1944; and *Stoic*, eight, north of Penang, Malaya, on 3 June 1944. These submarines were capable of carrying eight ground mines, deposited via their 21-inch torpedo tubes.[30]

S-class Submarine Minelayers
(217 feet, 715/990 tons, 14.75/9 knots, 6 O/42 E ship's complement)

Name	Comm.	No. Mines	Disposition
Sea Rover (P218)	7 Jul 43	8/ 8 Mk 2	Sold Oct 1949 and broken up at Faslane, Scotland in Jun 1950
Stoic (P231)	29 Jun 43	8/ 8 Mk 2	Sold Jul 1950 and broken up at Dalmuir, Scotland
Surf (P239)	18 Mar 43	8/ 16 Mk 2	Sold 28 Oct 1949 and arrived Faslane, Scotland Jul 1950 for breaking up[31]

Battle Honours earned by surface- and submarine-minelayers are listed in the following chapter, along with those of the Australian and Netherlands navies.

MINELAYING AIRCRAFT

The scope of this book precludes much detail about the operations of minelaying aircraft. However, it's important to note that aircraft planted 54,194 mines, and that 500 planes went missing during these operations. Identified below are the types of aircraft employed, and their standard mine loads (number and types of mines they could carry). U.S. Naval aircraft laid relatively few mines, and only quantities are provided.[32]

Royal Navy Fleet Air Arm

Aircraft	Mine Load, No. and Type
Swordfish, Albacore, Barracuda II, Avenger II	1-2,000lb (Mk 1-4 & Mk 6)

Royal and Dominion Air Forces

Aircraft	Mine Load	Aircraft	Mine Load
Beaufort I	1-2,000lb (Mk 1-4 & Mk 6)	Manchester I	4-2,000lb (Mk 1-4 & Mk 6)
Catalina PBY	4 mines/4,000 lbs maximum load	Marauder	2-1,000lb (Mk 5 & Mk 7)
Halifax III	4-1,000lb (Mk 5 & Mk 7)	Mosquito XIB	2-2,000lb (Mk 1-4 & Mk 6)
Hampden I	1-2,000lb (Mk 1-4 & Mk 6)	Stirling I	6-2,000lb (Mk 1-4 & Mk 6)
Lancaster III	6-2,000lb (Mk 1-4 & Mk 6)	Ventura (number and type of mines carried by RNZAF unknown)	3,000lbs bombs, six 325lb depth charges, 1 torpedo
Liberator I (VLR)	6-1,000lb (Mk 5 & Mk 7)	Wellington IC	2-2,000lb (Mk 1-4 & Mk 6)
Liberator II (LR)	6-1,000lb (Mk 5 & Mk 7)	Wellington X	2-2,000lb (Mk 1-4 & Mk 6)

U.S. Army Air Forces

Aircraft	Mine Load	Aircraft	Mine Load
B24 Liberator	8-Mks 13, 26, 36	B29 Superfortress	12-Mks 13, 26, 36
B25 Mitchell	2-Mks 13, 26, 36		

U.S. Navy Aircraft

Aircraft	#Mines Laid	Aircraft	#Mines Laid
PBY-5 Catalina	2,548	PB4Y-2 Privateer	186
PB2Y-3 Coronado	12	PV-1 Ventura	18
PB4Y-1 Privateer	101	TBF Avenger	320[33]

4

Royal Australian and Royal Netherlands Navy Minelayers

Photo 4-1

Heavy cruiser HMAS *Australia* under way during the 1930s.
Naval History and Heritage Command photograph #NH 52684

The Royal Australian Navy's Mine Warfare Forces in World War II consisted of a single auxiliary minelayer, HMAS *Bungaree* (M29)—a converted merchant vessel, which laid 9,289 mines in Australian and New Zealand waters during the war—and Minesweeping Groups of requisitioned craft based at Australian port cities as shown below:

Sydney	Minesweeping Group 50	Fremantle	Minesweeping Group 66
Melbourne	Minesweeping Group 54	Darwin	Minesweeping Group 70
Hobart	Minesweeping Group 60	Brisbane	Minesweeping Group 74
Adelaide	Minesweeping Group 63	Newcastle	Minesweeping Group 77[1]

The heavy cruiser HMAS *Australia* (D84) also merits mention. Although minelaying was not one of her usual duties, she undertook such in early November 1941.

AUSTRALIA MINES ANCHORAGES IN KERGUELEN ISLANDS, SOUTH ATLANTIC, IN NOVEMBER 1941

Since April 1940, German raiders had been operated in the South Atlantic and Indian Oceans, preying on Allied ships sailing alone. Believing that the passengers and crews of ten ships known to have been lost on the East Indies Station (or were long overdue) might have been put ashore, Vice Adm. Sir Ralph Leatham, RN, sent the light cruiser HMS *Neptune* to investigate the Kerguelen Islands. Inspection in October 1940 of these French islands, located far south in the Indian Ocean, found no evidence that they were in general use by raiders.[2]

As raider activity continued, HMAS *Australia* arrived at Columbo, Ceylon, on 21 October 1941. That same day the Admiralty informed Leatham, the station's commander, that the German raider *Atlantis* was known to have been at Kerguelen for about four weeks from mid-December 1940. There were also reports that raiders *Komet* and *Pinguin* had obtained stores from a supply ship there in March 1941. The Admiralty message also suggested that the Crozet Islands, farther west, might also prove to be a haven of these ships. Leatham proposed that *Australia* be attached to his command and be dispatched to search the islands. The Admiralty concurred and recommended that she lay ground mines in likely anchorages, to prevent their use by the enemy.[3]

The heavy cruiser immediately loaded a number of A Mk 1 mines that same day, 21 October, from a lighter in Columbo Harbor, and sailed the following morning for British Mauritius. The island, located in the Indian Ocean about 1,200 miles off the southeast coast of the African continent, was a source for fuel before continuing south. *Australia* reached the Kerguelens on 1 November. Over the next four days, as shore parties and ship's aircraft searched the islands, she mined Gazelle Basin (6 mines), Long Island South (4 mines), Island Harbor (2 mines) and Tucker Strait (6 mines). The Crozet Group was searched on the 6th and 7th of November, also without result, and *Australia* sailed for Durban, South Africa; arriving on the 11th.[4]

It was later learned that the Kerguelen Islands had last been used by German raiders in March 1941. The mines may have ultimately borne fruit, had not raider activity in the Indian Ocean virtually ceased after the sinking of *Kormoran* on 19 November 1941, by the light cruiser HMAS *Sydney* off the west coast of Australia. The location of these mines was a close-held secret, and it was not until September 1944 that a "Q" message was sent, defining the dangerous areas. No steps were taken to dispose of the mines after the war. As late at 1959, the French Naval Authorities sought information as to their location, which the Admiralty supplied.[5]

AUXILIARY MINELAYER HMAS *BUNGAREE*

Photo 4-2

Australian auxiliary minelayer HMAS *Bungaree* (M29). During the war, she laid mines at New Zealand, Australia, and nearby islands.
Australian War Memorial photograph 300494

The 357-foot *Bungaree* was a former coastal cargo ship completed in May 1937 by Caledon Shipbuilding and Engineering Co. Ltd, Dundee, Scotland, for operation by the Adelaide Steamship Company in Australian waters. Following requisition by the Royal Australian Navy in October 1940, conversion immediately began in Sydney, which involved turning her cargo holds into large mine magazines. Two sets of rails were installed on the mining deck to transport mines (all moored contact types) aft to her stern, for deployment in the water.

HMAS *Bungaree* was commissioned on 9 June 1941 at Garden Island, Sydney (under the command of Comdr. Norman Calder, RAN).

Two boilers providing steam to a reciprocating engine (2,500 hp), coupled to a single shaft, propelled the ponderous 7,494-ton ship at a top speed of only 11 knots. Carrying 423 mines when fully loaded, she was essentially a slow-moving ammunition dump, subject to attack and destruction of ship when in enemy waters. This reality for her and other minelayers gave rise to the moniker "nightraider." Her armament consisted of two 4-inch guns, and for anti-aircraft defense, two 2pdr guns, and eight Vickers water-cooled .303-inch machine guns.[6]

ROYAL AUSTRALIAN AIR FORCE MINELAYERS

Photo 4-3

A U.S. PBY-5A Catalina patrol bomber on a beach airstrip at Ulithi, Caroline Islands. These "Black Cats" (being painted black to better remain undetected at night), were the same type the Royal Australian Air Force used for stealthy nighttime missions to lay two or four mines at a time, in Japanese ports and harbors.
Naval History and Heritage Command photograph #309720

Royal Australian Air Force PBY-5 Catalinas (of Squadrons No. 11, 20, 42 and 43) flew 1,130 sorties between April 1943 and March 1945, planting a total of 2,512 mines.

NETHERLANDS EAST INDIES NAVAL MINELAYERS

Netherlands East Indies Naval Forces (at Batavia)
Vice Adm. Conrad E. L. Helfrich, RNLN

Vice Adm. Conrad E. L. Helfrich, RNLN, commanded the ABDA (American-British-Dutch-Australian) naval fleet in its unsuccessful attempt to protect the Dutch East Indies from Japanese attack. Following the Battle of the Java Sea on 27 February 1942, surviving ABDA light warships and small vessels were instructed to disperse to safety. Some escaped to Ceylon (today Sri Lanka) and Australia; others were intercepted and sunk by Japanese forces south of Java. With only his submarines remaining, Helfrich realized that continued naval defense of Java was impossible. Accordingly, on 2 March, he left with his staff in four Dutch Catalina flying-boats for Colombo, Ceylon. Following arrival there, he set up new headquarters as commander of Dutch sea, air, and land forces in the southwestern Pacific.[7]

Ten Dutch minelayers (two submarines and eight ships) were a part of the Royal Netherlands Naval Forces on 7 December 1941, when the Japanese attacked Pearl Harbor. Based at Soerabaja (now Surabaya), on Java's northeastern coast bordering the Madoera Strait, the minelaying submarines comprised Submarine Division 4. The surface minelayers were units of the Mine Service, which also included minesweepers. On that particular day, the two submarines were in the South China Sea. Three of the minelayers were also away from Soerabaja; *Gouden Leeuw* was at Tarakan Island off northeastern Borneo; *Pro Patria* at Palembang, Sumatra; and *Willem van der Zaan* at Lingga Island off Sumatra.[8]

Only HNLMS *O-19* and HNLMS *Willem Van Der Zaan* would escape the Netherlands East Indies. Submarine *O-20* was scuttled by her crew on 9 December 1941, after being damaged by a Japanese destroyer, and minelayer *Prins van Oranje* was sunk by enemy forces on 10 January 1942. The crews of the remaining six minelayers scuttled their ships to prevent capture by the Japanese.

- HNLMS *O-20*   Scuttled 9 December 1941
- HNLMS *Prins van Oranje*   Sunk 10 January 1942
- HNLMS *Pro Patria*   Scuttled 15 February 1942
- HNLMS *O-19*   Escaped 1 March 1942
- HNLMS *Willem Van Der Zaan*   Escaped 1 March 1942
- HNLMS *Bangkalan*   Scuttled 2 March 1942
- HNLMS *Rigel*   Scuttled 2 March 1942
- HNLMS *Soemenep*   Scuttled 2 March 1942

- HNLMS *Gouden Leeuw*   Scuttled 7 March 1942
- HNLMS *Krakatau*   Scuttled 8 March 1942

Summary information about the minelayers of the Netherlands East Indies Naval Forces follows; AA, HP, and MG are abbreviations for anti-aircraft, horsepower, and machine gun.

HNLMS *O-19*-class Minelaying Submarines
(264 feet, 982/1491 tons, 9/19.5 knots, 40 ship's complement)

The two minelaying submarines, built with a partial double-hull and a snorkel system, could carry forty mines. Their other armament consisted of eight 21-inch torpedo tubes (14 torpedoes), one 3.5-inch deck gun, two 40mm Bofors anti-aircraft guns, and one 12.7mm machine gun. Submarine length, surfaced and submerged displacement and speed, and manning is given below the heading. The submarines' commissioning dates, their final disposition, and the identities of their commanding officers on 7 December 1941, follow:

O-19	3 Jul 39	Lt. Comdr. Frederik Johan Adolf Knoops, RNLN	Escaped from Java 1 Mar 1942, Destroyed by explosives 10 Jul 1945 after grounding on Ladd Reef in the South China Sea
O-20	28 Aug 39	Lt. Comdr. Pieter Gerardus Johan Snippe, RNLN	Scuttled 9 Dec 1941 off Kota Baharu, after depth-charged by destroyers *Ayanami* and *Yugiri*, and further damaged by gunfire from destroyer *Uranami*[9]

HNLMS *Willem van der Zaan*-class Minelayer
(246 feet, 1407 tons, 15.5 knots, 92 ship's complement)

Two 4.7-inch guns, four 40mm AA, four 12.7mm MG, 120 mines. Triple-expansion-engines, 2,200 HP, two shafts.

Willem Van Der Zaan	21 Aug 39	Lt. Comdr. Gijsbertus Petrus Küller, RNLN	Rebuilt and commissioned as a frigate 13 Nov 1950, sold for scrapping on 6 Oct 1970[10]

HNLMS *Prins van Oranje*-class Minelayers
(230 feet, 1291 tons, 15 knots, 121 ship's complement)

Armed with two 75mm guns, two Vickers 40mm AA, two 12.7mm MG, and 150 mines. *Gouden Leeuw* had sonar from early 1942 with depth charges; *Prins van Oranje* had neither. Triple-expansion engines, 1750 HP, two shafts.

Gouden Leeuw	24 Feb 32	Lt. Comdr. Eugenius Johannes Cornelis van der Horst, RNLN	Scuttled by her crew on 7 Mar 1942 at Soerabaja

Prins van Oranje	2 Feb 32	Lt. Comdr. Anthonie Catharinus van Versendaal, RNLN	Sunk on the night of 10 Jan 1942, while trying to escape Tarakan, by Japanese patrol boat *38* and the destroyer *Yamakaze*[11]

HNLMS *Pro Patria*-class Minelayer
(154 feet, 612 tons, 10 knots, 61 ship's complement)

One 3-inch gun, two 12.7mm MG, 60 mines. One triple-expansion engine, 650 HP, one shaft.

Pro Patria	23 Aug 23	Lt. Louis Guiot, RNLN	Scuttled by crew on 15 Feb 1942 at Palembang, Sumatra, to prevent capture by the Japanese[12]

HNLMS *Krakatau*-class Minelayer
(213 feet, 982 tons, 15.5 knots, 105 ship's complement)

Two 3-inch guns, four 12.7mm MG, 150 mines. Triple-expansion engine, 2500 HP, one shaft.

Krakatau	11 Dec 24	Lt. Johan van Haga, RNLN	Scuttled by crew on 8 Mar 1942 off Madoera, Netherlands East Indies, to prevent capture by the Japanese[13]

HNLMS *Rigel*-class Auxiliary Minelayer
(221 feet, 1631 tons, 13 knots, 63-67 ship's complement)

During peacetime, used as a yacht for the governor-general of the Netherlands East Indies. Commissioned as a minelayer on 15 October 1939. Two 3-inch guns, two 12.7mm MG, 150 mines. Triple-expansion engines, 1400 HP, 2 shafts.

Rigel	6 Nov 31		Sunk at Batavia (Jakarta) on 2 Mar 1942, salvaged by the Japanese in March 1944 and partially repaired. Found in Jakarta in 1945 and returned to the N.E.I. government. Transferred to the Indonesian navy in 1951 and renamed *Dewakembar*[14]

HNLMS *Soemenep*-class Auxiliary Minelayer
(118 feet, 227 tons, 10 knots, 26 ship's complement)

Former tug *Zoutregie*. Following conversion to an auxiliary minelayer, recommissioned on 24 June 1940. Two 12.7mm MG, 26 mines. Triple-expansion engine, 400 HP, one shaft.

Soemenep	1931	Lt. Tobias Jellema, RNLN	Scuttled at Soerabaja on 2 Mar 1942 to prevent capture by the Japanese[15]

HNLMS *Bangkalan*-class Minelayer
(141 feet, 397 tons, 10 knots, 26 ship's complement)

Former tug *Willem van Braam*, rebuilt as survey vessel *Hydrograaf* and later renamed *Bangkalan* in 1935. Converted to an auxiliary minelayer in February 1942 at Naval Dockyard Soerabaja. One 75mm gun, two 12.7mm MG, 30 mines. Triple-expansion engine, 360 HP, one shaft.

Bangkalan	1926	Lt. Herman Nicolaas Sax, RNLN(R)	Scuttled by crew at Soerabaja on 2 Mar 1942 to prevent capture by the Japanese. Salvaged by the Japanese and commissioned.[16]

NAVAL AIR SERVICE

PBY-5 Catalinas of the Royal Netherlands Navy laid twenty-four mines as part of the blockade of Japan.[17]

BATTLE HONOURS

The Royal Australian Navy minelayer *Bungaree*, Royal Netherlands minesweeper *Abraham Crijnssen* and minelaying submarine *O-19*, and fifteen Royal Navy minelaying submarines earned Battle Honours during the war.

PACIFIC 1942-43

HNLMS *Abraham Crijnssen*   HMAS *Bungaree*

MALAYA 1942-45

HNLMS *O-19*	HMS *Tantivy*	HMS *Tradewind*
HMS *Porpoise*	HMS *Taurus*	HNS *Trenchant*
HMS *Sea Rover*	HMS *Templar*	HMS *Trespasser*
HMS *Stoic*	HMS *Thorough*	HMS *Truculent*
HMS *Tally-Ho*	HMS *Thule*	HMS *Tudor*[18]
HMS *Tantalus*		

5

Fall of Hong Kong and Singapore

The Japanese had only about 5,000 troops with very little artillery support…their troops were ill-equipped and not used to night fighting…their aircraft were for the most part obsolete and their pilots very mediocre.

—Maj. George Trist, RCA, adjutant of the Winnipeg Grenadiers, recalling information provided by Maj. Gen. Christopher Maltby, British Indian Army, regarding enemy strength, shortly after the arrival of the Royal Rifles and Winnipeg Grenadiers battalions at Hong Kong on 16 December 1941.[1]

The Royal Navy recognized in 1938, when making plans against the possibility of a war in Europe, that there were insufficient resources to maintain an adequate fleet in the Far East while guarding home waters and the Mediterranean against the expected onslaught from Germany and Italy. Japan's activities in China had given warning of her plans for expansion in the Far East. However, due to the paucity of fleet units, provisions were made that in the event Japan joined the Axis Powers, Singapore and Australasia would be defended, while only token resistance would be offered in the event of an attack on Hong Kong. Even this insufficient measure would be dependent upon transfer of the Mediterranean Fleet to the Far East, leaving the defense of the former theater in the hands of the French. The collapse of France in June 1940 precluded any possibility of naval assistance from the Mediterranean and, in fact, the China Station was stripped to provide ships for that theater.[2]

From 1865-1941, commander in chief, China, was the title of the admiral in charge of the Royal Navy's vessels and shore establishments in China. The command was known as the China Station. In response to increased Japanese threats, the China Station was merged with the East Indies Station in December 1941 to form the Eastern Fleet.

In 1941, commander in chief, China, Vice Adm. Sir Geoffrey Layton, RN (based in Singapore) facilitated a series of conferences to consider mutual defense problems in the theater. Anglo-Netherlands-

Australian interests were the main topic; America, then uncommitted, sent observers. Cooperation between the British and Netherlands East Indies fleets resulted from these discussions, agreement was reached on defensive mining plans, and arrangements were made for the supply of British mines to the Netherlands authorities.[3]

BRITISH MINEFIELD DEFENSES AT SINGAPORE

Local plans were made for the laying of defensive minefields in the approaches to Singapore and Hong Kong, and adequate stocks of mines were accumulated for this purpose, as well as for offensive mining by submarines or by the minelayer HMS *Adventure*, then a unit of the China Squadron. Separate arrangements were made by Australia for defensive mining in Australasia (a region of Oceania, encompassing Australia, New Zealand, neighboring islands in the Pacific, and New Guinea).[4]

When war broke out in Europe in September 1939, the *Adventure* had already been recalled to Britain and HMS *Seal* (N37), one of the three minelaying submarines on the China Station, followed her in October. Accordingly, old S-class destroyers of the Hong Kong and Singapore Local Defense Flotilla, and a Hong Kong river boat pressed into minelaying duties, were tasked with sowing the defensive fields.[5]

Photo 5-1

British minelayer HMS *Adventure* (M23) under way; date and location unknown. Naval History and Heritage Command photograph #NH 63246

In anticipation of such employment, the destroyers HMS *Scout* and *Tenedos* had sailed from Hong Kong on 24 August 1939 to form the local minelaying force at Singapore. (HMS *Thanet* and *Thracian* were retained at Hong Kong for the same purpose, and the fifth ship of the flotilla, HMS *Stronghold*, on passage from the United Kingdom, reached Singapore on 22 September.) On their arrival at Singapore about 28 August, *Scout* and *Tenedos* were converted for minelaying. This involved removing the after 4-inch gun, torpedo tubes, and depth charge racks and control units to enable mine rails to be fitted on port and starboard sides. Twenty mines were carried in each set of rails.[6]

On 2 September, *Scout* and *Tenedos* each loaded forty British Mk 14 moored mines at the Singapore Armament Depot. At the outbreak of war, Mine Depots were in operation or were being completed at Malta; Haifa; Trincomalee, Ceylon (now Sri Lanka); Singapore; Hong Kong; and in Australia—with storage facilities at Kilindini, Kenya; Durban, South Africa; and Freetown, Sierra Leone.[7]

Photo 5-2

British "S" class destroyer in the Dardanelles, circa 1919-1920.
Naval History and Heritage Command photograph #NH 63470

The Mk 14 mines, fitted with fourteen Hertz horns (named after German Heindrich Hertz' discoveries in contact mechanics) proved to have some shortcomings. It was found that a relatively hard impact by a vessel on a horn might on occasion fail to fracture the glass tube inside, and thereby fail to detonate the 320 or 500lb charge. The mines were also liable under certain circumstances to be countermined (exploded) by the detonation of an adjacent mine. For these and other reasons, it was decided to replace the Hertz horns with eleven switch horns, resembling spikes. When a switch horn was moved in any direction, it completed the electrical circuit from a battery to the detonator. Fitted with switch horns, and a 320, 450, or 500lb explosive, the modified ordnance was designated the Mk 17. The modified mine, which would become the standard British contact mine of World War II, could be laid in waters up to 500 fathoms deep.[8]

Photo 5-3

British Mk 17L mine on a Mk 17 anchor
Collection of Rob Hoole

Plans for the defense of Singapore included laying three minefields: No. 1 across the western entrances to the Johor Strait and covering the entrances to Selat Sembilan; No. 2 between Pulau Sebarok and the western side of St. John Island, covering the southern approach to

Keppel Harbor; and No 3. between the eastern side of St. John Island and Siglap, covering the eastern approaches to the harbor. *Scout* and *Tenedos* laid field No. 1 between 4-8 September; sowing 544 mines at a depth of four feet, in fifteen separate lines. Following this operation, *Scout* was restored to her normal destroyer configuration, leaving *Tenedos* to complete the remaining fields.[9]

Map 5-1

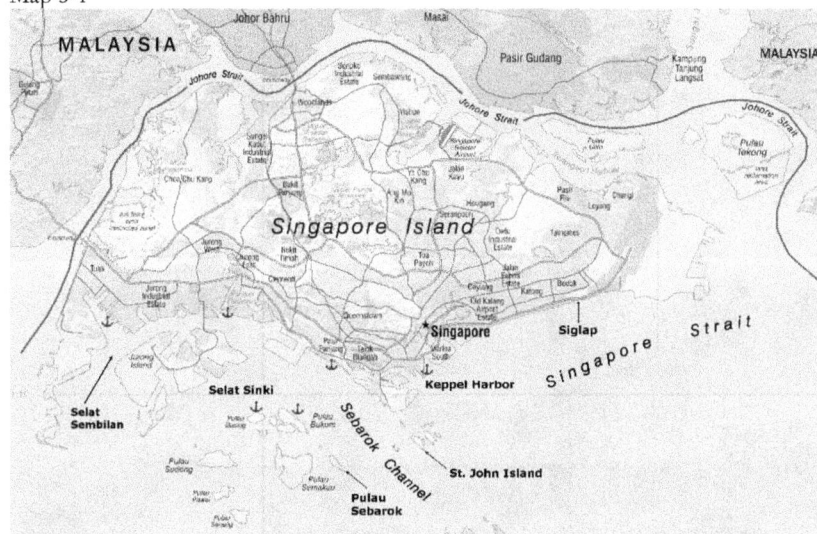

Singapore

Tenedos sowed the two smaller fields between 11-17 September, six lines (228 mines) in minefield No. 2, and seven lines (266 mines) in No. 3. The mines were once again set to a depth of four feet. The Admiralty Notice to Mariners of 16 September 1939 identified the danger area associated with field No. 1, and the Notice of 3 October those associated with fields No. 2 and 3. Neither provided any instructions to ships regarding entry into Singapore, and it was not until 2 December that Selat Sinki was designated in writing as the only safe entrance.[10]

As the Norwegian MV *Höegh Transporter* approached the harbor on 3 October, through the narrow gap existing between St. John Island and the southeastern limit of field No. 3, she struck a mine, and sank off the Outer Shoal Beacon. A similar incident occurred on 13 November, when the British liner SS *Sirdhana* was sunk in almost the same location. Instructions to shipping entering Singapore had presumably been broadcast locally, as both these losses were attributed to the failure of the ships' masters to obey local regulations.[11]

The independent minefields at Singapore were now complete, and the planned laying of supplementary controlled fields had to await the arrival from Hong Kong of the indicator loop minelayer HMS *Redstart* (commanded by Lt. Comdr. Alfred Frank Jackson, RN) at the end of November. *Tenedos* was reconverted for a general duties role, and not used again as a minelayer. When operations restarted in early 1941 to reinforce and relay the original fields, HMS *Stronghold* was employed for that purpose.[12]

Photo 5-4

Postcard of HMS *Redstart* (M62), circa 1939

BRITISH COLONY OF HONG KONG

The British Colony of Hong Kong encompassed Kowloon Peninsula, and a series of islands on the south coast of China, bordered to the north by Guangdong province. Separating Hong Kong Island from the city of Kowloon (formerly a Chinese military fort and later a walled city) on the mainland was the mile-wide waters of Victoria Harbor. Great Britain had acquired Hong Kong Island in 1842 and developed its principal city, Victoria, into a major port for trade with China. The acquisition in 1899 of the New Territories on the mainland, provided a 368-square-mile buffer zone to protect the island from direct artillery attack from the north. For self-defense, the colony was garrisoned with British and Indian troops, and coast defense batteries were constructed.[13]

Map 5-2

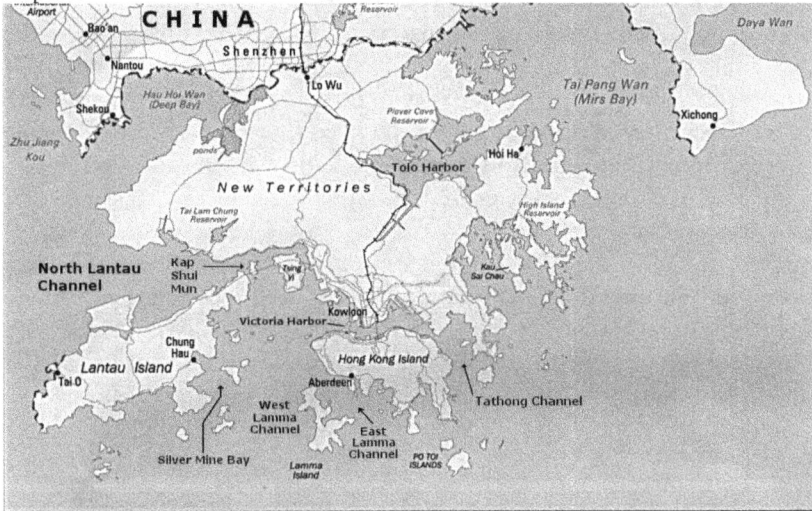

Hong Kong

ROYAL NAVY'S PRESENCE IN HONG KONG

The earliest association of the Royal Navy with Hong Kong came in 1841 when, arriving aboard HMS *Sulphur* in what became known as Victoria Harbor, Commodore Sir Gordon Bremer claimed the sparsely-inhabited island as a colony. Over the next fifty years, the Royal Navy gradually increased its presence at Hong Kong, building dockyards and other installations, including HM Victualling Yard to provision ships. A succession of naval ships served temporarily as administrative bases until the sail and steam-powered troop ship HMS *Tamar* was "hulked" to serve as RN headquarters in Hong Kong. Arriving off Victoria City on 11 April 1897, she anchored by the Naval Dockyard, while the base administration establishment that would bear her name was being built ashore. Thereafter, she served as the depot ship for the base.[14]

TAM MARTE, TAM PACE
(Alike in war and peace)

MINE DEFENSES AT HONG KONG

To counter any incursion of German naval forces in the Pacific and the possibility of Japan joining them as an ally, two independent minefields were laid at Hong Kong in 1939; one in the North Lantau Channel to cover the northern approaches to the island, and another closing the West Lamma Channel. Controlled minefields were installed in the two remaining entrances, with eight loops guarding all the southern and eastern approaches to the port. *Thanet* and *Thracian* carried out the mining, aided by the *Mao Lee* and *Mao Yeung*. The former vessel had been requisitioned in August for conversion to a mine carrier, as there were no shore loading facilities for the minelayers. The latter was a Hong Kong ferry boat, fitted out for minelaying with a capacity of 100 mines and commissioned on 28 August with a crew drawn from the river gunboat HMS *Robin* (T65).[15]

Photo 5-5

Panoramic photograph by Lum Long of Hong Kong Harbor, circa 1931-1933, with the Royal Navy Dockyard at the left.
Naval History and Heritage Command photograph #NH 94179

Redstart later put down a controlled field near Tathong Point and a similar field in the East Lamma Channel before sailing for Singapore on 17 November. The laying of the independent minefields was delayed because Commodore Arthur Malcolm Peters, RN (in charge of the Naval Establishments at Hong Kong) believed it inadvisable to lay them, and impede shipping using the port, until there was evidence of German U-boat activity on the station. However, following a message from the Admiralty that vigilance in anti-submarine measures was necessary in Far East waters, commander in chief, China, Adm. Sir Percy Noble, RN, ordered that the independent minefields off Hong Kong be laid forthwith.[16]

The planned fields of moored mines were designed to prevent U-boats from entering on the surface by way of the North Lantau and West Lamma Channels. *Thanet, Thracian,* and *Mao Yeung* were readied for operations, and the areas were mined between 25-31 October, with the *Mao Lee* providing resupply of mines. Having sown 278 Mk 14 mines in three lines in the North Lantau Channel, *Mao Yeung* was returned to her owners. *Thanet* and *Thracian* each laid two lines in West Lamma Channel for a total of 240 mines. *Thracian* later sowed an additional sixty mines across the entrance to Shap Long Bay, and twenty in the North Lantau field.[17]

Upon completion, the destroyers were reconverted for general duties. The minefields were declared in a Notice to Mariners, and no further mining was done until 1941. On completion of the controlled fields, a Notice to Mariners cancelling the previous one, provided information about all the dangerous areas, and advised that port entry was confined to the Tathong Channel.[18]

REINFORCEMENT OF THE SINGAPORE MINEFIELDS

Between 2-5 January 1941, following a determination that the fields of moored mines at Singapore were deteriorating due to rough weather and general corrosion of the ordnance, *Stronghold* laid 186 new mines in six deep and shallow lines within the original area. The remaining two fields were reinforced to seaward of those laid in 1939.[19]

Photo 5-6

British River Passenger Ship SS *Kung Wo* (built in Hong Kong for Yangtze River service of Indo-China Steam Navigation Co.) at Shanghai in the 1920s. Requisitioned for service as a minelayer, she was bombed and sunk 14 February 1942 by Japanese aircraft. Naval History and Heritage Command photograph #NH 77128

Following the arrival of the necessary mines from UK sources at Singapore, *Stronghold* carried out the work in field No. 3 between 3-8 March 1941, laying 418 mines in eleven deep and shallow lines. *Kung*

Wo, a converted British Yangtze river steamer, reinforced field No. 1 on 25-26 March, laying a total of 224 mines in four lines. She had been commissioned at Singapore on 7 March.[20]

LOSS OF HMS *BUFFALO* TO A MINE

Disaster struck Singapore on 5 April 1941, when the mooring vessel HMS *Buffalo* was sunk by a mine in field No. 3. She had left Singapore Road that morning to attempt the salvage of a Bristol Blenheim light bomber, which had crashed into the sea south of Siglap Obelisk. Embarked aboard the mooring vessel were a dockyard mooring party, an RAF salvage party, and an officer and a signalman from the light cruiser HMS *Dauntless*. A senior surveying officer directing *Buffalo* from a boat, provided faulty navigation, which led her into danger. The after part of the ship sank almost immediately. The fore part remained afloat for almost seventeen minutes before capsizing, taking with it several injured men, and those attempting their rescue. Total casualties were 30 dead and 25 injured.[21]

INCREASED JAPANESE AGGRESSION

In July 1941, Japan occupied the southern part of Indochina (Vietnam, Cambodia, and Laos). It was later learned that this was in preparation for an invasion of the Dutch East Indies. The United States, Great Britain and the Dominions immediately froze all Japanese assets, as a warning against further aggression and to help prevent the accumulation by Japan of war materiel; the Netherlands soon followed suit. Over the next few months Britain took steps to strengthen her defenses in the Far East. Measures taken included laying additional minefields off Hong Kong and Singapore, and making plans for Dutch minelayers in the Netherlands East Indies to assist in closing the southern approaches to Singapore on the outbreak of war.[22]

The Admiralty also transferred HMS *Teviot Bank* from Home Waters, to augment the small minelaying force at Singapore and to strengthen a minefield recently laid off the east coast of Johor. The mining of these waters had resulted from concerns about the possibility of enemy landings on Johor, at the southern end of the Malay Peninsula, across the Johor Strait from Singapore Island. HMS *Stronghold* visited Johor in February and July, laying 456 mines. *Teviot Bank* made one more lay in December 1941.[23]

JAPANESE ATTACKS ON HONG KONG/SINGAPORE

On 8 January 1941, the Japanese 25th Army under the command of Lt. Gen. Tomoyuki Yamashita launched an amphibious assault on the northern coast of British Malaya. Japanese troops landed at Kota Bharu and moved down the eastern coastline of the Malay Peninsula. This assault was made in conjunction with landings at Pattani and Songkhla in Thailand, for troop movement south across the Thailand-Malayan border to attack the western portion of Malaya. Shortly after Japanese forces landed on Kota Bharu, seventeen G3M Nell bombers of the Imperial Japanese Navy (flying from Thu Dau Mot in southern Indochina) began bombing strategic sites in Singapore.[24]

Yamashita, the 'Tiger of Malaya,' would march his troops for nine weeks through supposedly impenetrable jungle to capture Singapore, Britain's 'Gibralter of the East' on 15 February 1942. Yamashita later assumed command of the defense of the Philippines in the fall of 1944. After the war, he was hung for war crimes on 23 February 1946, for permitting his troops to commit brutal atrocities and other high crimes between 2 October 1944, and 2 September 1945.[25]

As Japanese forces attacked Pearl Harbor, Hong Kong, Thailand, and Malaya (the Malay Peninsula and island of Singapore) almost simultaneously, the Allies increased mine defenses at Singapore. On 7 and 8 December 1941, HMS *Teviot Bank* and HNLMS *Willem van der Zaan* began at once laying mines around the Singapore Strait. The Durian Strait was sown with 1,158 deep and shallow mines, with a narrow channel left down the center for shipping traffic; about 370 shallow mines were laid at either end of the Rhio Strait to close it; and the Koendoer Passage was fouled by a single line of 27 shallow mines. *Teviot Bank* also laid a line of 253 shallow mines east of Johor and, assisted by *Kung Wo*, put down an anti-submarine field (706 mines) in the eastern approaches to the Strait. At Hong Kong, S-class destroyers laid small fields at Port Shelter and in the Shui Mun Pass.[26]

"BLACK CHRISTMAS" THE FALL OF HONG KONG

We were greatly concerned to hear of the landings on Hong Kong Island which have been affected by the Japanese. We cannot judge from here the conditions which rendered these landings possible or prevented effective counter attacks upon the intruders. There must however be no thought of surrender. Every part of the island must be fought and the enemy resisted with the utmost stubbornness.... The eyes of the world are upon you. We expect you to resist to the end. The honour of the Empire is in your hands.

—Excerpt of a Most Secret message dated 22 December 1941, sent from the First Lord of the Admiralty, urging the British Governor of Hong Kong, Sir Mark Aitchison Young, to fight on, following the Japanese invasion of Hong Kong.[27]

On 8 December 1941, the Imperial Japanese Army's 38th Infantry Division (well-trained, battle-hardened troops) attacked Hong Kong from mainland China, supported by air strikes on the colony. As a precaution against just such a southward invasion across the New Territories, the British had built a fortified line inland from the city of Kowloon—a series of defensive positions linked together by paths. The so-called "Gin Drinkers Line" (which took its name from Gin Drinkers Bay, the southern anchor of the fortified line) was intended to hold off an attack until the Royal Navy arrived with relief. However, the line was never sufficiently garrisoned. There were only three British Army battalions (the 2nd Battalion, Royal Scots in the west, the 2/14th Battalion, Punjab Regiment in the center and the 5/7th Battalion, Rajput Regiment in the east) manning the line the night of 9 December, when it was easily penetrated by the Japanese.

Approximate Strength of Hong Kong Defense Forces, 8 December 1941

British	3,652	Hong Kong Volunteer Defense Corps	2,000
Canadian	1,982	Auxiliary Defense Units	2,112
Indian	2,254	Nursing Detachment	136
Local Colonial	2,428	Total	14,564[28]

Twelve thousand British, Canadian, and Indian forces, supported by the 2,000 members of the Hong Kong Volunteer Defense Forces, attempted to oppose the rapidly advancing enemy but were heavily outnumbered by twice that number of troops. Racing through the New Territories and Kowloon, Japanese forces would cross Victoria Harbor on 18 December.[29]

As Japanese troops moved against the New Territories, HMS *Thracian* laid mines in Kap Shuin Muni between Green Island and Tsing Yi. This action was undertaken to close a gap in the western entrance to the harbor, which had been left open to allow freedom of traffic between Macau and Hong Kong. *Thracian* came under air attack during the operation but was undamaged. On completion, she took up a patrol in the Lamma, Lantau and Lema islands area to intercept enemy shipping attempting to carry out landings. *Scout* and *Thanet*, the other two destroyers assigned to the Local Flotilla, were sent to Singapore via Manila for escort duties after Japan's initiation of war.[30]

Thracian continued her patrol on the 10th amidst air attacks, took part in attacks on craft attempting to land Japanese troops on Lamma Island, and assisted in the evacuation of personnel from Stonecutters Island. Over the next five days, when not engaged in patrol activities, she entered HM Dockyard at Victoria, for conversion to a destroyer role. On 12 December, she was dispatched with two motor torpedo boats to the foot of Devil's Peak to evacuate rearguard troops from Kowloon. During this operation, one MTB was hit and taken in tow. After landing the Army personnel at Aberdeen, *Thracian* returned to the dockyard to continue conversion, during which Hong Kong was under air attack and fire from enemy land units in Kowloon.[31]

On the 15th, *Thracian*, now a fully functional destroyer, stood out of harbor to resume her patrol. She was dispatched after nightfall to carry out an attack on enemy ships embarking troops in Kowloon Bay for an assault landing on Hong Kong Island. During this operation, *Thracian* grounded during high speed passage near Lamma Island and ruptured a fuel tank. The ship's company were able to control serious flooding, but the destroyer developed a list. Nevertheless, *Thracian* engaged enemy shipping in Kowloon Bay the next day and sank several small vessels.[32]

Thracian arrived at Aberdeen on 16 December for repairs. Inspection revealed major structural damage requiring dry docking but, owing to bomb damage during persistent attacks, the dock gates could not be opened, making the completion of necessary work impossible. With one fuel tank out of commission, sufficient fuel could not be carried to make passage to Hong Kong. Thus, Comdr. Arthur Luard Pears, RN (Retired), her commanding officer, decided to offload all stores and portable equipment ashore, and scuttle (purposefully sink) his ship to avoid its capture by the Japanese. *Thracian* sailed from Aberdeen the following day with only a skeleton crew aboard and deliberately ran aground between Repulse Bay and Deepwater Bay. She was subsequently salvaged by the Japanese and taken to Japan.[33]

Photo 5-7

View of Okosuka, Japan, after the end of World War II, showing *Thracian* (ex-British destroyer) at left. She had been captured at Hong Kong, BCC, in December, 1941. Naval History and Heritage Command photograph #NH 83526

HMS *REDSTART* SCUTTLED IN DEEPWATER BAY

Scuttling of the more diminutive *Redstart* followed. The loop minelayer (under the command of retired Comdr. Henry Charles Sylvester Collingwood Selby, RN, since October 1940) had arrived at Singapore from Hong Kong on 17 January 1941. She spent the next several months laying and relaying mines and, at completion, had left for Hong Kong, arriving there on 12 April. During the second, and into the third week in December, despite the recurring Japanese air attacks on Hong Kong, *Redstart* managed to survive. This was likely due in part, to her crew having painted the ship brown and green on 12 December for camouflage purposes. The minelayer spent her last few days operating between Aberdeen harbor and Deepwater Bay.[34]

Redstart was in Deepwater Bay, close in to shore, on 19 December when, at about 1030, Selby received orders to scuttle the ship. The sea cocks were opened and *Redstart* sank as her crew abandoned and proceeded to Aberdeen harbor for further orders. Selby and a contingent of sailors went up into the hills to join the fight on land. Two days later, he was wounded and taken prisoner. Selby spent the rest of the war in Hong Kong prison camps—first North Point, then Argyle Street and Sham Shui Po—before being repatriated in late 1945.[35]

FINAL DAYS

On 18 December, as thick fog and smoke shrouded the harbor and the north side of Hong Kong Island, Lt. Gen. Takashi Sakai ordered Maj. Gen. Ito Takeo to prepare his 38th Infantry Division to at once invade Hong Kong from their positions in the captured New Territories. By 2200, the first wave of over 3,500 soldiers set off across the Lye Mun Channel. Landing on the northeast corner of the island, they seized Taikoo Dockyard and captured the minesweepers HMS *Taitam* and *Waglan* under construction in the yard. At midnight, the second wave of 4,000 soldiers began crossing the channel, and by early morning, a Japanese foothold on the island was secure. By that evening, IJA troops held all key points on the island.[36]

As the Japanese closed in, the river gunboat HMS *Tern*, boom defense vessel HMS *Barlight* and three other ships were also scuttled to avoid capture. A mixed naval force of about two hundred, including men from *Thracian*, were sent to assist Canadian Army units by taking over defense of the Central Ordnance Munitions Depot (known as the "Little Hong Kong ordnance base area"). On 25 December 1941, "Black Sunday," the governor of Hong Kong, Sir Mark Aitchison Young, surrendered to the Japanese.[37]

HMS *THRACIAN* CASUALTIES

Nineteen *Thracian* crewmen died aboard ship, or ashore after scuttling the minelayer. Another twenty-two perished while prisoners of war, half when the 444-foot *Lisbon Maru* was torpedoed by the submarine USS *Grouper* (SS-214) on the night of 30 September 1942. The unmarked POW ship was carrying 800 Japanese troops, heading home after having taken part in the capture of Hong Kong, and 1,816 British prisoners-of-war being transported to slave labor camps in Japan. The POWs included members of the Royal Navy, Royal Artillery, 2nd Battalion Royal Scots and 1st Battalion Middlesex Regiment. During the twenty-five hours it took the *Lisbon Maru* to sink, the Japanese made no attempt to rescue their prisoners. Instead, they locked them in the ship's holds to drown and machine-gunned those able to get off the vessel.[38]

At some point, some soldiers from either the Royal Scots or the Royal Artillery were apparently able to reach the bridge, kill the Japanese soldiers still aboard, and release the remaining prisoners from the holds. As British servicemen slipped over the side into the water, many were taken under rifle and machine-gun fire from surrounding Japanese vessels (a destroyer and three or four landing craft). Some islands about four miles distant appeared to offer the only salvation. After several hours of struggling against the chop, the stronger swimmers started to reach the islands. Once contacted, local Chinese fishermen launched as many junks and sampans as were available, and quickly retrieved 977 British prisoners from the China Sea. A summary of *Redstart* and *Thracian* casualties follows.[39]

HMS *Redstart*

Date	Name	Rating	How Lost
24 Dec 41	William Young	Stoker Petty Officer	Killed
25 Dec 41	John Cremen	Act/Leading Stoker	Killed

HMS *Thracian*

DOW: Died of Wounds   MPK: Missing Presumed Killed

16 December 1941

Name	Rank/Rating	How Lost
Wilfred Trehake	Steward PO	DOW after bombing

19 December 1941

Eric Godfrey Burton	Stoker 2c	Killed, bombing
Frederick Butlin	Ordinary Seaman	Killed, bombing
John Daniel Foster	Ordinary Seaman	Killed, bombing
James Hyland	Stoker 1c	Killed, bombing
John Oliver Jenkinson	Ordinary Seaman	Killed, bombing
Henry J. W. Parsons	Stoker Petty Officer	Killed, bombing

Clifford Reynolds	Able Seaman	Killed, bombing
Henry Walter Slark	Able Seaman	Killed, bombing

25 December 1941

Herbert Thomas Deakin	Stoker Petty Officer	MPK, bombing
George Egan	Able Seaman	Killed
Henry Dollar Greig	Ordinary Seaman	MPK, bombing
George Harrison	Ordinary Seaman	MPK, bombing
Robert Henderson	Ordinary Seaman	MPK, bombing
Robert John Keith	Ordinary Seaman	MPK, bombing
James Lowden Kendall	Stoker 1c	MPK, bombing
John Kingham	Ordinary Seaman	MPK, bombing
Harvey Loveday Porrett	Able Seaman	MPK, bombing
Frederick G. Thomas	Stoker 1c	MPK, bombing

9 April 1942

Leslie John Gardiner	Chief Stoker	Died as a POW

27 September 1942

Harold Cullum	Petty Officer	Died as a POW

2 October 1942

Albert W. J. Archer	Stoker 1c	Died as a POW (*Lisbon Maru*)
Arthur Edward Birch	Leading Seaman	Died as a POW (*Lisbon Maru*)
David Kenneth Bliss	Able Seaman	Died as a POW (*Lisbon Maru*)
Francis Cassin	Able Seaman	Died as a POW (*Lisbon Maru*)
John Raphael Crangle	Able Seaman	Died as a POW (*Lisbon Maru*)
Charles Edward Fage	Able Seaman	Died as a POW (Camp Amagasaki, Osaka)
Richard Howell Finch	Leading Seaman	Died as a POW (*Lisbon Maru*)
George Hill	Able Seaman	Died as a POW (*Lisbon Maru*)
Alexander Lees	ERA 3c	Died as a POW (*Lisbon Maru*)
Bartholomew O'Sullivan	Stoker Petty Officer	Died as a POW (*Lisbon Maru*)
Thomas John Stone	Leading Stoker	Died as a POW (*Lisbon Maru*)
John Campbell Wilson	Lead Telegraphist	Died as a POW (*Lisbon Maru*)

12 October 1942

Montague Pearson	Joiner 5c	Died as a POW

17 October 1942

Melville Maurice Morgan	Stoker 1c	Died as a POW

20 October 1942

John Duffy	Petty Officer	Died as a POW

27 October 1942

Ralph Schofield	Able Seaman	Died as a POW

30 October 1942

George Leonard Adams	Petty Officer	Died as a POW

10 March 1943

Henry Davies	Able Seaman	Died as a POW

2 June 1943

Robert Edward Halfyard	Leading Seaman	Died as a POW

14 February 1945

Ernest Burgin	Able Seaman	Died as a POW[40]

FALL OF SINGAPORE

On 10 December 1941, the mighty battleship HMS *Prince of Wales* and her consort the battlecruiser HMS *Repulse* were sunk by Japanese aircraft when they attempted to intercept landings in British Malaya. Following the evacuation of Penang on 16 December, the destroyer HMS *Express* laid eighteen A Mark I/IV ground mines in the approaches to that port on the 24th. Once again, seaward defenses mattered little, as Japanese troops advanced overland. British defense forces retired to the island of Singapore on 31 January. The enemy landed there on 8 February, and a week later, Singapore capitulated on 15 February.[41]

EVACUATION OF MILITARY/CIVILIAN PERSONNEL

On 11 February, Rear Adm. Ernest John Spooner, RN, the senior naval officer remaining at Singapore, decided to abandon the colony and evacuate all naval personnel employing Eastern Fleet and local defense units. The cruiser HMS *Durban* (D99), auxiliary patrol ship HMS *Kedah* (FY035), and destroyers HMS *Jupiter* (F85), and HMS *Stronghold* (H50) arrived at Singapore at 0300 on 12 February from Batavia. They sailed before daylight the same morning escorting the refrigerated cargo liner MV *Empire Star* and the cargo ship MV *Gorgon*, carrying the greater part of shore-based naval personnel. These ships, and other tankers, merchant ships, and auxiliaries (collectively carrying thousands of servicemen and civilian evacuees) proceeding southward from Singapore, were heavily attacked by aircraft throughout the day. The convoy reached Tandjong Priok, the port for Batavia (now Jakarta, Indonesia, then the capital of the Dutch Indies on the island of Java), on 13 February.[42]

One day earlier, the destroyer HMS *Scout* and auxiliary patrol ships *Darvel* and *Kinta* had arrived at Batavia with RAF personnel evacuated from Singapore. *Stronghold* was later sunk on 2 March after being in action with the Japanese cruisers *Maya*, *Arashi*, and *Nowaki*, accompanied by two destroyers. She had been escorting the Dutch SS *Zaandam* with refugees from Tjilatjap bound for Fremantle, when she had become detached from the faster merchant ship. The British survivors (reportedly 50) were rescued by the captured Dutch SS *Sigli* and later transferred to the Japanese cruiser *Maya*.[43]

Japanese naval and air forces operating in the seas around Malaya destroyed many of the ships that left Singapore in the final days before the surrender on 15 February. Rear Admiral Spooner, his chief staff officer, the captain of the dockyard, and nearly all remaining naval personnel left Singapore in various auxiliary craft and yachts on 13 February. As enemy air attacks on shipping fleeing Singapore continued, the minelayer *Kung Wo*, among other vessels, was sunk on 14 February. Parties of evacuees and survivors accumulating on various islands and in Sumatra, endeavored to proceed overland to the West Coast. Some, including Spooner, died ashore of thirst, starvation, malaria, exhaustion or other causes.[44]

HMS *Circle* and HMS *Medusa*, which had been requisitioned by the Royal Navy in 1939 and converted to auxiliary minesweepers, escaped to Australia just before the fall of Singapore. They became HMAS *Medea* and HMAS *Mercedes*, respectively.[45]

Of the seven Royal Navy mine warfare ships identified below, only HMS *Sin Aik Lee* escaped to safety from Singapore; and it was only temporary. She arrived at Batavia on 13 February but was later lost on 1 March 1942 when she encountered two Japanese destroyers in Soendra Strait while trying to make Tjilatjap in company with the auxiliary minesweeper HMS *Rahman* and the motor launch HMS *ML-1063*. The other two vessels were sunk as well.[46]

Date	Ship	Minelayer Disp. Tons/ Year Built	Disposition
14 Feb 42	*Kung Wo*	4,636/1921	Sunk by aircraft bombs near Lingga Archipelago, Singapore area
Minesweepers			
14 Feb 42	HMS *Changteh* (Lt. (E) Holm, RNR)	244	Left Singapore 13 February with 40 RAF personnel, bombed and sunk en route to Indragiri
14 Feb 42	HMS *Li Wo* (T/Lt. Thomas Wilkinson RNR)	707/1938	Sunk by Japanese surface craft after leaving Singapore 13 February
16 Feb 42	HMS *Fuh Wo* (T/Lt. B. Shaw, RNVR)	954/1922	Beached and scuttled by crew on Banka Island
17 Feb 42	HMS *Jerak* (Lt. H. C. Butcher, SSRNVR)	208	Left Singapore PM 13 February, to mark the entrance to Durian Strait, sunk by aircraft bombs in Singapore area on the 17th
1 Mar 42	HMS *Sin Aik Lee* (Lt. J. M. Brander, SSRNVR)	198/1928	Arrived at Batavia from Singapore on 13 Feb 42, sunk on 1 March
February-March 42	HMS *Tapah* (Comdr. G. E. W. W. Bayly, SSRNVR)	208/1926	Left Singapore PM 12 February, to mark north end of the Durian Strait swept channel, believed lost by enemy action[47]

LI WO ENGAGES JAPANESE CONVOY IN BATTLE

Lieutenant Wilkinson's valour was equaled only by the skill with which he fought his ship. The VICTORIA CROSS is bestowed upon him posthumously in recognition both of his own heroism and self-sacrifice, and of that of all who fought and died with him.

—Third Supplement to *The London Gazette* of Friday, the 13th of December, 1946

On 14 February, HMS *Li Wo* (formerly a passenger steamer on the upper Yangtze River) was on passage to Batavia. Aboard her were eighty-four officers and men, including one civilian, mainly survivors of His Majesty's Ships which had been sunk, and a few from units of the Army and the Royal Air Force. Her armament was one 4-inch gun, and two machine guns. After leaving Singapore the previous day, she had beaten off four air attacks, one in which fifty-two planes took part, and had sustained considerable damage. Late in the afternoon, two enemy convoys came into view, the larger one under escort by Japanese naval units, including a heavy cruiser and some destroyers.[48]

Lt. Thomas Wilkinson, RNR, gathered his ad hoc crew together and told them that rather than try to escape, he intended to engage the convoy and to fight to the last, in the hope of inflicting some damage upon the enemy. In making this decision, which drew resolute support from all those assembled, Wilkinson knew that his ship faced certain destruction, and that his own chances of survival were small.[49]

With battle ensign hoisted, HMS *Li Wo* engaged the enemy, using her machine guns with effect against the crews of ships within range. A volunteer crew manning the 4-inch gun, hit and set on a fire a transport. After a little over an hour, with his ship critically damaged and sinking, Wilkinson decided to ram the transport, which had been abandoned by her crew. *Li Wo*'s gallant fight ended when, her shells spent, and under heavy fire from the enemy cruiser, Wilkinson ordered abandon ship. He himself remained on board, and went down with her. There were but some ten survivors, who were later made prisoners of war.[50]

Wilkinson, who was 43 years old, and a temporary lieutenant in the Royal Naval Reserve, was awarded the Victoria Cross posthumously. Other members of ship's company also received awards for valor. Sub-lieutenant Ronald G. G. Stanton was the *Li Wo*'s only surviving officer.

Victoria Cross

T/Lt. Thomas Wilkinson, RNR   Commanding Officer

Distinguished Service Order

T/Sub Lt. Ronald George Gladstone Stanton, RNR   First Lieutenant

Conspicuous Gallantry Medal

A/Petty Officer Arthur William Thompson   Gun layer for 4-inch gun

Distinguished Service Medal

Leading Seaman Victor Spencer   Port machine gunner
Able Seaman Albert Spendlove[51]   Member of 4-inch gun crew

FALL OF SINGAPORE FACILITATES LOSS OF THE NETHERLANDS EAST INDIES

The forces at the disposal of the Allies in the ABDA area (outside Singapore) were never likely to be able to stop the Japanese advance, as Japan could easily allocate stronger naval and air forces without incurring undue risks elsewhere. In the circumstances, the only possible course for the Allies was to hold on with what forces they would muster in order to deny the N.E.I. to the enemy as long as possible and inflict on him as many casualties as possible. It was inevitable in so doing that we should expose ourselves to having our weak forces in the area overwhelmed but this result was on the whole preferable to abandoning it without firing a shot.

Singapore was however, the key to the situation, and once it fell, it was certain that the N.E.I. would be speedily over run. Singapore was the only place in the area which was a self-contained fortress, and might be expected to hold out in isolation.

—Vice Adm. Sir Geoffrey Layton, RN, diary entries.[52]

6

Fall of the Netherlands East Indies

The manner of the Japanese advance resembled the insidious yet irresistible clutching of multiple tentacles. Like some vast octopus it relied on strangling many small points rather than concentrating on a vital organ. No one arm attempted to meet the entire strength of the Abda fleet. Each fastened on a small portion of the enemy and, by crippling him locally, finished by killing the entire animal...

The Japanese spread their tentacles cautiously, never extending beyond the range of land-based aircraft unless they had carrier support. The distance of each advance was determined by the radius of fighter planes from airfields under their control. This range was generally less than 400 miles, but the Japanese made these short hops in surprisingly rapid succession. Amphibious operations, preceded by air strikes and covered by air power, developed with terrifying regularity. Before the Allies had consolidated a new position, they were confronted with a system of air bases from which enemy aircraft operated on their front, flanks and even rear.

—Naval historian Samuel Eliot Morison artfully describing the strategy that Japanese military forces employed in their conquest of the Netherlands East Indies.[1]

The main objective of the ABDA (American-British-Dutch-Australian) command established at Soerabaja, Java, on 15 January 1942, was to maintain control of the "Malay Barrier." This term referred to a notional line running down the Malayan Peninsula, through Singapore and the southernmost islands of the Netherlands East Indies. Their task was to stop the advance of Japanese forces, which in late December had begun establishing themselves in the southern Philippines—at Davao on Mindanao, and to the southwest at Jolo, an island in the Sulu Archipelago within the Mindanao group—from which points they could invade the Netherlands East Indies. Davao and Jolo provided the enemy airfields from which to launch attacks and set up parallel drives southward, one down through the Molucca Sea and another down the Makassar Strait.[2]

Map 6-1

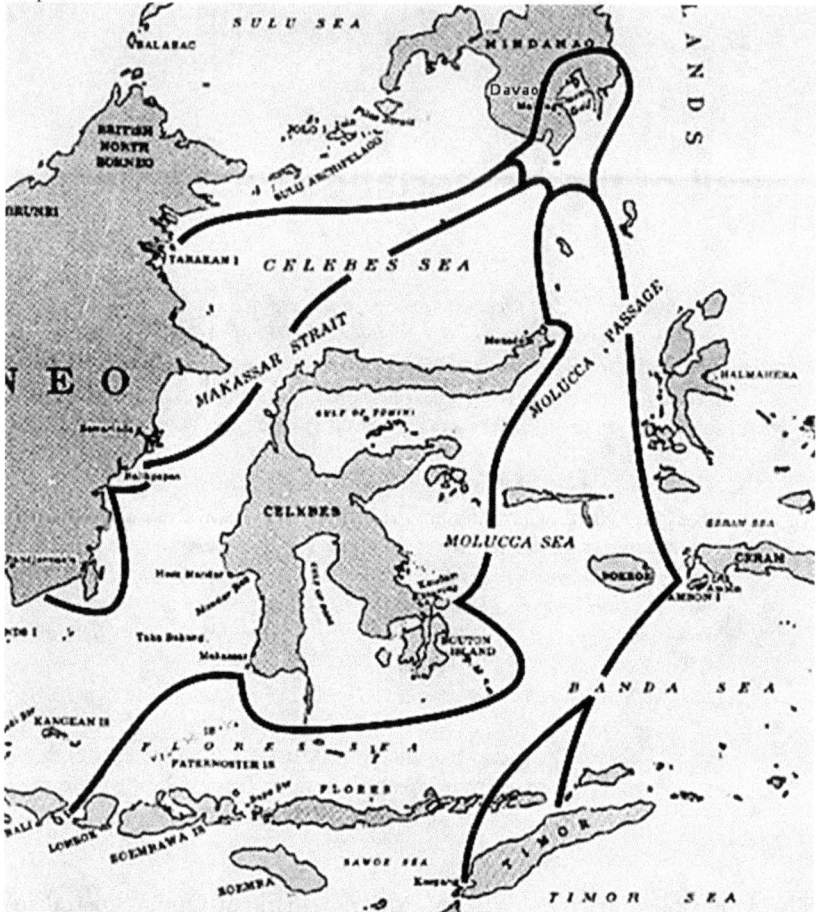

The movements of Japanese forces from Davao, Mindanao, southward to Java in the Netherlands East Indies resembled the tentacles of an octopus.

DEFEAT IN THE BATTLE OF THE JAVA SEA

The relatively short, but hard fought Allied defense of the Netherlands East Indies came to a bitter end with the defeat of the ABDA Striking Force by Japanese naval forces in the Battle of the Java Sea fought on 27 February 1942 and the ensuing Battle of Sunda Strait. The latter series of skirmishes lasted through 1 March. In late morning that day, Vice Adm. Conrad E. L. Helfrich, RNLN, formally announced that the Governor General of the Netherlands East Indies had dissolved the ABDA naval command. Vice Adm. William A. Glassford, commander, U.S. Naval Forces, Southwest Pacific, had earlier ordered all American ships to Exmouth Gulf, Australia, and the British commander, Rear Adm. Arthur Palliser, ordered the Royal Navy units there as well.[3]

In anticipation of the likelihood of evacuating Java, Rear Adm. Pieter Koenraad, the commander of the Soerabaja Naval Base, had directed the 2nd Minesweeper Division on 17 February to be ready to leave for Australia upon receipt of the coded signal KPX. Yet, as personnel began demolishing the base, there was still no guidance to escape. Orders finally came on 6 March. Surviving ships were to withdraw to Australia, or Colombo, Ceylon, and vessels unable to make the voyage were to be destroyed, to prevent their capture by the enemy.[4]

However, since Japanese forces effectively controlled the sea and the air around Java, escape seemed so improbable that many perceived any attempt to do so would be suicidal. Lt. Comdr. J. R. L. Lebeau (commander, 2nd Minesweeper Division) called a meeting of the commanding and executive officers of the four ships of his division, and told them they could make their own decision about trying to escape. He also gave commanding officers the option to destroy their ships and not try to escape—contradictory to the order KPX.[5]

Photo 6-1

Dutch minesweeper HNLMS *Abraham Crijnssen*, date and location unknown. Courtesy of www.uboat.net

Only three of the four minesweepers left the harbor on 6 March. Lt. Comdr. J. P. A. Dekker sank HNLMS *Pieter de Bitter* with explosive charges alongside the cruiser quay at Soerabaja, an action for which he was court martialed after the war. HNLMS *Jan van Amstel* (Lt. C. de Greeuw) and HNLMS *Eland Dubois* (Lt. H. de Jong) stood out of port and sailed for the Gili Islands. HNLMS *Abraham Crijnssen* departed thereafter, sailing independently.[6]

During the preceding days, Lt. Comdr. Anthonie van Miert, *Abraham Crijnssen*'s commanding officer, and his crew had been busy making preparations to escape Java. The ship's exterior was painted in a disruptive pattern, and covered with nets and tree branches to create the appearance of a tropical island. With the assistance of his executive officer (Lt. A. D. H. Heringa), van Miert had gone around to the other ships in the division as well as to the minelayer *Gouden Leeuw* seeking volunteers to join the *Crijnssen*. He then held an "All Hands" meeting, at which he announced his intention to attempt an escape, and to permit anyone who did not want to remain aboard to leave. About half of his crew departed the ship, including all of the Indonesian sailors.[7]

Abraham Crijnssen sailed at 2130 on 6 March with navigation lights extinguished, portholes covered, and yacht *Urania 2* in tow. She also was bound for the Gili Islands, which lay to the south and southeast of Madoera Island off the northeastern coast of Java. The minesweeper proceeded along the south coast of Madoera and while in passage through the Madoera Strait, cut loose the yacht in early morning darkness on 7 March. Later that morning, van Miert found the other two minesweepers lying at anchor off Gili Radja. He took aboard some fuel from the *Eland Dubois*, and then continued on to an anchorage at Gili Genteng, where *Crijnssen* was further camouflaged. This decision proved to be prudent, as *Dubois* and *Amstel*, completely uncamouflaged, were sighted that day by a Japanese aircraft.[8]

Map 6-2

Route taken by HNLMS *Abraham Crijnssen* to reach the Indian Ocean from Soerabaja, Java, before proceeding southeast to the northwest coast of Australia.

Since *Dubois* had only a partial crew, was experiencing boiler problems, and had insufficient fuel to make it to Australia, the decision was made on 8 March to scuttle her and transfer those aboard to *Amstel*. *Amstel* was camouflaged with foliage from shore, and set off that evening in an easterly direction. She was discovered in the Madoera Strait at 2330 by the Japanese destroyer *Arashio* and sunk by gunfire, with 21 of the more than 80 men killed, including the captain of *Eland Dubois*. *Amstel* survivors were later picked up by another Japanese destroyer.[9]

Crijnssen had likely avoided the *Arashio* by leaving the Gili Islands area a day earlier. Following enhancement of his ship's "island appearance," van Miert had departed Gili Genteng on the evening of 7 March. Proceeding eastward, he sailed south of Sapoedi Island, then turned northward between Goa-Goa and Karang Takat reef, before changing to a northeastern course en route to the Kangean Islands. In early morning on the 8th, the minesweeper arrived off the north shore of Kangean Island and anchored there. Once again, the branches and leaves were replaced as much as possible, with the crew paying extra attention to the local plant species. Thus began, a schedule of remaining at anchor under camouflage by day and sailing by night.[10]

Crijnssen departed at 1845 that evening, and sailing southeast between Pageroean and Sekala, began the dangerous passage of the Bali Sea in the direction of Soembawa Island. Between 2300 and 2330, an unidentified silhouette was spotted and course altered. The following morning, the minesweeper arrived at Poto-Paddu Bay, on the north shore of Soembawa. While crewmembers went to work refreshing the camouflage, a party went ashore in the motor dingy to collect drinking water and information from the inhabitants. From delegates of the local sultan and a representative of a Dutch interisland commercial shipping company, they learned that there were no Japanese troops on the island, and that it had been four days since enemy planes had been sighted. After stocking up on water and coconuts, the ship left after sundown.[11]

HNLMS *Abraham Crijnssen* arrived at the entrance to the Alas Strait at 2215 on 9 March, transited it at 13.5 knots, and immediately reduced speed to 10 knots to conserve fuel. Sloughing alone across the open Indian Ocean was unsettling, but finally (at 0800 on the 13th), the Northwest Cape of Australia came into view. Although critically short on fuel, the minesweeper continued southward along the coast and reached Geraldton at noon on 15 March. Despite minimal armament, a maximum speed of 15 knots, and bunkerage for only 110 tons of fuel, she had completed her seemingly implausible, lengthy, solo journey.[12]

CREWMEMBERS AWARDED CROSS OF MERIT

On 20 February 1941, the Dutch government in exile in London had instituted several new awards for bravery. Amongst the new decorations was the "Cross of Merit" (Kruis van Verdienste) an award for "working in the interest of the Netherlands while faced with enemy actions and distinguishing oneself through valor and resolute behavior."

For his courage and ingenuity, Lt. Comdr. Anthonie van Miert, RNLN, received the Cross of Merit in September 1942. Nine other crewmen of the *Abraham Crijnssen* received the same honor in November 1943.[13]

ESCAPE OF THREE AUXILIARY MINESWEEPERS

While van Miert had earlier been waiting for orders to leave Soerabaja, three 80-ton auxiliary minesweepers of the 4th Minesweeper Division had sailed on the afternoon of 3 March. Two days earlier, Rear Admiral Koenraad had ordered the division commander, Lt. Comdr. J. J. C. Korthals Altes, to evacuate his ships to a safe port, or scuttle them and send their crews to Tjilatjap for evacuation. Altes decided the division would take its chances at sea. When "topped off," the diminutive ships carried about 800 gallons of fuel, with a range of about 1,180 miles.[14]

When the ships' crews learned of the plan to attempt a long voyage through enemy-controlled waters, many men deserted. Although the preface HNLMS (His/Her Netherlands Majesty's Ship) signifies that *Merbaboe*, *Rindjani* and *Smeroe* were commissioned ships, they were in fact, single-engine craft capable of only 10 knots, and fitted with but two machineguns for self-protection. Altes gave the remaining sailors the choice of leaving Java, or remaining and taking their chances ashore. The commanding officers of five of the six auxiliary minesweepers chose to stay. The remaining ship's officers were assigned to *Merbaboe*, *Rindjani* and *Smeroe*. Lt. Johannes Pieter Rotgans, RNLN, formerly the commanding officer of HNLMS *A* (a 140-foot minesweeper scuttled on 1 March 1942, after suffering damage a day earlier during a Japanese air raid on Soerabaja) was assigned command of the group.[15]

Minesweeper	Commanding Officer
HNLMS *Merbaboe* (HMV10)	Luitenant ter zee der 3e klasse P. A. H. Roozen
HNLMS *Rindjani* (HMV11)	Luitenant ter zee der 3e klasse Henri Jedeloo
	Luitenant ter zee der 3e klasse L. Kuiper
HNLMS *Smeroe* (HMV12)	Luitenant ter zee der 3e klasse Masselink[16]

[Authors' note: The authors are unsure whether the commanding officers cited in the table were aboard the auxiliary minesweepers during the voyage to Australia. The rank of these officers was equivalent to that of an ensign in the USN, and a midshipman in the RN or RAN. Equivalent ranks have been used elsewhere in the book.]

The relatively new 74-foot vessels had been constructed in 1941 as government dispatch boats to transport people and small amounts of cargo, and later converted to minesweepers. The other units of the division—*Ardjoeno*, *Kawi*, and *Salak*—were slightly older (1940-1941) former dispatch boats. They were stripped of fuel, supplies, and water, and scuttled. *Merbaboe*, *Rindjani* and *Smeroe* then each had approximately 1,200 gallons of fuel and 925 gallons of water, with food for a month. The extra fuel in drums on deck supported a voyage of 1,300-2,000 miles, depending on speed.[17]

VOYAGE TO AUSTRALIA

Map 6-3

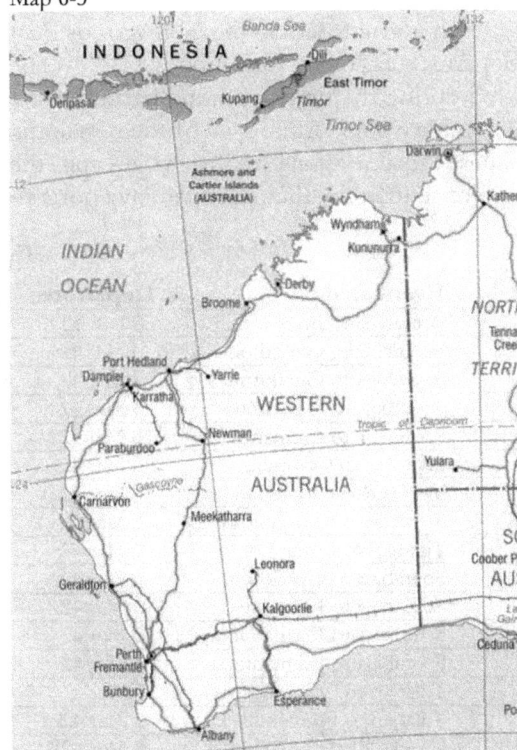

Western Australia

Owing to reported Japanese activity in the Madoera Strait, the group went out the Westwater Channel leaving Soerabaja the afternoon of 3 March and began passage along the north coast of Madoera. At daybreak, the minesweepers found a place to hide in close to shore off the northeast coast of Madoera, until nightfall. Sailing at 1800 that evening, 4 March, the group continued eastward, passed north of the Kangean Islands, and then set a course east-southeast bound for the Paternoster Islands. Also known as the Balabalagans, this island chain lay north of Soembawa.[18]

On 6 March, the group passed through the Paternoster Islands and through the Alas Strait and into the Indian Ocean. In early evening the following day, the minesweepers set a course for Broome, Australia. Arrival at their destination at 2030 on 10 March brought to an end a remarkable week-long journey by the remaining members of the 4th Minesweeping Division. The intrepid little ships subsequently moved down the coast to Fremantle.[19]

AFTERMATH

Only a few remnants of the Dutch Naval forces in the Netherlands East Indies escaped to Fremantle, Australia, or to Colombo, Ceylon. The bulk of the fleet was destroyed during the first or second battles of the Java Sea, or the Battle of the Sunda Strait. Almost all of the remaining forces were sunk by Japanese forces while trying to escape the Netherlands East Indies, or were scuttled by their crews in Java ports to prevent capture by the enemy.

Ship/Submarine	Evacuated to	Departure
cruiser *Tromp*	Soerabaja to Fremantle	23 Feb 42
submarine *K-8*	Soerabaja to Geraldton	3 Mar 42
submarine *K-9*	Soerabaja to Geraldton	2 Mar 42
submarine *K-11*	Soerabaja to Colombo	3 Mar 42
submarine *K-12*	Soerabaja to Fremantle	6 Mar 42
submarine *K-14*	Java Zee to Colombo	3 Mar 42
submarine *K-15*	Veeckensbay to Colombo	6 Mar 42
submarine *O-19*	Straat Sape to Colombo	7 Mar 42
minelayer *Willem van der Zaan*	Tjilatjap to Columbo	1 Mar 42
minesweeper *Abraham Crijnssen*	Soerabaja to Geraldton	6 Mar 42
auxiliary minesweeper *Merbaboe*	Soerabaja to Fremantle	3 Mar 42
auxiliary minesweeper *Rindjani*	Soerabaja to Fremantle	3 Mar 42
auxiliary minesweeper *Smeroe*	Soerabaja to Fremantle	3 Mar 42
supply ship SS *Zuiderkruis*	Tjilatjap to Columbo	27 Feb 42
supply/depot ship MV *Janssens*	Tjilatjap to Fremantle	3 Mar 42
tanker MV *Petronella*	Veeckensbay to Colombo	8 Mar 42[20]

7

Attack on Pearl Harbor

FROM: SECNAV *[Secretary of the Navy]*
TO: ALNAV *[All Navy]*

WHILE YOU HAVE SUFFERED FROM A TREACHEROUS ATTACK YOUR COMMANDER IN CHIEF HAS INFORMED ME THAT YOUR COURAGE AND STAMINA REMAIN MAGNIFICANT. YOU KNOW YOU WILL HAVE YOUR REVENGE. RECRUITING STATIONS ARE JAMMED WITH MEN EAGER TO JOIN YOU.

THE SECRETARY OF THE NAVY

Waves of fighters and bombers launched from all six of Japan's first-rate aircraft carriers—*Akagi, Hiryu, Kaga, Soryu, Shokaku* and *Zuikaku*—struck Pearl Harbor on 7 December 1941. With over 420 embarked aircraft, these ships constituted, by far, the most powerful carrier task force ever assembled. Their main targets were battleships and aircraft carriers; fortunately, the Pacific Fleet's carriers were at sea. Secondary targets were airfields. Of first priority were Hickam Field, Wheeler Field, and Ford Island at Pearl Harbor, followed by elsewhere on Oahu, Kanoehe Naval Air Station, Bellows Field, and Ewa Marine Corps Station. The carriers were a part of Vice Adm. Chuichi Nagumo's Striking Force, positioned 275 miles north of Pearl Harbor. His goal was to wipe out the major part of the Pacific Fleet at Pearl Harbor and destroy all the military aircraft on Oahu. The Force had sortied from Tankan Bay in the Kurile Islands on 26 November, and set a course for Pearl Harbor across a part of the North Pacific generally avoided by merchant shipping. Nagumo's orders were to abandon the mission if detected, or should diplomacy work an unanticipated miracle. However, only one ship (a Japanese freighter) was encountered during the transit which was characterized by bad weather and rough seas.[1]

Photo 7-1

Japanese naval aircraft preparing to take off from an aircraft carrier (believed to be the *Shokaku*) to attack Pearl Harbor the morning of 7 December 1941. Plane in the foreground is a Zero fighter. This is probably the launch of the second attack wave. U.S. Navy photograph #80-G-71198, now in the collections of the National Archives

There was some concern among the Japanese about the PBY Catalina patrol planes on Oahu, which had the range to find the enemy fleet and track it for hundreds of miles. The mission of the PBYs was scouting for submarines and enemy positions with an additional role of search and rescue. Unbeknownst to the Japanese, commander in chief Pacific Fleet, Adm. Husband E. Kimmel, USN, was reluctant to use the aircraft for extensive patrol, it being difficult to get parts for them. He calculated that if war with Japan broke out he would need those aircraft to patrol in advance of his fleet. As a result, very few PBYs were used for patrol, and search sectors were to the south of Oahu in the direction of the nearest Japanese possessions. None of the Catalinas were assigned to patrol northward, the direction from which the Japanese Fleet actually came. War Plan Orange (the U.S. strategy for dealing with a possible war with Japan) assigned patrolling to the Army. But on Oahu, there were only twelve B17 Flying Fortresses that were capable of long range patrols. Thus, neither the Navy nor the Army was maintaining patrols adequate to detect the approaching Striking Force.[2]

A patrolling Catalina did find a Japanese midget submarine just off the entrance to Pearl Harbor at 0700 on the morning of the attack. The

PBY dropped its depth bombs and sent a coded message to its base at 0715, but by the time this message was decoded and passed on to Admiral Kimmel, enemy bombs had already begun falling.[3]

INITIAL ENEMY CONTACT

Photo 7-2

Coastal minesweeper USS *Condor* (AMc-14) in 1941.
Bureau of Ships collection photograph 19-N-24615 in the U.S. National Archives

Contact with the submarine was first made at 0342 by the coastal minesweeper *Condor* (AMc-14), carrying out a routine sweep of a mile-wide area from the Pearl Harbor entrance out to the 100-fathom curve. As the officer of the deck and quartermaster of the watch peered through binoculars in an attempt to pierce the darkness ahead, they sighted the periscope of a submerged submarine off the entrance buoys of the harbor in a restricted defensive sea area where U.S. submarines were prohibited from operating submerged. A mere seventy-seven feet in length, *Condor*—the former fishing vessel *New Example*—was slightly shorter than the diminutive submarine (later identified as a Japanese type-*A* midget), and was not ideally suited to deal with it. She was pulling sweep gear, which severely limited her maneuverability, had no depth charges, and was armed only with a single .50-caliber machine gun. *Condor* accordingly informed the *Ward* (DD-139), patrolling nearby, of this contact by visual signal. The destroyer instituted a search and at, about 0637, sighted the periscope of a submarine apparently trailing the general stores issue ship *Antares* (AKS-3), as it approached the harbor entrance to Pearl Harbor.[4]

Ward witnessed a PBY (from squadron VP-14) circle and drop what appeared to be, two smoke pots near the object. At 0645, the destroyer commenced her attack, firing one salvo each from her No. 1 and 3 guns, followed by depth charges on the submarine. At 0651, the destroyer sent a radio message to the commandant, Fourteenth Naval District: "We have dropped depth charges upon sub operating in defensive sea area." The commanding officer, however, after reflecting that this message might not be interpreted as showing a surface submarine contact, two minutes later sent the supplementary message: "We have attacked fired upon and dropped depth charges upon submarine operating in defensive sea area." This message was received by the Bishop's Point radio station, relayed to the officer in charge, Net and Boom Defenses, Inshore Patrol, and delivered by the communications watch officer, Fourteenth Naval District, to the duty officer, who notified his chief of staff. However, because the message moved with excruciating slowness upward through the chain of command, the base commander didn't receive a report of the submarine until just shortly before Japanese carrier aircraft initiated an unprovoked attack on the U.S. Pacific Fleet inside the harbor.[5]

Sixty-one years later, on 28 August 2002, two deep diving submersibles operated by the Hawaii Undersea Research Laboratory found the midget submarine sunk by the *Ward*. The wreckage lies in 400 meters of water about five miles off the mouth of Pearl Harbor. Five of the midgets had been transported by *I* type "mother" submarines and launched near the entrance to Pearl Harbor the night before the attack. At least one of the 2-man submarines penetrated the harbor, where it was sunk by the destroyer *Monaghan* (DD-354). Another drifted around to the east coast of Oahu and was captured there the day after the attack. In 1960, the fourth was discovered on the sea floor off the harbor entrance. The final submarine remains unaccounted for to date.[6]

FOURTEENTH NAVAL DISTRICT MINESWEEPERS
Among the more than ninety ships of the Pacific Fleet present at Pearl Harbor that morning were twenty-three mine warfare vessels. The four smallest were district craft assigned to the Inshore Patrol of the 14th Naval District Local Defense Force. *Condor* (AMc-14), *Cockatoo* (AMc-

8), *Crossbill* (AMc-9), and *Reedbird* (AMc-30) were former fishing vessels. Fitted with minesweeping gear and designated coastal minesweepers, they were responsible for sweeping the approaches to Pearl Harbor.[7]

Commandant, 14th Naval District/Commander, Hawaiian Sea Frontier
Rear Adm. Claude C. Bloch, USN
14th Naval District Local Defense Force
Inshore Patrol: Capt. C. H. Morrison
XAMc Division: Charles B. Carlon

Ship	Length (ft.) Disp. (tons)	Built/Acq. by Navy	Commanding Officer
Reedbird (AMc-30)	87/223	1935	Charles B. Carlon
ex-*Fearless*		18 Nov 40	
Cockatoo (AMc-8)	88/185	1936	
ex-*Vashon*		23 Oct 40	
Condor (AMc-14)	77/195	1937	Ens. M. H. Habbell, USNR
ex-*New Example*		28 Oct 40	
Crossbill (AMc-9)	81/213	1937	Robert W. Costello[8]
ex-*North Star*		31 Oct 40	

War Plan-Rainbow No. 5—issued by the Chief of Naval Operations on 26 May 1941 in anticipation of America's probable entry into war in the European and Pacific theaters—was the catalyst for the Navy's acquisition of civilian and commercial vessels such as these. Under this plan (destined to be the basis for American strategy in World War II), there would be early deployment of United States forces to the eastern Atlantic, and to either or both the African and European continents, followed by offensive operations to defeat Germany and Italy. A strategic defensive was to be maintained in the Pacific until success against the European Axis Powers permitted a major transfer of fleet units to the Pacific for an offensive against Japan.[9]

Since the Navy had insufficient forces to provide the commanders of local defense forces with the ships and craft necessary to carry out their responsibilities, it assigned them mostly "X" vessels instead. Virtually non-existent at the outbreak of the war, these vessels (designated by the letter X preceding their normal ship classification) were to be acquired from sources other than the Navy or Coast Guard. When directed, commanders were to lease or purchase privately-owned and commercial vessels and prepare them for military service. In preparation for this possibility, naval district representatives began scouring their local waterfronts to identify candidate craft for conversion. The designation for coastal minesweepers was XAMc. On the day of the attack on Pearl Harbor, the four ex-fishing vessels acquired by the 14th Naval District for service as coastal minesweepers comprised XAMc Division, a component of the Inshore Patrol.[10]

In a related action, the U.S. Pacific Fleet was reorganized in April 1941, with some additional changes made on 31 October 1941 to provide for all phases of type, intertype, and Fleet training, concurrent with performance of certain required patrol and escort duties. Type training refers to training conducted by the same type ships, and intertype by more than one type of ship training together. Fleet training brings the components of a fleet together to train as an integral group. At the time of the Japanese attack, the mine warfare ships home ported at Pearl Harbor (other than the District craft) were assigned to either the Minecraft Battle Force or the Base Force.[11]

The assignments of the ships on 1 October 1941, and the identities of their commanding officers on that date are provided in the table below. In some cases, a change of command had taken place after that day, and a different officer commanded a particular ship on the day of the attack. Mine Squadron One was a component of the Minecraft Battle Force, and Squadrons Two and Four of the Base Force. The ships present at Pearl Harbor on 7 December 1941 are identified by asterisks. Mine Divisions 5 and 6 (of Mine Squadron Two) were absent, as were two ships of Mine Squadron Four. *Boggs, Chandler, Hovey,* and *Lamberton* were operating with the cruiser *Minneapolis* (CA-36) about twenty-five miles south of Oahu. *Dorsey, Eliot, Hopkins, Long,* and *Southard* were with the cruiser *Indianapolis* (CA-35) off Johnston Atoll, several hundred miles to the southwest. *Robin* was towing a barge to Johnston Atoll, and *Kingfisher* was on station duty at Tutuila, Samoa.[12]

Minecraft Battle Force: Rear Adm. William R. Furlong, USN
* *Oglala* (CM-4) - Flagship: Comdr. E. P. Speight

Mine Squadron One

Mine Division 1 Comdr. John F. Crowe Jr.	Mine Division 2 Comdr. Ross P. Whitemarsh
* *Pruitt* (DM-22) F Lt. Comdr. William G. Beecher Jr.	* *Gamble* (DM-15) F Lt. Comdr. Donald Allen Crandell
* *Preble* (DM-20) Lt. Comdr. Charles F. Chillingworth Jr.	* *Breese* (DM-18) Lt. Comdr. Herald Franklin Stout
* *Sicard* (DM-21) Lt. Comdr. William Christian Schultz	* *Montgomery* (DM-17) Lt. Comdr. Richard Allen Guthrie
* *Tracy* (DM-19) Lt. Comdr. George Richardson Phelan	* *Ramsay* (DM-16) Lt. Comdr. Gelzer Loyall Sims

Base Force: Rear Adm. William. L. Calhoun, USN
Mine Squadron Two: Comdr. George F. Hussey Jr.

Hopkins (DMS-13) – Flagship: Lt. Comdr. R. W. Clark

Mine Division 4 Comdr. W. H. Hartt Jr.	Mine Division 5 Comdr. S. H. Hurt	Mine Division 6 Comdr. E. D. Gibb
* *Perry* (DMS-17) F Lt. Comdr. L. H. Miller	*Southard* (DMS-10) F Lt. Comdr. J. B. Cochran	*Dorsey* (DMS-1) F Lt. Comdr. R. M. MacKinnon
* *Trever* (DMS-16) Lt. Comdr. D. M. Agnew	*Long* (DMS-12) Lt. Comdr. W. S. Veeder	*Elliot* (DMS-4) Lt. Comdr. C. D. Reynolds
* *Wasmuth* (DMS-15) Lt. Comdr. J. L. Willfong	*Chandler* (DMS-9) Lt. Comdr. H. H. Tiemroth	*Lamberton* (DMS-2) Lt. Comdr. W. J. O'Brien
* *Zane* (DMS-14) Lt. Comdr. L. M. LeHardy	*Hovey* (DMS-11) Lt. Comdr. J. E. Florance	*Boggs* (DMS-3) Lt. Comdr. D. G. Roberts

Mine Squadron Four

Mine Division 10	Mine Division 11
* *Bobolink* (AM-20) Lt. (jg) J. L. Foley	* *Rail* (AM-26) Lt. Comdr. F. W. Beard
* *Grebe* (AM-43) Lt. Comdr. E. D. McEathron	*Robin* (AM-3) Lt. D. G. Greenlee Jr.
Kingfisher (AM-25) Lt. Comdr. C. B. Schiano	* *Tern* (AM-31) Lt. W. B. Pendleton
* *Vireo* (AM-52) Lt. Comdr. F. J. Ilsemann	* *Turkey* (AM-13) Lt. Comdr. T. F. Fowler[13]

The letter "F" in the table, denotes Division Flagship

MINE FORCE FLAGSHIP USS *OGLALA* (CM-4) SUNK

I was on the deck of my Flagship, U.S.S. OGLALA (CM4) and saw the first enemy bomb fall on the seaward end of Ford Island close to the water. This one did not hit the planes parked there. Another fell immediately afterwards in the same vicinity and caused fires near the water. U.S. planes were on the ground nearby and later flames flared up from structures at that end (south end) of the island.

The next bombs fell alongside or on board the seven battleships moored at "F" moorings on the east side of Ford Island.

—From a report by Commander, Minecraft, Battle Force, Rear Adm. William R. Furlong, USN, regarding the loss of his flagship, the USS *Oglala*, to a torpedo during the attack on Pearl Harbor.[14]

Photo 7-3

View from the Ten-Ten Dock the morning of 7 December 1941 looking toward the
Pearl Harbor Navy Yard's drydocks. In the foreground is the capsized USS *Oglala*
(CM-4), with the light cruiser USS *Helena* (CL-50) farther down the pier.
U.S. Navy Photograph, now in the collections of the National Archives.

Shortly before 0800 on the morning of 7 December, the bridge watch
on every ship in the harbor was watching the station signal tower. When
the tower hoisted the Preparatory pennant at 0755, all of the ships would
hoist Prep simultaneously. When the tower hauled down Prep at 0800,
all the ships would observe colors—by simultaneously hoisting the
American flag and union jack aboard each vessel. Prep was in the air,
when the roaring engines of approaching torpedo planes and dive
bombers broke the stillness of the Sunday morning. Witnessing bombs
falling on Ford Island from his flagship *Oglala*, moored at Ten-Ten
Dock, Rear Adm. William B. Furlong, commander, Minecraft, Battle
Force, ordered the signal "All Ships in Harbor Sortie [get under way as
soon as possible]" hoisted aloft.[15]

Furlong then witnessed Japanese aircraft, flying 50-100 feet above
the water, drop what appeared to be three torpedoes or mines in the
channel between *Oglala* and the seaward end of Ford Island. He judged
they were either torpedoes or mines, and not bombs, because no
explosion-induced water plumes went up from them. Plumes over 100
feet high were being produced by bombs hitting close aboard the

battleships.   A plane then dropped a torpedo close aboard *Oglala*'s starboard beam.   The torpedo ran under *Oglala* and struck the cruiser *Helena* (CL-50), moored inboard of her alongside the pier, and exploded. The distance between the ships was about eight feet and the depth of the water about forty-two feet.[16]

The explosion lifted up floor plates in *Oglala*'s fireroom, and ruptured her hull.   At the same time, several Japanese planes strafed her decks with machine gunfire, and two or three minutes later a dive bomber dropped a bomb, which fell between *Oglala* and *Helena* and exploded.   The detonation created a geyser of oil and water, but no flash, flame or smoke was observed, indicating that the bomb had a delay-action fuse which exploded the bomb under water.   As the fireroom started to flood rapidly, personnel had time to secure the fires, and abandon the space.   The engine room also began to flood, as a result of the torpedo, and more rapidly, following the bomb explosion. By 0830, *Oglala* had a 5-degree list to port, and was settling by the stern.   No electrical power was available for pumping and it was obvious that the ship could not be kept afloat much longer.[17]

Photo 7-4

Rear Adm. William R. Furlong, USN, commander, Minecraft, Battle Force, 7 December 1941. He became commandant of the Pearl Harbor Navy Yard, 25 December 1941. Naval History and Heritage Command photograph #NH 54300

As anti-aircraft batteries aboard the ships opened fire, Furlong hailed two small contractor tugs working with dredges across the channel from the minelayer to obtain assistance in moving *Oglala* aft of *Helena* so that the cruiser might sortie, if able. By 0900, she was berthed astern of the cruiser, secured to the pier with all available mooring lines. Nevertheless, her list and trim continued to increase. Furlong ordered all hands to abandon ship shortly after 0900. Japanese planes were strafing the ship, and *Oglala*'s gun crews kept up anti-aircraft fire until the ship's list had increased to 20 degrees and the men on the machine guns were sliding off the deck and the angle was too steep for others to remain on the deck and serve the 3-inch gun. The machine guns were slid off the top of the deckhouse to the pier, and set up there.[18]

Oglala's list increased to about 40 degrees to port, her mooring lines parted and her upper deck rail was so far over, it appeared that she might sink at any moment. Furlong ordered the gun crews to abandon, and he left with them. The minelayer capsized and sank just before 1000, resting on her port side alongside Ten-Ten Dock.[19]

Commander, Minecraft, Battle Force later described, in a report written at 1100 that morning, the heroic actions of *Oglala*'s gun crews and signalmen prior to abandonment of the ship:

> The guns' crews manned their battle stations promptly and stood to their guns during bombing and strafing as if at target practice, keeping up a continuous fire at enemy planes during the bombing and strafing. The signal force manned their bridge stations and sent signals during the action; one to sortie and one to the *NEVADA* warning her of mines during which time the bridge was struck by machine gun bullets. The man on the [boiler] fires when the fireroom was flooding very promptly turned off the oil fires and no one suffered oil burns.[20]

After Mine Division Two sortied from Pearl Harbor upon orders from Furlong, he reported for duty to Adm. Husband E. Kimmel, commander in chief Pacific Fleet, with his staff.[21]

MINE DIVISION TWO, BERTH D-3, MIDDLE LOCH

It is estimated that about 70 planes were used in the attack. The dive bombers seemed to be very slow. They were equipped with two machine guns forward and one machine gun aft. Most bombs were released from a horizontal position. About 30 high altitude bombers were observed. These planes flew in formation and were painted white on their underside, which made them blend in perfectly with the high alto cumulus clouds. This was highly effective camouflage. All anti-aircraft fire was below these planes.

The anti-aircraft battery of this vessel is considered inadequate. It is composed of two .50 cal. machine guns and one 3" 23 cal. AA [anti-aircraft] gun. The following landing force equipment was used to augment the anti-aircraft battery; and although its effectiveness is doubtful it served as a means of satisfying the offensive spirit of the crew, 3" 30 cal. Lewis machine guns, 3 Browning automatic rifles.

—Lt. Alexander Coxe Jr., USN, executive officer USS *Breese* (DM-18). *Breese* was the division guard ship (responsible for monitoring radio traffic) and Coxe the nest duty officer.[22]

Map 7-1

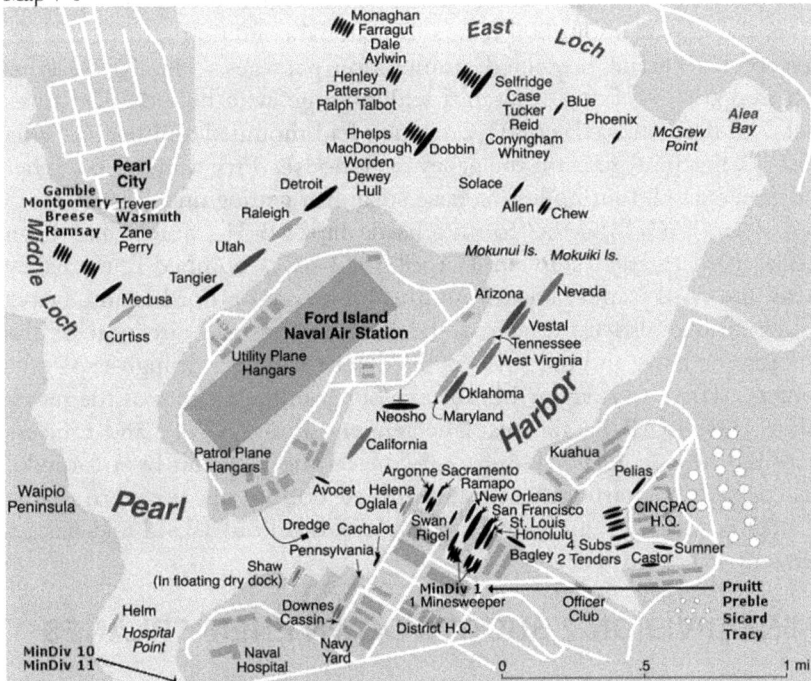

Locations of U.S. Navy ships at Pearl Harbor on 7 December 1941

Mine Division 2—*Gamble* (flagship), *Montgomery*, *Breese*, and *Ramsay*—was moored in a nest in Middle Loch, south-southwest of Pearl City, with the order of the ships from the north, listed as above. At 0755, two dive bombers approached Ford Island from the west at an altitude of 200 feet, and bombed the sea plane hangar and adjacent gasoline tanks on its west end. Aboard *Breese* and the other ships, many actions began happening concurrently. General alarms were sounded, calling crews to battle stations; material condition Affirm was ordered set to increase ship compartmentation (to better contain fires and/or flooding in the event of battle damage); anti-aircraft batteries were manned; and as planes began coming in and attacking battleships and Naval Air Station, Ford Island, preparations were begun to get under way.[23]

Breese also dispatched boats to the Pearl City landing to pick up returning men. At 0757, her starboard .50-caliber machine gun opened fire, manned by the gangway watch who was a qualified machine gunner. The ship's 3"/23 gun opened at 0805, using fuse settings 3 to 12 seconds, pre-set, and her commanding officer, Lt. Comdr. Herald Franklin Stout, USN, then dismissed the crews of the machine guns to attend to getting under way.[24]

Aboard *Gamble*, .50-caliber machine guns opened on planes passing over the nest at about 800 feet, and material condition Affirm was set except for certain protected ammunition passages. Her 3"/23 guns opened at 0759, firing at aircraft within range, with time delayed fuses set 3 to 8 seconds. By 0805, crewmen had mounted and begun firing .30-cal. machine guns on the galley deck house. Five minutes later, her engineers lit all four boilers to raise steam for getting under way.[25]

Utah (AG-16)—the former battleship BB-31, which had been reclassified a target ship on 1 April 1932—was bombed immediately after the Ford Island hangar. As fires broke out on Ford Island, heavy black smoke also came across the harbor from the direction of the submarine base, and the sky was filled with anti-aircraft high-explosive bursts. There were many bombs exploding in the vicinity of the repair ship *Medusa* (AR-1), the seaplane tender *Curtiss* (AV-4), and cruisers moored next to *Utah*. During this melee, the *Utah*, hit by a torpedo, listed badly to starboard and then turned bottom up; the stern of the *Curtiss* was on fire; and a PBY took off from Ford Island and passed directly overhead *Breese*.[26]

MINE DIVISION 2 SHOOTS DOWN ENEMY PLANES

Two planes are to be credited to ships of Mine Division TWO. Other A.A. fire may have struck the planes, but .50 cal. and 3"/23 fire from mine Division TWO finished them. There was no distinguished action by any individual, but all men showed courage, zeal and initiative. They did not hesitate in their action in face of strafing attack from planes. Their conduct under fire was commendable.

—Lt. Comdr. Donald A. Crandell, USN, commanding officer,
USS *Gamble* (DM-15), flagship of Mine Division Two.[27]

At 0913, a projectile from *Breese*'s 3"/23 struck a dive bomber, winging over after an attack on the *Curtiss*, just aft of the pilot's cockpit. The plane caught fire and broke apart in the air, the forward section with engine landed on the north shore of Waipio Peninsula and burned for some time. Tracer fire for the direct hit had been easy to follow at the short range involved. At 0925, a Japanese plane impacted the water about 1,000 yards off *Gamble*'s port beam. It was believed shot down by BM2c W. L. Roberts, *Gamble*'s port .50-caliber machine gunner, and GM3c H. W. Joos, on the starboard .50-cal.[28]

ORDERS TO GET UNDER WAY/MIDGET-SUB SUNK

Photo 7-5

Japanese type *A* midget submarine sunk by the destroyer USS *Monaghan* (DD-354) in Pearl Harbor on 7 December 1941. The submarine's hull shows the effects of depth charges and ramming. The upper background of the photograph was overpainted for censorship purposes.
Naval History and Heritage Command photograph #NH 54302

The minelayers had received a signal from the tower at 0825, to get under way immediately. A submarine was reported in the harbor at 0826, and about 0830 a sighting was made aboard *Breese* of two small submarines in the North Channel. She could not take them under fire because of her interior berth in the nest of minelayers. The destroyer *Monaghan* (DD-354), however, proceeded down the channel at high speed. She rammed the leading submarine, which apparently had just fired a torpedo at the seaplane tender *Curtiss*, which missed. *Monaghan* dropped two depth charges, the first directly upon the leading submarine and the second in the approximate position of the second which was no longer visible. About ten seconds later, a midget submarine surfaced upside down, before immediately sinking.[29]

At 0855, *Ramsay* got under way, the first unit of Mine Division Two to stand out. Earlier, as her liberty party was returning from shore leave in *Montgomery*'s boat, they had been strafed by torpedo planes, which also fired three torpedoes into *Utah* and one into *Raleigh*. Now as *Ramsay* proceeded out of the harbor, her gunners took an enemy plane within effective range under fire with .50-caliber and destroyed the aircraft. This action was claimed by the gun crew and also substantiated by several observers. Upon clearing the entrance buoys, the minelayer took up anti-submarine patrol duties.[30]

With *Ramsay* clear of her side, *Breese* left Middle Loch at 0917 and proceeded out of the harbor. At 0930, a lookout reported a periscope off the Coal Docks but it was not seen from the bridge nor picked up on sonar. After clearing the channel entrance, *Breese* took up offshore patrol duties in sector three. At 1108, following a report by a motor torpedo boat of a periscope, the minelayer dropped two depth charges in the spot indicated, but with no result. However, less than a half hour later, she picked up the sounds of a submarine on sonar at 1135. The second of two depth charges dropped in an ensuing attack brought up an oil slick and some debris. No further sound was heard but on the second pass over the target, contact was made when "pinging" from pulsed sonar produced return echoes. A subsequent attack was made with four deep-set depth charges with no tangible result. Meanwhile, several other destroyers carried out attacks in the same locality.[31]

Gamble left *Montgomery*'s side at 0930 and proceeded outbound. Seven minutes later, as Japanese planes attacked near the main channel, she anchored temporarily, astern of *Medusa*, at 0955. Once under way again, *Gamble*'s gunners shifted the .30-cal. machine guns to the top of the pilot house. As the minelayer cleared the entrance channel at 1021, her commanding officer, Lieutenant Commander Crandell, ordered eight depth charges armed, and she commenced an anti-submarine patrol off the Pearl Harbor entrance.[32]

Gamble gained sound contact on a submarine at 1204 and dropped three depth charges. At that time, she bore 162° T from Diamond Head Light, 2.5 miles distant. At 1255, following receipt of orders from CincPac to join friendly forces, *Gamble* proceeded west at 20 knots. At 1628, a sighting was made of a smoke bomb off the ship's port bow. Three minutes later, a submarine surfaced and *Gamble* fired one 4-inch gun round, which was short and to the left of the target. Upon the submarine displaying U.S. colors, the minelayer ceased firing, and the sub then submerged and fired a recognition red smoke flare.[33]

Photo 7-6

Destroyers USS *Gamble* (DD-123) and USS *Breese* (DD-122), circa 1919, long before their service as minelayers.
Naval History and Heritage Command photograph #NH 94956

Photo 7-7

Vice Adm. William F. Halsey, USN (center), with members of his staff on the bridge of the aircraft carrier USS *Enterprise* (CV-6).
Navy photograph #80-G-464485, in the collections of the National Archives

Continuing on a westward course, *Gamble* sighted the *Enterprise* (CV-6) at 1732 and exchanged calls. Instructed by commander, Aircraft, Battle Force, Vice Adm. William F. Halsey, USN, to join the aircraft carrier, the minelayer did so, taking station as third ship with two other plane guard destroyers.[34]

Montgomery, the last of Mine Division Two to depart Middle Loch, got under way at 1017, and an hour later, established an anti-submarine patrol in Sector One off the entrance to Pearl Harbor. Her experiences during the preceding air attacks were similar to the other minelayers with one exception. After civilians reported two Japanese swimming around a downed plane near the Pearl City dock, the ship's motor whaleboat was sent to investigate. The boat crew found the pilot still afloat. He was motioned to get into the boat several times, but refused to obey, and instead reached inside his jacket. At this action, a seaman first class shot him, and he slipped beneath the surface.[35]

MINE DIVISION 1 AT PEARL HARBOR NAVY YARD

Photo 7-8

Minelayers USS *Sicard* (DM-21), USS *Tracy* (CM-19), USS *Preble* (DM-20) and USS *Pruitt* (DM-22) nested together at a berth, 30 July 1939.
Naval History and Heritage Command photograph #NH 81004

At the time of the attack on Pearl Harbor, the four ships of Mine Division One—*Pruitt*, *Preble*, *Sicard*, and *Tracy*—were in the Navy Yard Repair Basin, east of Ford Island. *Pruitt*, the flagship, was moored at berth 18, undergoing routine overhaul, with *Sicard* and the fleet tug *Ontario* (AT-13) moored to port in that order. All machinery and armament were disabled for overhaul; and there were only a duty section and duty officer on board, with the remainder of her crew berthed at the Receiving Ship Barracks at the Navy Yard.[36]

At 0753 a group of ten low flying planes came in from a southerly direction and bombed the hangars on the north end of Ford Island. Red "meatballs" (Japanese insignia) were plainly visible on the wings of all the planes. At 0801, dive bombers and torpedo planes attacked "battleship row," followed by a high-altitude level bombing attack on the battle wagons. Attacks continued in the area of the Navy Yard until about 1130 with the heaviest attack on Pearl Harbor by dive bombers and horizontal bombers at 0907. Three low-flying Japanese fighter planes were shot down in the vicinity of *Pruitt* apparently by small caliber weapons.[37]

The initial surprise of the attack passed quickly and aboard *Pruitt*, duty section personnel began arming themselves with small arms in the ready locker—.30-caliber machine guns, Browning automatic rifles, service rifles, and service pistols. As low-flying, attacking planes were taken under fire, fire hoses were laid out and crewmen not armed were organized in two fire-fighting parties, one forward and one aft. Steel helmets and gas masks were distributed, assembly of .50-caliber machine guns was begun, and steps were taken to have 3"/23 anti-aircraft guns hoisted on board. Navy Yard personnel distributed small caliber ammunition and a party of men were sent from the ship to the ammunition depot to obtain 3"/23 anti-aircraft ammunition.[38]

Men who could be spared from *Pruitt* were sent to the cruiser *New Orleans* (CA-32) and destroyer *Cummings* (DD-365) in nearby berths to assist in the operation of their anti-aircraft batteries. There was no damage to the *Pruitt*. The only casualty was RM3c George Richard Keith, USN. He was sent from the Receiving Barracks, where the crews of ships in overhaul were berthed and messed, to assist the *Pennsylvania* (BB-38). Keith was killed aboard the battleship, as were also one *Sicard* and three *Tracy* crewmembers.[39]

Mine Force Personnel Casualties at Pearl Harbor

Ship	Killed Aboard	Casualty
Pruitt (DM-22)	*Pennsylvania* (BB-38)	RM3c George Richard Keith, USN
Sicard (DM-21)	*Pennsylvania* (BB-38)	S2c Warren Paul Hickok, USN
Tracy (DM-19)	*Pennsylvania* (BB-38)	S1c John Arthur Bird, USN RM3c John Wallace Pence, USN F2c Laddie John Zacek, USN[40]

Sicard, moored starboard side to *Pruitt*, was also totally disabled as to main and auxiliary machinery, boilers, and gun batteries. By 0815, condition of readiness Affirm was set as well as possible, considering that numerous air ports and hatches defied closure due to work in progress. Two fire parties were formed, and by 0820 two .30-cal. machine guns were mounted and serviced with the meager amount of small arms ammunition left aboard.[41]

As Japanese planes came in from the southwest to dive bomb targets on Ford Island (followed by waves of horizontal and dive bombers whose objectives were ships at Ten-Ten dock, Ford Island, and battleships moored in the vicinity of Ford Island), *Sicard* opened fire on planes within range. Some hits were observed on fuselage and tail surfaces, but no apparent damage done to the aircraft. At 0827, the minelayer's machine guns ceased firing, having expended 300 rounds.[42]

Twenty men were sent to the *Cummings* to handle ammunition and four gunner's mates to *New Orleans* to aid in preparing her batteries for firing. A party of ten men had previously been detailed to go to *Pennsylvania* from the Receiving Barracks to assist in damage control. A muster of *Sicard* crewmembers, upon the return of all working parties aboard ship, identified the absence of Sea2c Warren Hickok, killed aboard the *Pennsylvania*. Japanese air attacks concluded at 1135, upon the withdrawal of planes carrying out a horizontal bombing raid.[43]

Preble and *Tracy* were moored in berth B-15 undergoing, like their division mates, overhaul. The order of the ships from the pier outboard were *Tracy*, *Preble*, and *Cummings*. All guns and ammunition were removed from the minelayers, and their engineering plants were dismantled. *Cummings*, moored to the starboard side of *Preble*, opened fire on enemy planes at 0800. At 0820, *Preble* sent a party of men to the destroyer to supplement her gun crews and ammunition party. These sailors were recalled from the *Cummings* before she got under way at 1040. Aboard *Preble*, machine guns and 3"/23 guns were reassembled and mounted, with ammunition obtained from the Marine Barracks, and the Naval Ammunition Depot, Lualualei, Oahu—to which the ship's ammunition had been sent.[44]

Aboard *Tracy*, Ens. L. B. Ensoy, USN, and the other members of the watch on deck first observed dive bombers, coming from the north, attack battleships and Ford Island. This attack was immediately followed by those of horizontal bombers and dive bombers on the same objectives plus ships in the dry dock. At 0805, torpedo planes came in from an easterly direction and launched torpedoes against the battleships at Ford Island. At the first attack, *Tracy* was closed up as much as possible, and her crew started to break out and assemble .30- and .50-caliber machine guns. Fire hoses were connected but not laid out, and steel helmets and damage control gear were issued to personnel.[45]

At 0820, men from the *Tracy* were sent to the *Cummings* alongside, to assist at gun batteries and in the ammunition train. The commanding officer of the destroyer noted in his report in the attack that day:

> During the action, the ammunition supply crews of the ship were augmented by approximately thirty men from the destroyers *Preble* and *Tracy* which were moored inboard, undergoing overhaul and whose antiaircraft armament was disabled. Although unfamiliar with the arrangements for ammunition supply these men enthusiastically and creditably participated in the action. They returned to their own ships prior to getting underway.[46]

By 0825, three .30-caliber Lewis guns, and very limited ammunition, were ready for use aboard *Tracy*. Concurrently, approximately fifteen men were sent to the *Pennsylvania*. Some of the men helped to fight fires in the drydock, while others went aboard the battleship to assist with the ammunition train. While setting fuses in the after starboard gun compartment, Fireman Second Class Laddie Zacek was killed by a bomb hit above this space. The same blast also killed Seaman First Class John Bird and Radioman Third Class John Pence, and *Pruitt*'s Radioman Third Class George Keith.[47]

Tracy's commanding officer, Lt. Comdr. George R. Phelan, USN, arrived on board at 0915, to find the two .50-caliber machine guns had been mounted and parties were attempting to borrow ammunition from other ships present. As a group of dive bombers attacked out of the sun, two drums of ammunition appeared miraculously and fire was opened. Following a salvo by the *Cummings*, one plane pulled out over the submarine base and crashed in flames in the vicinity of Hospital Point. Intermittent action against stray planes lasted until about 1000, by which time all officers and men had reported aboard, and more ammunition was becoming available.[48]

Cummings got under way at 1040. Prior to doing so, she returned all the *Tracy* men. Following crew muster aboard the minelayer, approximately ten *Tracy* personnel were sent to help fight fires on the battleship *California* (BB-44). By 1100, it became apparent that the Japanese had withdrawn. An hour later, *Tracy* sent four gunner's mates qualified in mines to the West Loch ammunition depot.[49]

MINESWEEPERS PRESENT AT PEARL HARBOR

In addition to the four ex-fishing boats—*Reedbird*, *Cockatoo*, *Condor*, and *Crossbill*—of the Inshore Patrol, whose duty was make daily mile-wide sweeps from the Pearl Harbor entrance out to the 100-fathom curve, there were ten other minesweepers present at Pearl Harbor. The ships of Mine Division Four—*Perry*, *Trever*, *Wasmuth*, and *Zane*—were moored in a nest at buoys D-7 and D-7-S in Middle Loch, with their bows toward Pearl City. Their order from port to starboard was *Trever*, *Wasmuth*, *Zane*, and *Perry*. Like the minelayers, they were old "flushdeck" destroyers, but were currently employed as minesweepers.[50]

Six of the eight *Lapwing*-class minesweepers of Mine Division Ten and Eleven were also in port. When the attack developed, the *Bobolink* was moored in a nest at the westerly end of the old Coal Docks, seaward around Hospital Point from the Navy Yard, with three other auxiliary minesweepers. *Vireo* and *Turkey* were inboard of her, and *Rail* outboard. *Tern* was moored alongside the north end of Ten-Ten Dock, undergoing upkeep by the tender *Argonne* (AG-31), with all machinery dead, and receiving steam, water and electricity from the dock. *Grebe* was in the Repair Basin at the Navy Yard with Mine Division One.[51]

Although purposefully built as minesweepers, these 187-foot *Lapwings* (like the antiquated destroyers converted to minelayers and minesweepers) were of World War I vintage.

AFTERMATH

Rear Adm. William R. Furlong was appointed commandant of the Pearl Harbor Navy Yard after the Japanese Attack on Pearl Harbor, tasked with the salvaging and repairing of the sunken ships. His former flagship *Oglala* was refloated and, following extensive repairs, was placed in "ordinary" status. She was later reclassified an Internal Combustion Engine Repair Ship (ARG-1). The ships comprising Minecraft, Battle Force, became a part of Train Squadron 6, whose mission was the establishment of advance bases, training of landing force units, screening fleet movements, and lastly, mining and minesweeping. Mine Squadron 6 was split up, with part of the ships assigned to the Alaskan Command and the others sent to the South Pacific. (In World War II, Train Squadron referred to a squadron of logistics support ships.) Mine Squadron 1 initially maintained the Offshore Patrol in the Hawaiian area, while also performing myriad chores for the Fleet.[52]

In April 1942, a Pacific Fleet reorganization established Service Squadron 6, under Captain G. D. Hull, which included minecraft. Not until 15 October 1944, were the now rapidly expanding and widely scattered mine forces separated from the overgrown Service Force and given type status as Minecraft, Pacific Fleet.[53]

8

Second Attack on Pearl Harbor/
Mining of French Frigate Shoals

I just threw up my hands and said it might be a good idea to remind everybody concerned that this nation was at war.

—Capt. Joseph J. Rochefort, USN, describing his reaction upon learning that his warnings to the Pacific Fleet and Fourteenth Naval District headquarters in Hawaii had been ignored. Rochefort's Combat Intelligence Unit decoded Japanese messages ordering a second air attack on Pearl Harbor on the night of 4/5 March 1942.[1]

In the early morning on 5 March 1942, the U.S. Army's Air Defense Command Information and Control Center (ICC) at Fort Shafter on Oahu received a call on VHF radio from a fixed-radar site at Kokee on Kauai. Two unknown aircraft were approaching from the southwest, headed for Oahu. The ICC handed over the track to the Opana radar site (located just inland from the north shore of Oahu), which picked up the aircraft twenty miles east of Kauai. Jean Fraser, a WARD on duty, plotted the approach path of the aircraft.[2]

Less than five weeks earlier, 104 WARDs of the recently formed Women's Air Raid Defense had moved into quarters at Fort Shafter on 1 February 1942, and taken over plotting duties on all four 6-hour shifts. The group would soon be designated WARD Detachment, Company A, 515th Signal Aircraft Warning Regiment (Special). Later that month, new SCR-271 radar sites at Mount Haleakala on Maui, at Kokee on Kauai, and at Pahoa on Hawaii began operations, creating additional plotting positions.[3]

The importance of the Women's Auxiliary Air Force (WAAF) in Britain's air defense centers was well known before the war, but congressional opposition in 1941 had blocked the establishment of an American equivalent. Undeterred, Brig. Gen. Howard C. Davidson, commander of the 14th Pursuit Wing and Air Defense Command, had appealed to the War Department for an emergency executive order creating a WAAF-like organization for Hawaii. Executive Order 9063, authorizing such action, was approved on Christmas Day.[4]

Unbeknownst to Fraser (other WARDs on duty, and Army Air Force officers arriving in the ICC at Fort Shafter), the planes were preparing to execute Operation K. This codeword referred to a Japanese plan to bomb the Ten-Ten Dock at Pearl Harbor; so-called because it was 1,010 feet long. The dock could berth the largest ships in the Pacific Fleet for maintenance and repairs, and was therefore of strategic importance. Its destruction would force damaged carriers or battleships to retire to the continental United States for repairs.[5]

U.S. Navy code breakers had previously intercepted and decoded enemy radio communications surrounding the mission. However, for some reason, Naval headquarters had not acted on the team's discovery. No ships were dispatched to French Frigate Shoals, 500 miles northwest of Oahu, to intercept Japanese submarines scheduled to refuel the planes during their long flight to Pearl Harbor from Wotje Atoll in the Marshall Islands, or to attack the flying boats while refueling on the water.[6]

Photo 8-1

Japanese Kawanishi H8K2 ("Emily") flying boat in the Central Pacific, 2 July 1944. This plane was the successor to the K1 type aircraft used for Operation K. National Archives photograph #80-G-241258

Fortunately, the U.S. Army radar operators and message handling personnel on duty that night were vigilant. Air Defense Commanders, upon being informed of the two incoming contacts from the northwest, scrambled four Curtiss P40 Warhawks from Wheeler Field. They also sent five Navy PBY Catalinas to search for the Japanese aircraft carriers, from which the planes must have been launched. The fighter planes failed to make visual contact on the moonless, rainy night, but searchlights on the ground stabbing into the darkness did illuminate a large four-engine flying boat.[7]

Lt. Hisao Hashizume was piloting *Y-71*, the lead Kawanishi H8K1 "Emily" and Ens. Shosuke Sasao the second aircraft (*Y-72*). They and their crews had no more success locating Pearl Harbor on that foul night than did the interceptors in trying to find them. Honolulu and the naval base were "blacked out," lights extinguished, for just such an air raid. Each plane carried two 550-pound bombs, which the pilots, unable to find the naval base, dropped blindly. One pair exploded harmlessly off the entrance to Pearl Harbor, the other, one-fifth mile from Roosevelt High School, on the outskirts of Honolulu.[8]

Map 8-1

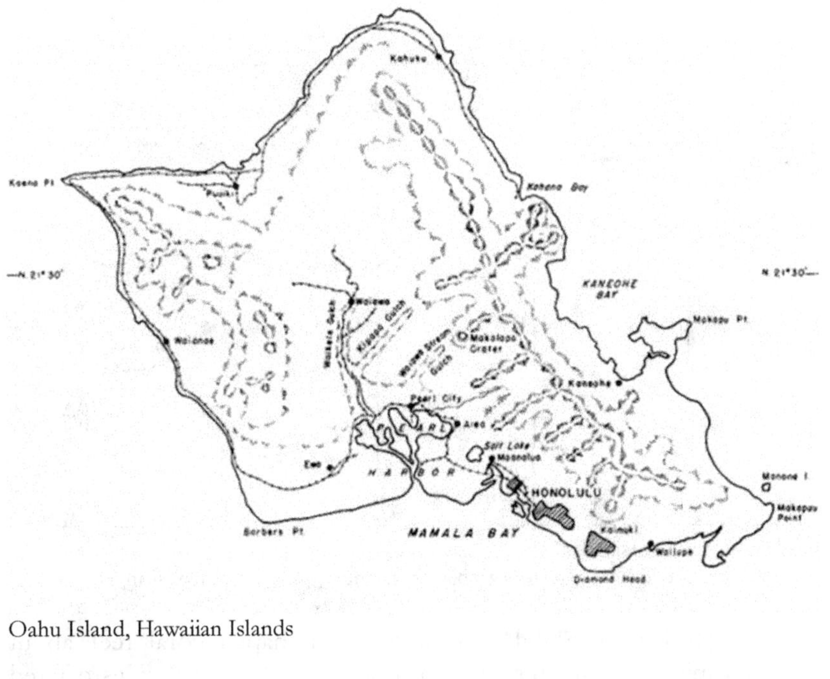

Oahu Island, Hawaiian Islands

The two planes independently turned back to the southeast for the long return flight back. Sasao made it to Wotje Atoll as per the mission plan, while Hashizume—having suffered some damage to his plane on takeoff from the refueling point—bypassed Wotje and flew onward to Jaluit, another atoll in the Marshall Islands where better maintenance capabilities were available. Both planes returned safely, their crews exhausted from the long ordeal of the flight.[9]

MINELAYERS SENT TO FRENCH FRIGATE SHOALS TO DENY CONTINUED USE BY ENEMY SUBMARINES

The end of March found Mine Division 1—*Pruitt, Breese, Sicard,* and *Tracy*—making ready for sea. Fuel tanks and storerooms were filled and, on 28 March, the ships' crews loaded mines at West Loch. The last mines were aboard by 1400, and the ships then stood out of the harbor, swung south and west around Barber's Point, and set a course for French Frigate Shoals. Navy planners had concluded that no Japanese long-range aircraft, including "Emily" flying boats, had the endurance to make it to Pearl Harbor from Japanese-held territory. A study of the ocean areas northwest of Hawaii, identified the shoal as an ideal location for a seaplane to land and receive fuel from a submarine.[10]

Map 8-2

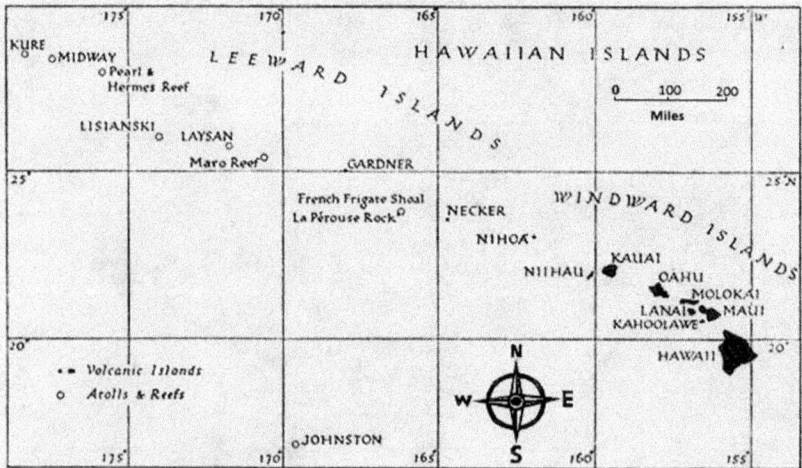

Outlying Hawaiian Islands.
www.lib.utexas.edu/maps/historical/pacific_islands_1943_1945/hawaiian_islands.jpg

French Frigate Shoals was a crescent-shaped coral reef about eighteen miles in diameter. Its north and east (leeward) sides offered shelter to ships from winds and currents. The south and west sides were open allowing access into the atoll. (Tern Island in the northwest corner of the reef was later developed by the U.S. Navy as a seaplane base. A ship channel was dredged through the barrier reef, and a seaplane runway was cleared in adjacent waters. The dredging's were used to increase the area of the island to permit the construction of a landing field, 3,100 feet long and 250 feet wide.)[11]

Photo 8-2

French Frigate Shoals airfield on Tern Island, 13 December 1961; then the site of a U.S. Coast Guard LORAN transmitting station.
Naval History and Heritage Command #NH 91212

The supposition that French Frigate Shoals served as a refueling point proved correct. Five submarines from the Japanese 6th Fleet had supported Operation K. Three large scout plane-carrying submarines—*I-15*, *I-19* and *I-26*—carrying aviation fuel had made their way to the shoals. The fourth submarine, *I-9*, was positioned between Wotje Atoll, in the Marshall Islands, and the refueling point, to provide navigational support; and the fifth one, *I-23*, to the south-southwest of Keahole Point, Hawaii, to provide weather reports and, if necessary, an air-sea rescue. The latter submarine disappeared while on this mission and was not heard from again.[12]

MINEFIELDS SOWN AT FRENCH FRIGATE SHOALS

Commander, Mine Division 1, had orders to lay three fields at the shoals to prevent any additional use of the sheltered waters by the Japanese. The first mine rolled off *Pruitt*'s fantail at 1233 on 30 March 1942. At the completion of the field a couple of miles westward of La Perouse Pinnacle, *Pruitt*, *Preble*, *Sicard*, and *Tracy* returned to Pearl Harbor for more mines. On 1 April, while *Tracy* was at the West Loch ammunition depot to load eighty-five Mk 6 mines, Lt. Comdr. John L. Collis relieved Lt. Comdr. George R. Phelan as her commanding officer.[13]

Mine Division 1 arrived in the mining area at 0930 on 3 April and completed laying field B (340 mines, northwest of La Perouse) in three hours. The ships returned to Pearl Harbor the afternoon of 4 April, and spent the next two days loading for the final operation. Arriving back in the area the morning of 8 April, a reference boat was positioned to assist with the mining. After stationing it at a buoy marking the southeast corner of field A, the ships maneuvered to set up their approach for a mining run. The field, five miles southwest of La Perouse, was completed at noon, and the boat recovered.[14]

The division arrived back at Pearl Harbor the afternoon of 9 April. *Preble* joined the offshore patrol; *Pruitt* and *Sicard* moored at the fuel dock; and *Tracy* entered the Navy Yard for a restricted availability period to receive echo ranging equipment (sonar) and other alterations.[15]

ASSIGNMENT QUEEN

The following day, commander, 14th Naval District and Hawaiian Sea Frontier, Rear Adm. David W. Bagley, designated *Sicard* for a special mission. Lt. Col. Alfred R. Pefley, USMC, and Comdr. Charles B. Momsen, USN, provided details, and that evening the ship began loading light cargo for a group of Marines/Naval personnel that were to embark. (Momsen, assigned to the Naval District staff, was the inventor of a submarine escape underwater breathing apparatus later known as the "Momsen lung.") Loading continued during the forenoon on 11 April. By early afternoon, all equipment and personnel were aboard, including: a 3-inch/23-caliber gun, fifty-six oil drums, distiller unit, radio gear, ammunition, dry stores, lumber, steel plate, and the personal effects of Gunner Michael Peskin, USMC, eight other Marines, three radiomen and a pharmacist's mate.[16]

After clearing port, at 1405, *Sicard* set a westerly course to pass seventeen miles south of Kalua Rock. The weather was extremely bad, and a heavy swell and gusty wind caused the ship to roll as much as 45 degrees. These conditions persisted. La Perouse Pinnacle was sighted in early afternoon on 13 April, and *Sicard* began steering various courses and speeds to clear the minefield.[17]

Map 8-3

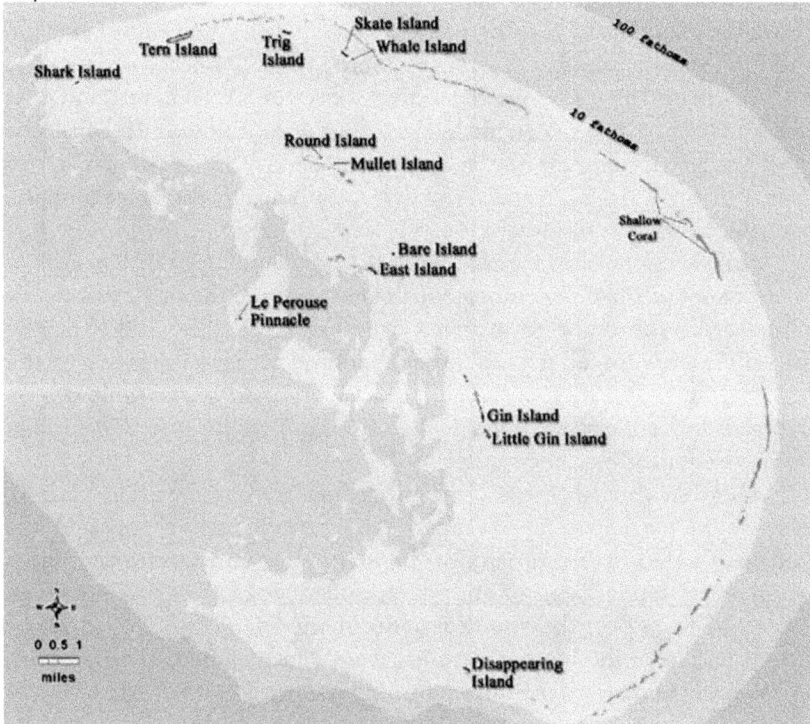

French Frigate Shoals Atoll with principal islands.
www.papahanaumokuakea.gov

After anchoring 1,800 yards from East Island, the motor whaleboat was launched, with the executive officer and a landing party aboard to conduct a reconnaissance of the island. No place suitable for a landing was found other than a rickety boat dock on the northwest side. Reaching it involved 45-60 minutes travel through a choppy head sea, and another 10-15 minutes through shoals and coral heads. A shack reported to be on the island was found, infested with goony birds and littered with trash. Over the next three days, all personnel and equipment were landed despite persistent bad weather and seas, which necessitated bailing out the boat after every trip to East Island. On the morning of 17 April, Gunner Peskin and his party went ashore, and *Sicard* set a return course for Pearl Harbor. (It's unclear from research materials, what their mission was, but it may have been to establish a manned reconnaissance outpost.)[18]

PREBLE SENT TO PEARL AND HERMES REEF

In a separate action, *Preble* was dispatched to Pearl and Hermes Reef. Her destination lay 1,300 miles northwest of Honolulu and 87 miles east-southeast of Midway Atoll. The third northernmost atoll in the Hawaiian Islands had been named for two English whaleships (the *Pearl* and the *Hermes*), which had wrecked there in 1822. *Preble*'s orders were to investigate suspected enemy activity, destroy all facilities, and take prisoners if possible. Arriving at the atoll the morning of 19 April, she took shore installations under fire with her 4-inch guns, while two aircraft from VMF-222 carried out dive-bombing attacks in conjunction with the naval bombardment.[19]

Following the shelling and bomb drops, a landing party was sent to investigate the atoll. The motor whaleboat carrying the men was unable to make the shore due to treacherous reefs and hidden sand bars, and the sailors were forced to ford the last hundred yards to the beach. After first searching and then completing the destruction of the buildings on the island, the party returned to the ship and reported having found nothing but dead gooney birds.[20]

The following day, the same procedure was used at Kure Island, absent the participation of aircraft. A search ashore revealed the remains of two fires estimated to be less than two months old, and a grapnel with ten fathoms of line which evidenced only a short exposure. A lean-to found on the southeast end of the island was destroyed by fire, as had been the buildings on Pearl and Hermes Reef. Upon return of the party aboard, *Preble* set a return course for Pearl Harbor.[21]

9

Build-up of Station Defenses/
Forces at Midway Atoll

Our citizens can now rejoice that a momentous victory is in the making. Perhaps we will be forgiven if we claim we are about midway to our objective.

—Adm. Chester W. Nimitz, June 1942, following the U.S. Navy's victory over the Japanese in the Battle of Midway, 3-7 June 1942.

After a short period of upkeep and crew recreation at Pearl Harbor, *Preble* took up local duties. On 11 May 1942, Marine Gunner W. E. Hemingway and a detail of twelve enlisted men reported aboard for transportation with six tons of stores to French Frigate Shoals. The ship arrived there the following day, and the Marines left the ship. The group they were replacing, Gunner Peskin and the men assigned to him, came aboard at 1600 for transport back to Pearl Harbor. *Preble* stood out in early evening, but a short time later received orders to return to the shoals to guard against possible enemy air and submarine activity.[1]

Around this same time, Comdr. Joseph J. Rochefort—the officer in charge of communications intelligence processing at the Pearl Harbor Naval Intelligence Center—and Lt. Comdr. Edwin T. Layton, the intelligence officer for the Pacific Fleet, identified Midway Island and Dutch Harbor, Aleutian Islands, as specific Japanese objectives for an attack. American cryptanalysts then discovered the date cipher used in Japanese message traffic. After examining previously intercepted messages, they predicted an attack on Midway on 4 June. Commander in Chief, Pacific Fleet, Adm. Chester W. Nimitz, USN, used this estimate to plan American countermeasures.[2]

Midway, located 1,141 nautical miles west-northwest of Honolulu, is actually a coral atoll, six miles in diameter, comprised of three islands: Sand, Eastern and Spit. The atoll's name is said to come from its location midway between San Francisco and Tokyo. It was first discovered in 1859, and the United States, recognizing its strategic importance, claimed it eight years later, in 1867, when Capt. William

Reynolds, commanding the screw sloop-of-war *Lackawanna*, raised the American flag over the atoll. Under Navy direction, the construction of facilities for Catalina seaplanes began in March 1940, and Naval Air Station, Midway was commissioned on 18 August 1941. Despite its small size, the station hosted an airstrip on Eastern Island, as well as a hangar for seaplanes and other facilities on larger Sand Island.[3]

Photo 9-1

Midway Atoll, 24 November 1941. Eastern Island, then the site of Midway's airfield, is in the foreground. Sand Island, location of most other base facilities, is across the entrance channel.
Official U.S. Navy photograph 80-G-451086 from the National Archives

Nimitz had visited Midway in early May 1942 to inspect its defenses and confer with the local commanders. As the Japanese threat became more imminent, he dispatched more ground and air forces to the atoll, crowding Eastern Island with U.S. Marine Corps, Navy, and Army Air Force planes. On 29 May, the seaplane tender *Ballard* (AVD-10) and PT boats of Motor Torpedo Boat Squadron 1 arrived at Midway to bolster seaward defenses, and four YPs (ex-San Diego tuna clippers) arrived to augment local defense forces.[4]

Yard patrol craft (YP) was the Navy's designation for civilian vessels acquired by commanders of Naval Districts and Sea Frontiers, armed with guns and if large enough, depth charges, and sent off to war to serve as patrol vessels. In anticipation of an impending air battle, the YPs were positioned near some small islands southeast of the station to refuel PBY Catalinas and rescue downed aviators. *YP-284* (ex-*Endeavor*) was allocated to Lisianski, *YP-290* (ex-*Picaroto*) to Laysan, *YP-345* (ex-*Yankee*) to Gardner Pinnacles, and *YP-350* (ex-*Victoria*) to Necker. Other local defense force units were stationed near Midway or at other nearby lesser islands, reefs or shoals:

Midway Island	*PT-20, PT-21, PT-22, PT-24, PT-25, PT-26, PT-27,* and *PT-28*
Kure Island	*PT-29, PT-30*, and four small patrol craft
French Frigate Shoals	*Thornton* (AVD-11), *Ballard* (AVD-10), *Clark* (DD-361), and *Kaloli* (AOG-13)
Pearl and Hermes Reef	*Crystal* (PY-25) and *Vireo* (ATO-144)[5]

The seaplane tenders *Ballard* and *Thornton*, like the *Preble* and other minelayers and minesweepers, were converted world War I vintage "flush deck" destroyers.

LOSS OF *YP-277* AT FRENCH FRIGATE SHOALS

> The *Preble attempted to steam in close to pick up possible survivors, but they were deep in the minefield and it was impossible. A PT boat was dispatched from the island and managed to navigate the minefield and pick up two seriously injured survivors. They were transferred to the* Preble *where they were treated by corpsmen.*
>
> —USS *Preble* War History describing the loss of the 116-foot yard patrol vessel USS *YP-277* on 23 May 1942 to an American minefield at French Frigate Shoals.[6]

During preparations to organize Midway for battle, YPs (from a group of fourteen ex-tuna clippers that had arrived at Pearl Harbor from San

Diego on 17 May) ferried aviation fuel and did whatever else was asked of them to assist in this effort. One of these vessels, the *YP-277* (ex-*Triunfo*)—part of a group of four YPs transporting provisions, parts, and fuel to Motor Torpedo Boat Squadron 1—became the first U.S. Navy ship lost to an American minefield in the Pacific. The squadron PT boats had broken down at French Frigate Shoals, 487 nautical miles northwest of Honolulu, while en route to Midway. Arriving at the atoll in the early morning of 23 May, the YPs stood in toward the anchorage near La Perouse Pinnacle which jutted up between the tips of a crescent-shaped reef. The *YP-239* (ex-*Challenger*) led the other vessels—*348*, *277*, and *237*—along a nautical track that it believed would skirt a defensive minefield laid seven weeks earlier to prevent enemy submarines from using the area as a refueling point for flying boat raids on Oahu.[7]

Photo 9-2

Night Action off Tulagi by Richard DeRosset, depicts the destruction of USS *YP-346* (the former San Diego tuna clipper *Prospect*) by the Japanese light cruiser IJN Sendai off Guadalcanal on 8 September 1942.

While proceeding toward the anchorage, the third ship in the column, *YP-277*, exploded a mine that blew her stern off and set her deck cargo of high octane fuel ablaze. Boatswain's Mate First John R. Bruce, the commanding officer of the vessel astern, brought the *YP-237* (ex-*Anna M.*) to within seventy-five feet of her, and quickly lowered a boat. Then, in spite of the danger presented by the field and exploding gasoline drums and ammunition aboard the stricken vessel, he entered the area to search for survivors. Bruce rescued Seamen Second Class

John Elijah Callin and George Alonzo Hazzard, and recovered the body of another crew member. For his courageous action, he was awarded the Navy and Marine Corps Medal.[8]

Preble was patrolling two to three miles seaward of the northern end of the minefield, when at 1935 a sighting was made of a column of white smoke and flames about twelve miles distant in the vicinity of the southern end of the field. In the fading light with visibility further impaired by rain squalls, it was not possible to determine the source of the fire. Believing that one of the PT boats that had just arrived at the shoals might have entered the field from the south and exploded a mine, *Preble* headed toward the smoke and flames at 20 knots on a course to clear the field to seaward. Earlier, a PT boat had informed *Preble* that they had incorrect information regarding the location of the minefield. Ten of the eleven boats of Motor Torpedo Boat Squadron 1 were accounted for—three having been sighted at French Frigate and seven reported to be at Necker. (As it turned out, the missing boat was *PT-23*, which had broken a crankshaft the first day on the squadron's transit from Pearl Harbor to Midway, and had to turn back.)[9]

At about 1945, *Preble* sighted *YP-348* and *YP-239* between the fire and La Perouse Pinnacle, and *YP-237* to seaward of the flames. The minelayer informed the YPs of the correct location of the field, and safe courses to steer to avoid it. Ens. A. H. Bryant, USNR, the commanding officer of *PT-42*, proceeded at high speed from inshore the minefield toward *YP-277*, which was enveloped in flames. A short time later, Bryant came alongside *Preble* and transferred to her the two survivors which John Bruce had rescued from the water.[10]

The following morning, *PT-42* escorted *YP-348* and *YP-237* into the lagoon at French Frigate Shoals; *YP-234* was already anchored there. Later that morning, *Preble* launched a boat to investigate floating wreckage of *YP-277*. No bodies or survivors were found on what appeared to be the side and top of the pilot house or superstructure. This section of the vessel had apparently been blown clear by the explosion as it was unburned. *Preble* entered the lagoon and anchored south of East Island to transfer the survivors to a patrol plane for transport to the hospital at Pearl Harbor. In late afternoon, YPs *348* and *237* were sent to Necker Island to fuel the seven PT boats there. *Preble* remained on patrol at French Frigate Shoals until 29 May when, relieved of her duties by the seaplane tender *Thorton* (AVD-11), she set a course for Pearl Harbor.[11]

THE BATTLE OF MIDWAY

Before we're through with them, the Japanese language will be spoken only in hell.

—Remark by Vice Adm. William F. Halsey, USN, on 8 December
1941, upon entering Pearl Harbor that evening and surveying
the wreckage of the Pacific Fleet.

The presence of U.S. ships at French Frigate Shoals prevented the refueling there of Japanese flying boats to reconnoiter Pearl Harbor. Thus, the Japanese were unable to visually confirm the departure of two carrier strike forces (Vice Adm. William F. Halsey's Task Force 16, temporarily under the command of Rear Adm. Raymond A. Spruance, and Rear Adm. Frank J. Fletcher's Task Force 17) from Pearl Harbor. Japanese communications intelligence (COMINT) stations had learned of carrier movements in and out of Pearl Harbor by listening to increased air-ground radio chatter, and traffic analysis of "Urgent" American radio messages coming out of Pearl Harbor suggested at least one task force was at sea. Incredibly, these discoveries were withheld from the Japanese Midway Strike Force because of Adm. Yamamoto Isoroku's (commander in chief of Japan's Combined Fleet) strict radio silence restrictions.[12]

The Battle of Midway proved disastrous for the Japanese. The loss of four of their front-line aircraft carriers, together with 250 aircraft and some 100 of their best pilots, deprived them of the powerful striking force with which they had achieved their conquests to date, and with which they had planned to oppose American efforts to counterattack. As a result of this battle, Japanese expansion to the east was stopped and Midway Island remained an important American outpost. From this time forward, the balance of power in the Pacific shifted steadily to the Allies' side over the course of the war.[13]

Mine Warfare in the Aleutians

Weather colder and fogs more frequent. Seas getting higher.

—USS *Sicard* (DM-21) War Diary entry for 24 June 1942, describing
deteriorating weather as she and two division mates proceed
northward up American's rugged Pacific Northwest coast.[1]

No enemy surface forces located outside Kiska-Attu area since June Twelfth. [It]
now appears probable that air action in first week of campaign and results of
Midway victory plus our western air striking concentration had discouraged for the
present enemy advance east of Kiska.

—Message from Rear Adm. Robert A. Theobald to commander,
U.S. Pacific Fleet on 1 July 1942, regarding conditions in the
Alaskan Theater following the Battle of Midway,
and Aleutian Islands portion of the battle.[2]

In late afternoon on 19 June 1942, Mine Division 1 (less *Tracy*)—*Pruitt*,
Preble, and *Sicard*—stood out of Pearl Harbor. The division commander,
Capt. John F. Crowe Jr., USN, was aboard *Pruitt*. Designated Task
Group 6.8, the three ships were bound for Kodiak, by way of Seattle,
Washington, and the nearby Indian Island Ammunition Depot between
Port Townsend Bay and Kilisut Harbor. With Mokapu Point (the
southeastern-most feature on Oahu) to port, the destroyer minelayers
formed "column open order," spacing of 1,000 yards, came left to a
north-northeast course, and "rang up" the ordered speed of 16 knots.[3]

This formation was preferred for open ocean transits, because it
allowed all the ships to see each other, thereby facilitating visual
communications. The lead ship steered the ordered course and speed,
and served as the guide for the others following astern. The second ship
was typically positioned 4 degrees off the port quarter of the guide at
standard distance, and the third ship 2 degrees off the starboard quarter
of the guide at twice the standard distance. Additional ships alternated
between port and starboard positioning.

The three 314-foot, 1190-ton minelayers were all wellpast their prime, having been originally put into service as destroyers shortly after the end of World War I. The dates they were commissioned, and the identities of their present commanding officers are provided in the table.

Task Group 6.8 (CoMinDiv 1): Capt. John F. Crowe Jr., USN

Ship	Date Comm.	Commanding Officer
Preble (DM-20)	19 Mar 1920	Lt. Comdr. Harry Darlington Johnston, USN
Pruitt (DM-22)	2 Sep 1920	Lt. Comdr. Edwin Warren Herron, USN
Sicard (DM-21)	9 Jun 1920	Lt. Comdr. William Julius Richter, USN

BATTLE OF MIDWAY (3-7 JUNE 1942)

The three minelayers had been ordered to Kodiak to help strengthen the defenses of the U.S. forces based there, following the occupation by Japanese forces of Kiska and Attu Islands, during the Aleutian Islands portion of the Battle of Midway. Following neutralization of the U.S. Fleet at Pearl Harbor on 7 December 1941, Japan had rapidly expanded her empire into the South Pacific and Southeast Asia, and established a defensive perimeter of island positions to the southeast and to the south. This first phase of the war terminated in the seizure of the Dutch East Indies, whose resources Japan believed were necessary to sustain her power and position in the Western Pacific.[4]

In the second phase, Japan planned to seize additional outposts to guard the newly gained Empire against attack, to consolidate positions and to cut enemy supply lines. This included a plan to occupy Midway and the Aleutians in order to establish an outer defense line to the east and northeast of Japan. Occupation of these points would allow for the establishment of air coverage from these bases to a radius of 1,300 miles, sufficient for attacks against the Hawaiian Islands.[5]

A team of Pacific Fleet cryptanalysts broke Japan's top secret naval code on 15 May and were able to piece together Adm. Yamamoto Isoroku's plan to occupy Midway and the Aleutians. Informed of this intelligence, Admiral Nimitz had established Task Force 8 under Rear Adm. Robert A. Theobald on 21 May. Nimitz provided Theobald (who had been serving as his commander of Pacific Fleet destroyers) with a force of five cruisers, and units of Destroyer Division 11.[6]

Theobald would also have command of the sparse U.S. Navy, Army, and Canadian forces already based in Alaska. These forces consisted of the so-called "Alaskan Navy" commanded by Capt. Ralph C. Parker, USN; nearly two hundred planes (mostly Army bombers and fighters) under Maj. Gen. Simon B. Buckner Jr., USA; and the twenty

Navy Catalina PBY seaplanes of Patrol Wing Four. Having received his orders to prepare Alaska against a Japanese attack, Theobald left Pearl Harbor aboard the destroyer *Reid* bound for Kodiak, headquarters of the Alaskan Naval Sector.[7]

ALEUTIANS PORTION OF THE BATTLE OF MIDWAY

While fighting was taking place in the waters and skies around Midway Island, on 6 June, under cover of fog, Japanese forces had begun the occupation of Attu and Kiska whose only military installations were meteorological outposts. (These landings followed attacks on Dutch Harbor on the 2nd and 4th of June by enemy carrier-based aircraft.) Reports from these stations ceased the next day, and while it was not unusual for either station to miss broadcasting one or more weather updates, it was curious for both stations to miss all of them. On 10 June, after the fog had abated somewhat, a PBY Catalina seaplane reported substantial Japanese forces on both islands: four ships in Kiska harbor, one probably a cruiser and one a destroyer; and at Attu, a tent camp and numerous small boats and landing barges.[8]

Map 10-1

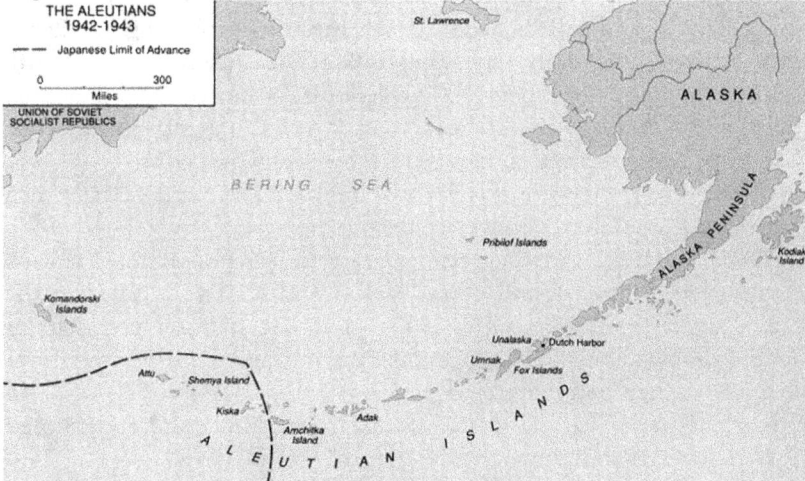

Japanese invasion forces landed at Attu and Kiska, at the westward end of the Aleutian Islands, in early June 1942. Kodiak Island lies at the eastern end of the chain.
Aleutian Islands: The U.S. Army Campaigns of World War II

The Aleutian portion of the Battle of Midway ended in mid-June 1942 with enemy forces occupying two islands of little value. Theobald's expectation of a fleet action had not materialized and the Aleutians battle became a contest of air power, as had Midway.

Reconnaissance flights during the latter part of the month revealed that while a majority of the Japanese ships had apparently withdrawn, the remaining strong landing forces on Kiska and Attu were digging in and establishing advance bases on the barren islands. Amid the stalemate in the Territory of Alaska, the Navy dispatched two additional flag officers to the theater. On 22 June, Rear Adm. W. W. Smith, who had been Nimitz's chief of staff, reported to commander, Task Force 8, embarked aboard the *Indianapolis*, and assumed command of the main body. Theobald and his staff then relocated to Naval Air Station, Kodiak. Four days later, Rear Adm. John W. Reeves Jr., formerly commander, Northwest Sea Frontier, relieved Capt. Ralph C. Parker as commander, Alaskan Sector.[9]

TRANSIT OF TASK GROUP 6.8 NORTHWARD

For the minelayers' first few days of the transit, the weather was generally fair, with good visibility. This began to change on 24 June, and the following day rough seas forced the ships to reduce speed to 12 knots. Tragedy struck in waters south of Cape Flattery, known to offer brutal offshore rides across building cross seas. Complex weather patterns, and wave sets (that finish their long ride across the fetch of the North Pacific, and reflect back off the massive, projecting headland) create an area of volatility. As *Sicard* took a 50-degree roll to starboard, Chief Gunner's Mate Robert H. Gray lost his footing on the galley deck house, and slid under the lifelines abaft No. 3 Gun, and into the sea. Gray had been serving as a member of the 4-inch gun crew.[10]

Three life rings were thrown in Gray's vicinity, and he was seen swimming toward one. However, despite being known to be a strong swimmer, the cold, dirty sea claimed him in the few minutes it took *Sicard* to swing around and come back for him. *Preble* joined in the search, which continued until 1835, when the division commander ordered the two ships to abandon their efforts, and form column order on *Pruitt*—which had continued on a northeast course at 5 knots. An understated *Sicard* war diary entry about the existing conditions, noted "Weather rough, visibility poor."[11]

The task group entered Puget Sound at 0821 the following morning, 26 July, passed Port Angeles at noon, and moored in late afternoon at Pier 41 in Seattle. Fueling and provisioning of *Preble*, *Pruitt*, and *Sicard* began at once. The ships got under way at 0800 on 29 July for the Indian Island Ammunition Depot, arriving there shortly after twelve noon. "Minemen" were sent ashore to inspect mines and prepare them for loading, while their shipmates aboard the minelayers performed upkeep and other work. The Mineman rating was first established on 12 October 1943 and disestablished 1 January 1948. The rating was then re-established 9 June 1948. Initially, Gunner's Mates who were assigned duties to work on mines wore a GM rating badge with the silhouette of a Mk 6 mine around the crossed guns (nicknamed "Keyhole" Gunner's Mates). These gunner's mates were commonly referred to as "mining men." The actual MN rating badge (shown here) was created when the rate was established in 1943.[12]

USN Enlisted Rating Insignia of World War II

Mineman 3rd Class (MN3c)	Mineman 2nd Class (MN2c)	Mineman 1st Class (MN1c)	Chief Mineman (CMN)

After being briefed on the results of the inspection, Captain Crowe advised the depot's inspector of ordnance and the operations officer for commander, Northwestern Sea Frontier (under whose command the ships would be operating) that four or five days would be required to thoroughly check the mines, disassemble and reassemble ones with deficiencies, and bring the resultant satisfactory ordnance aboard ship.[13]

The loading of mines and gun ammunition was completed on 3 July, and the ships sailed the following morning. *Sicard*, the last to leave the pier, joined *Pruitt* and *Preble* in a column formation and the group steered various courses and speeds conforming to the channel while heading out Puget Sound. A sighting was made of the Cape Flattery Light, three miles distant, at 1430, as the ships continued west at 17.5 knots, before turning to a northwest course. As was common in these waters, the weather was foggy and visibility poor.[14]

Photo 10-1

Ice Floes, Kodiak. Painting by Edward T. Grigware; 1943. In World War II, the United States' main base in the Aleutians was at Kodiak, Alaska.
Naval History and Heritage Command Accession #: 74-062-A

The transit was uneventful until early morning on 7 July, when at 0300 a report was received of enemy submarine activity. Aboard *Sicard*, No. 3 and 4 boilers were lighted, the propulsion plant split, and the crew called to General Quarters. (The term "split plant" refers to aligning a ship's support systems, engines, pumps, and other machinery so that two or more propulsion plants are available, each complete in itself. Each propulsion plant operates its own propeller shaft. If one plant were to be put out of action by battle damage, the other plant could continue to drive the ship ahead.) The ship secured from GQ at 0430, and an hour and a half later, Kodiak Island came into view. The three minelayers passed Chiniak Point abeam to port, and then entered the channel leading to Womens Bay. Mooring at Marginal Pier at 0907, they made preparations to receive fuel and, with no information regarding future operations, went into a modified upkeep status.[15]

Over the next several days, the crews of *Sicard*, *Pruitt*, and *Preble* performed upkeep and other ship's work, while the minemen inspected and overhauled additional mines awaiting at Kodiak. Upon arrival on 7 July, Captain Crowe had reported for duty to Rear Adm. Robert A.

Theobald, USN, commander, Alaskan Sector/North Pacific Force (Task Force 8). Theobald informed him that no mines would be laid until the North Channel had been wire-dragged, and a request was made that "Explorer" proceed immediately to Kodiak to carry out this action. The USC&GS ship *Explorer*, built in 1904, had been acquired by the Navy and commissioned USS *Explorer* on 3 June 1918. Returned to the Coast and Geodetic Survey on 31 March 1919, she eventually came into possession of the Army. Although now formally the survey ship USAT *Atkins* (FS 237), she was still routinely referred to as "Explorer."[16]

The mines at Kodiak were aboard lighters anchored in the bay. It was decided to use the storage shed at the end of Permanent Pier for the inspection, testing and reassembly process, and the lighters were moored alongside. In general, the same defects as in the mines at Indian Island were found. Work to correct these deficiencies progressed, and on 13 July, "Explorer" began wire-dragging operations in the vicinity of the location of the planned minefield. The field, designed to protect U.S. naval forces at Kodiak from attack by enemy submarines, was to be situated north and south of Humpback Rock, between Woody and Long islands.[17]

On 16 July, a Coast and Geodetic Survey officer came aboard the *Pruitt* to brief Captain Crowe. He reported that the main channel had been wire-dragged to a depth of 38 feet, and Leg Three of the minefield to 30 feet. Wire dragging by "Explorer" of Legs One and Two was then in progress, and completed on 18 July. *Sicard*, *Pruitt*, and *Preble* departed Womens Bay at 0830 on 19 July, bound for the mining area. Following a practice run, they planted Leg One at low slack water and then returned to port to load mines for Leg Two.[18]

They stood out of the bay the following morning and followed the same procedure; a practice run preceding planting Leg Two at low slack water. Leg Three was sown the following morning. The operation was completed with 575 mines in the water. The ships received sailing orders late that afternoon, and Mine Division 1 departed Kodiak that evening. Clearing the bay, *Pruitt*, *Preble*, and *Sicard* formed column, open order, and began the return trek to Pearl Harbor. Arriving there on 27 July, the minelayers moored at the Navy Yard to refuel.[19]

MINELAYING AT ADAK, ALASKA

Four minelayers were dispatched to the Aleutians in September 1942, to lay a larger series of fields at Adak (1,532 mines). Commander, Mine Division 1 left Pearl Harbor with *Pruitt* and *Sicard* (Task Group 6.5.2) on 16 September, after loading Mk 6 mines at West Loch. Upon arrival of the ships in the anchorage at Adak on 22 September, Captain Crowe reported to Brig. Gen. Eugene M. Lundrum, who was in charge of the Army forces ashore, and received instructions relative to the scheduled operations. Upon the arrival of *Ramsay* and *Montgomery* in early evening, Crowe assumed duties as commander, Task Group 8.12.[20]

Photo 10-2

Adak Harbor. Painting by William F. Draper; 1942. Dwarfed by snow-clad mountains, a Liberty Ship lies berthed at the pier as P38s and P40s practice maneuvers in the sky. Naval History and Heritage Command accession #88-189-AV

The responsibilities of the task group included mining, escorting and anti-submarine activities. Captain Crowe was charged with guarding the harbor, protecting shipping there, and acting as liaison officer to the Army. On 25 September, the four minelayers planted field Affirm A, after "Explorer" had wire-dragged the area.[21]

Also present at Adak were four PT boats of Torpedo Squadron 1 under the command of Lt. Clinton McKellar Jr., USN. On 24 October, PTs *24*, *27*, and *28* took receipt of some Mk 12 mines from *Montgomery* and *Ramsay*. They laid them in Umak Pass and in Kagalaska, to the east of Adak, the following day. This was the only time PT boats were so employed in the Pacific, but without success. Four of the mines exploded inside an hour and all within eight hours. The mines were variants of the submarine version Mk 12 Mod 0 bottom mines with a primitive "dip-needle" magnetic-firing mechanism. The Mk 12 Mod 2 mines may have exploded (self-destructed) after laying, due to magnetic anomalies in the region. They were likely a quick modification of the submarine mine, as they were never officially produced in quantity nor part of the long-term fleet stockpile of mines.[22]

Photo 10-3

Mk 12 Mine
Courtesy of Ron Swart

The final minelaying operation was completed on 8 November 1942. The month began with extremely heavy seas and high winds up to 60 knots, with gusts of 75 knots. The weather in the Aleutian Islands, especially toward the western part, is among the worst in the world. When sudden blasts of cold dense wind, called "williwaws," sweep down from snow and ice fields of coastal mountains, winds may increase to gale proportions; speeds of 100 knots are not uncommon. When these conditions occur, heavy seas and strong currents running through passes and channels near jagged island shorelines and shoals make navigation extremely hazardous. Rain is common even on good days and when it is not raining, there is normally fog. Moreover, it is a peculiarity of the area that fog and wind may persist together for many days at a time.[23]

After the weather abated somewhat, *Pruitt*, *Sicard*, and *Ramsay* loaded mines on 4 November and planted same. They loaded and planted another field on the 5th; loaded the next day; planted a field on the 7th; and upon return to port, loaded mines needed for the final operation on 8 November.[24]

MINE DIVISION ONE STAFF NAVIGATOR LAUDED

Comdr. Wilbur Haines Cheney Jr., USN, was awarded the Legion of Merit for his actions as Mine Division 1 mining officer and navigator. The medal citation attests to the challenges the Adak area presented the minelayers and their crews in carrying out operations:

The President of the United States of America takes pleasure in presenting the Legion of Merit to Commander Wilbur Haines Cheney, Jr., United States Navy, for exceptionally meritorious conduct in the performance of outstanding services to the Government of the United States as Division Mining Officer and also as Navigator of the U.S.S. *PRUITT* (DM-22), Flagship of Captain John F. Crowe, Jr., U.S. Navy, Commanding a Task Group of light mine layers engaged in the passages around Adak, Aleutian Islands in October and November 1942. He rendered extremely valuable assistance to the Task Group Commander in the preparation of the mines and in the navigation of the Task Group thereby contributing materially to the planting of highly efficient mine fields. As a result of his expert navigation in narrow, dangerous and poorly charted waters, in the face of extremely stormy weather, the Task Group most expeditiously and most accurately conducted the mine laying operations, resulting in complete blocking of the several bays and passages around Adak, thereby contributing greatly to the security of the U.S. Army forces occupying Adak in close proximity to the Japanese garrison at Kiska. His advice to the Task Group Commander pertaining to the final preparation of the mines and his advice relative to navigational matters in conducting the several approaches to the mining areas showed superb skill and were far beyond that normally to be expected.

U.S. ASSAULT LANDINGS/OCCUPATION OF ATTU

Under ferocious attack by enemy land, sea, and air, the two battalions to the fore were both almost smashed. We have barely been able to sustain this day. I arranged so the wounded and the ill in the field hospital were disposed of, the light ones by themselves, the serious ones by the medics. I made the civilian employees who were noncombatants each take up a weapon, form a unit, both army and navy combined, and follow the attack unit. We had them make a resolve [to die], lest we together suffer the shame of being taken prisoner while alive. It is not that there is no other way; I simply did not wish to sully the soldiers' last moments. We will carry out a charge with the heroic spirits [of those killed in battle].

—Col. Yamazaki Yasuyo, commander of the Japanese forces on Attu, describing his actions prior to ordering a Gyokusai (Banzai charge) by his remaining troops against U.S. Army assault forces.[25]

On 11 May 1943, units of the U.S. 17th Infantry (of Maj. Gen. Albert Brown's 7th Infantry Division) came ashore on Attu to retake the island from Japanese Imperial Army forces. *Pruitt* and *Ramsay* earned a battle star for their support of the fierce fighting ashore, carried out under brutal weather conditions, during the occupation of Attu (code named Operation LANDCRAB), as did also three destroyer minesweepers.[26]

Photo 10-4

Minelayer USS *Pruitt* (DM-22) leads landing craft from the attack transport USS *Heywood* (APA-6) toward their landing beaches in Massacre Bay, Attu, on the first day of the invasion, 11 May 1943. *Pruitt* used her radar and searchlight to guide successive waves of boats nine miles through the fog.
Naval History and Heritage Command photograph #NH 78232

Pruitt served as the control ship for the Southern Force landings at Massacre Bay on 11 May 1943, leading waves of assault boats from the transport area to the departure point from which the boats proceeded to beaches Blue, Yellow, and Rainbow. The landings at Massacre Bay proved difficult in the extreme due to heavy fog and vaguely charted shoal areas, which were only partially defined by incomplete information collected in 1934. *Ramsay* arrived with a small convoy at the bay on the 13th. She, along with the gunboat USS *Charleston* (PG-51) and destroyer minesweeper USS *Lamberton* (DMS-2), had escorted the transport ships USS *Grant* (AP-29) and USAT *Chirikof*—bringing additional Army troops—from Kuluk Bay, Adak.[27]

Aleutians Operation: Attu Occupation

Ship	Award Period	Commanding Officer
Pruitt (DM-22)	11-29 May 43	Lt. Comdr. Richard Claggett Williams Jr., USN
Ramsay (DM-16)	13 May-2 Jun 43	Lt. Comdr. Charles Helmick Crichton, USN
Chandler (DMS-9)	11 May-2 Jun 43	Lt. Comdr. Harry LeRoy Thompson Jr., USN
Elliot (DMS-4)	11 May-2 Jun 43	Lt. Comdr. Henry Mullins Jr., USN
Lamberton (DMS-2)	13 May-2 Jun 43	Lt. Comdr. Baxter Morrison McKay, USN

From 11-28 May 1943, the Americans fought the Japanese, as well as the weather, to take back the island. After two weeks of relentless warfare, American units managed to push the remaining Japanese defenders back to a pocket around Chichagof Harbor. In desperation, Col. Yasuyo Yamasaki, commanding a now much smaller force, decided to order a Banzai charge. In preparation, he burnt all his papers and issued orders that all wounded Japanese soldiers unable to take part in the assault be killed.[28]

The surprise attack broke through American front-line positions, and shocked rear-echelon troops (a mixed group of Army engineers, medical, and headquarters personnel), which found themselves engaged in hand-to-hand combat on the crest of a hill with Japanese soldiers. The fighting continued until only a few enemy remained. Twenty-eight Japanese were captured; 2,351 having been killed since the U.S. landings by combat, brutal conditions on the island, or by order of their commander. The Americans suffered 549 killed, 1,148 wounded, with an additional 1,200 soldiers debilitated due to severe cold injuries and 614 to disease (including exposure).[29]

Map 10-2

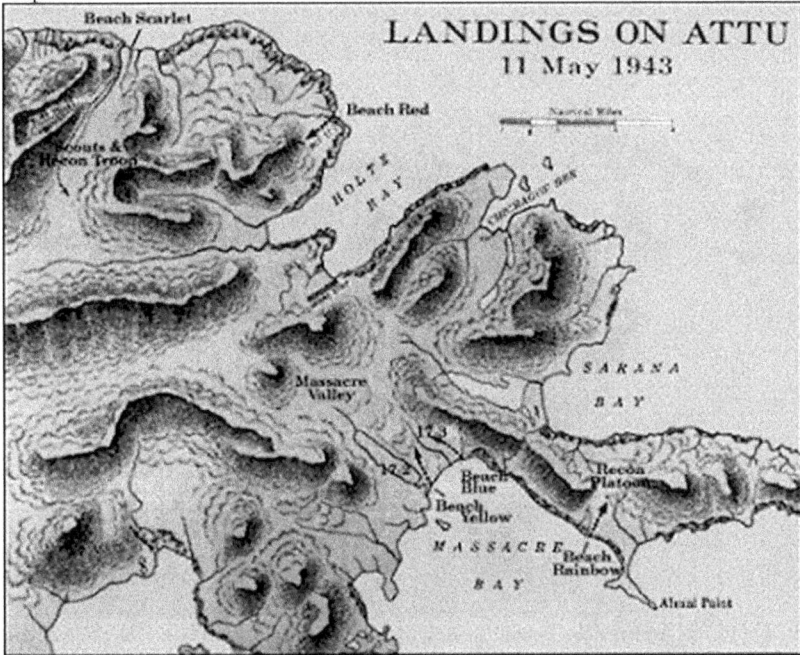

LANDINGS ON ATTU
11 May 1943

Western side of Attu Island in the Aleutians

ALLIED LANDINGS AT KISKA

The final stage in securing the Aleutians was retaking Kiska from the Japanese. A considerably larger combined Canadian-American force of thirty-three thousand assault troops was assembled to overcome the estimated 5,000-man garrison on the island. When the Allies went ashore on 15 August 1943, they found that the Japanese had evacuated the island on 28 July. The Allied occupation of Kiska marked the end of the Aleutian Islands Campaign.[30]

Photo 10-5

Kiska Harbor. A transport, probably the *Nissan Maru*, sinking and two other transports on fire after U.S. attack on the Japanese base on 16 June 1942, during the early days of the Aleutians Campaign (June 1942-August 1943).
National Archives photograph #80-G-11686

AFTERMATH

During the lengthy Aleutians Campaign, the elderly destroyer minelayers and minesweepers, growing rusty in their joints and hampered by excessive topside weight, took a beating in heavy seas generated by fierce winds. Some of these ships took part in the bombardment of Kiska and U.S. Army landings at Attu and Kiska, and continued escort and patrol duties before finally being pulled out in September 1943. *Sicard* collided with the destroyer *Macdonaugh* (DD-351) in heavy fog during the Attu landings, but was patched up and laying mines in the Solomon Islands six months later.[31]

11

Mine Division Two in the South Pacific

Photo 11-1

Pago Pago Harbor, Tutuila Island, Samoa, 1899.
Naval History and Heritage Command photograph #NH 1457

Within days of the attack on Pearl Harbor, Japan swallowed up Guam, Indochina and Thailand, and by Christmas had taken Wake Island and Hong Kong. Within two months, her forces had occupied Manila, Singapore, and British Malaya (now called Malaysia). In March 1942 the Allies lost Java and Burma, and Japanese armies were in the Owen Stanley Mountains of New Guinea, with the coast of Australia almost in sight. In May, Corregidor surrendered, the Philippines fell, and Japan invaded the Solomon Islands. There was no let-up in the progress of the Japanese; it appeared they would win the war if they could continue their march to Australia and New Zealand and overrun them.[1]

Thus, keeping open the 7,800-mile-long sea route from Panama to Sydney was a strategic imperative. The northern route, west-southwest from Hawaii, was already controlled by the Japanese operating from island bastions. Only the southern route, via South Sea Islands, was available for use. From Bora Bora, the midway point, shipping destined for Australia had to sail through or close to a number of the island groups. First came the Cook Islands, then the Samoa, Tonga, and Fiji groups, and finally, a thousand miles or so from the Australian coast, the New Hebrides group (today Vanuatu) and New Caledonia, forming the eastern rim of the Coral Sea. The utilization of these islands was critical to safeguarding the shipping route.[2]

One of the steps undertaken by the Navy to support the South Pacific supply line was the strengthening of American Samoa. The plan called for increasing facilities at the existing naval station at Tutuila and establishing two new advance bases to support land, sea, and air forces: on Upolu, an island under New Zealand mandate; and on Wallis, a French possession three hundred miles to the west. On both Upolu and Wallis, combined landplane and seaplane bases were to be built for the use of U.S. Navy and Marine Corps air units.[3]

Map 11-1

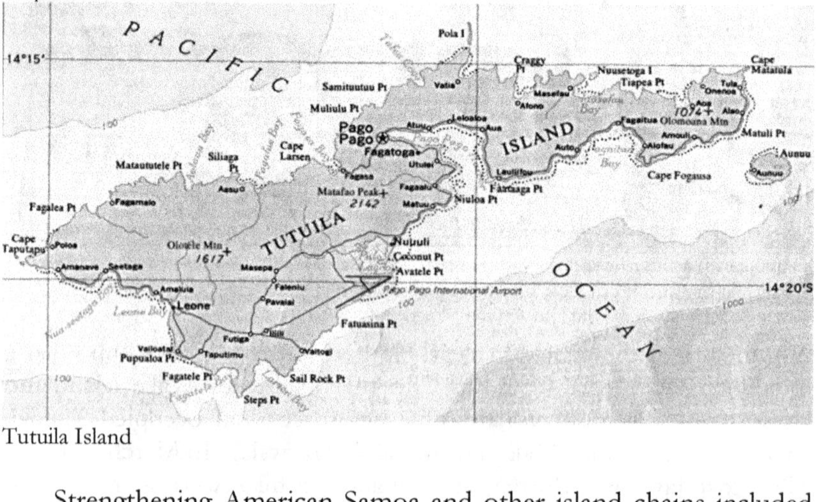

Tutuila Island

Strengthening American Samoa and other island chains included laying defensive minefields. On 22 February 1942, *Gamble* and *Ramsay* were en route from Pearl Harbor to the South Pacific, in company with the destroyer tender *Dobbin* (AD-3), the provisions stores ship *Bridge* (AF-1), and the chartered tanker *D. G. Schofield*. Commander, Mine Division Two (Comdr. Ross P. Whitemarsh, USN) was aboard *Gamble*. During the transit, *Ramsay* test fired her 4"/50 battery and 3"/23 anti-aircraft mount. Following their arrival at Pago Pago the morning of 4 March, *Gamble* and *Ramsay* reported to the naval governor for temporary duty, and minemen aboard the ships began preparatory work to the planned operations off Tutuila.[2]

Ramsay loaded twenty-three Mk 6 moored mines on 9 March, and laid them off Fagauso Point the next day. *Gamble* followed up this initial effort with a reinforcing line to existing minefield A on the 10th. The two minelayers next conducted a survey of other planned fields and over the next two weeks, laid fields C, D, and F. At completion, *Ramsay* left Tutuila at noon on 27 March to proceed independently to Apia, British Samoa. Later that day, *Gamble* joined the troop transport SS *President Garfield* en route to the same destination, providing her escort. *Garfield*, operated by the American President Lines, was subsequently purchased by the Navy, and commissioned as the attack transport USS *Thomas Jefferson* (APA-30) on 31 August 1942.[4]

Photo 11-2

"Apia facing the Entrance of Harbor" by Rear Adm. Lewis A. Kimberly, USN. Naval History and Heritage Command photograph #NH 42116

Following a stop at Apia, *Gamble* and *Ramsay* left the picturesque island on 31 March, bound for Suva, Fiji Islands. Arriving at Suva on 4 April, the two ships moored to King's Wharf. During forthcoming operations in Fijian waters, Comdr. Andrew D. Holden, RNZNR (commanding officer HMNZS *Matai*, and senior officer 25th Minesweeping Flotilla) assisted commander, Mine Division Two by providing the positions of Dan buoys and markers laid by his ship to establish the boundaries of the proposed fields. These operations were supported in part by delivery by the general stores issue ship USS *Castor* (AKS-1) of 480 Mk 6 moored mines to Suva.[5]

During the first operation on 7 April (in which the minesweeper USS *Kingfisher* AM-25 joined *Ramsay* and *Gamble* in laying mines), *Gamble* struck an uncharted coral pinnacle near Lovuka Island with her port propeller. Unable to operate it at speeds greater than 5 knots, and requiring shipyard repairs, she departed on 15 April for Pearl Harbor. Best obtainable transit speed was 18 knots, using her starboard engines with the port propeller stopped. *Ramsay* then operated alone until *Montgomery*, *Gamble*'s replacement, arrived at Fiji.[6]

A summary of ensuing operations in Fijian waters is provided below, with dates, areas mined, and numbers of mines laid in each area.

U.S. Defensive Mines Laid in Fijian Waters, April-May 1942

Date	Ship	Areas Mined	#Mines
7 Apr 42	*Gamble*	NW of Tivoa Island	80
	Ramsay	NW of Tivoa Island	80
	Kingfisher	NW of Tivoa Island	72
10 Apr 42	*Gamble*	Vicinity of Tivoa Island	80
	Ramsay	Vicinity of Tivoa Island	80
12 Apr 42	*Gamble*	Nuviui Island field	80
	Ramsay	Nuviui Island field	80
14 Apr 42	*Gamble*	Tuvarua Island field (47), Tivoa field (48)	95
	Ramsay	Tuvarua Island field (54), Tivoa field (41)	95
19 Apr 42	*Ramsay*	Rovondrau Bay	80
20 Apr 42	*Ramsay*	Nukumbutho Passage (10), Nukalau Passage (10)	20
21 Apr 42	*Ramsay*	Vicinity of Moturiki Island	87
23 Apr 42	*Ramsay*	Across Moturiki Channel (16), inshore of Naingani Island (69)	85
25 Apr 42	*Ramsay*	Off Naingani Island (83), Nananu Ira Passage (5), Malaki Passage (10), Nukurauvula Passage (12)	110
26 Apr 42	*Ramsay*	Natom Bu Ndrauivi Passage (10), Manava Passage (10)	20
28 Apr 42	*Ramsay*	Point Passage (42), Nyavu Passage (20), Kumbalau Passage (17)	79
1 May 42	*Ramsay*	Sausau Pass (21), Kia Island Pass (14), Mali Pass (50)	85
Total number of mines			1,308[7]

Preparations for each operation involved loading mines in port, sowing the designated area the following day, and then returning to port to load mines for the next operation. *Ramsay* was able on 25 April to lay more mines than she could carry, because a lighter from Nukulan brought her resupply. The term "Pass" in the preceding table refers to an entry into reefs, as opposed to "Passage" between islands.[8]

Map 11-2

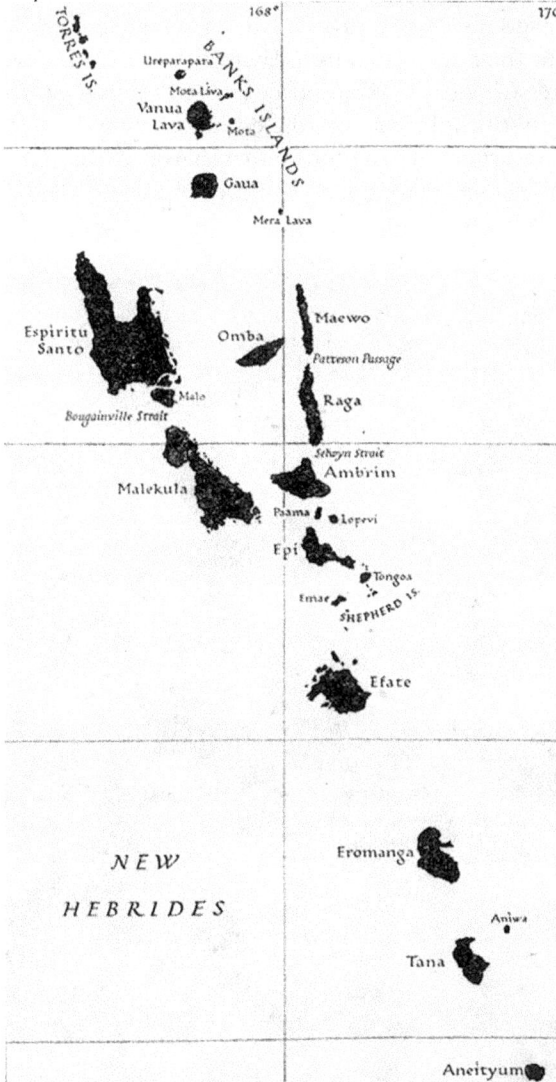

New Hebrides Islands (today Vanuatu)

THE NEW HEBRIDES

On 3 May 1942, *Ramsay* and *Montgomery* got under way from King's Wharf, proceeded out Mbenga Passage and headed west, bound for Port Vila, Efate Island, New Hebrides. Tucked inside a well-protected harbor on the southwest side of Efate, at the eastern end of Meli Bay, Port Vila was an important city and the site of an American and Australian air base. The two minelayers arrived at Efate on 5 May and anchored in Vila Harbor.[9]

On 10 May, *Ramsay* and *Montgomery* proceeded to Havannah Harbor at Efate, and carried out their first operation, laying a line of fourteen mines each in Milliard Channel. That afternoon, the two ships collectively sowed seventy-one mines across the entrance and westward of South Bay, before returning to the Fila inner harbor. In mid-afternoon on 12 May, *Montgomery* stood out, to proceed independently to Tongatapu, Tonga Islands.[10]

Map 11-3

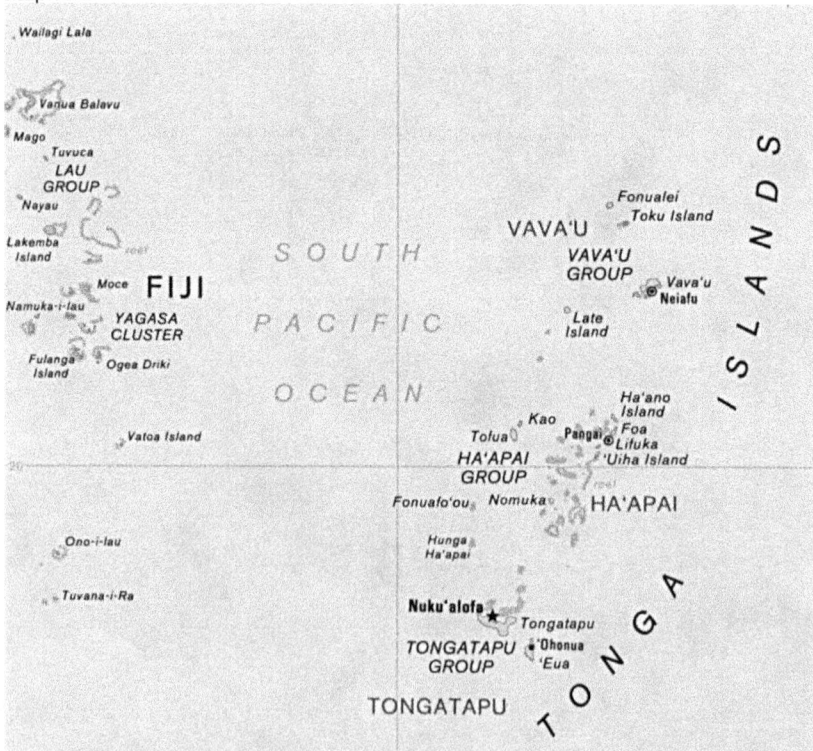

Fijian and Tonga Islands

TONGA ISLANDS

Montgomery arrived at Nuku'alofa Harbor, Tongatapu, on the 18th, following a stop at Suva, en route.   Two days later, she began investigating passages and reefs in the harbor and entrance for the possibility of laying mines. On 21 May, Lt. Comdr. John Andrews Jr., USN, relieved Lt. Comdr. Richard Allen Guthrie, USN, as commanding officer. Following completion of the survey, *Montgomery* laid eighty-six mines on the 24th, and eighty-one on 26 May. After lack of success in trying to sink a mine floating inside the reef during the latter operation, she patrolled offshore during the night to await daylight. At 0741 the following morning, rifle fire dispatched the potential "shipkiller."[11]

Photo 11-3

Men pushing U.S. Mk 6 moored mines along the dispersing track at Tongatapu, Tonga Islands, 8 June 1942. In the background is the destroyer tender USS *Whitney* (AD-4) and an oiler, either the USS *Cuyama* (AO-3) or the *Kanawha* (AO-1).
Naval History and Heritage Command photograph #NH 91170

After laying forty-nine mines on 2 June, unfavorable weather prevented operations on the 4th and 5th. Efforts resumed in following days, and on 10 June, *Montgomery* completed all mining operations at Tongatapu.

U.S. Defensive Mines Laid in Tongan Waters, June 1944

Date	#Mines	Date	# Mines	Date	#Mines
24 May 42	86	6 Jun 42	47	9 Jun 42	80
26 May 42	81	7 Jun 42	62	10 Jun 42	33
2 Jun 42	49	8 Jun 42	97	Total Mines	535[12]

Between 11 to 14 June, and 17 to 23 June, *Montgomery* assisted in planting buoys in Lahi passage. She departed Tongatapu on the 24th, in company with *Ramsay*, bound for Pearl Harbor. The two minelayers arrived in home port around noon on 3 July 1942.[13]

PRECEDING OPERATIONS BY *RAMSAY* AT EFATE

While *Montgomery* was engaged at Tongatapu, *Ramsay* had been finishing up mining efforts at Efate, and serving as a "general dogsbody." Mine warfare ships were frequently called upon to be "maids of all duties," and destroyer minelayers occasionally were tasked with combatant ship duties, as will be shown later in the book. Following the departure of *Montgomery*, *Ramsay* towed the coal hulk "Star of Russia" (pulled off the beach by the light cruiser HMNZS *Achilles*) to a berth next to the Liberty ship SS *Benjamin Franklin*. On 15 May, she proceeded to South Bay to provide a pilot to the French MV *Requin* to guide her through the South Bay minefield. Other duties included listening watches while at anchor in Fila Harbor, and patrols off the entrance to Meli Bay.[14]

U.S. Defensive Mines Laid at Efate, New Hebrides, May-June 1942

Date	Ship	Areas Mined	#Mines
10 May 42	*Ramsay*	Milliard Channel (14), South Bay (35)	49
10 May 42	*Montgomery*	Milliard Channel (14), South Bay (36)	50
28 May 42	*Ramsay*	Meli Bay	50
30 May 42	*Ramsay*	Ningut River (24), Marafong (10), Song (18), north Undine Bay (21)	73
9 Jun 42	*Ramsay*	south Undine Bay	80
11 Jun 42	*Ramsay*	Meli Bay	65
		Total Mines	367[15]

In preparation for her final lays, *Ramsay* assisted in unloading the Liberty ship SS *Paul Revere* on 1 June, and in mine assembly. In mid-afternoon on 11 June, her work completed, *Ramsay* left anchorage at Iririki Island in Meli Bay (near Port Vila), for Suva. After fueling to capacity there on the 13th, she resumed passage to the Tonga Islands, and entered Nuku'alofa Anchorage in early afternoon on 15 June to join *Montgomery*.[16]

12

Minelayers Sent Back
to the South Pacific

At the beginning of the month, we were operating under Service Force, Cincpac. We were told we had a small mining job to do in the South Pacific. We completed our mining job and we were then assigned to Comsopac and Comairsopac for service connected with the big push in the Solomons. That assignment completed we were further assigned to Comtaskfor 62 for further services in the supply of the Solomons. On this last assignment, we were hunted by cruisers and submarines and bombed by enemy bombers. The enemy came over pretty regularly three times a day. Of five escorts, two returned.

We are hardly the type for this sort of job. We have no torpedoes. Our minor caliber ack [anti-aircraft gunfire] is scarcely effective against anything but a torpedo plane attack. If we are to be used for other purposes than minelaying our capabilities would seem to confine us to ocean escort out of range of enemy bombers. We are excellent antisubmarine vessels for ocean escort because we can carry more depth charges than any type afloat by utilizing the entire length of our mine tracks. If we continue our present duties we will not be of any service to anyone in a very short time.

—USS *Tracy* (DM-19) August 1942 War Diary entry, describing being assigned to a variety of operational commands in succession, due to a critical shortage of newer, more capable combatant ships.[1]

On 22 July 1942, *Gamble*, *Breese*, and *Tracy*, then operating in Hawaiian waters, were assigned to Task Group 2.7, under Comdr. Ross P. Whitemarsh. *Gamble* and *Breese* were units of Mine Division Two, and *Tracy* of Mine Division One. While awaiting tasking to carry out their specialty, the ships had been employed in ocean escort of intraisland convoys, and offshore patrol duty at Pearl Harbor. Following receipt that morning of orders to the South Pacific for mining operations, *Gamble* moved from her berth at Pearl Harbor to Naval Ammunition Depot West Loch. Once *Breese* had cleared the pier, *Gamble* breasted in. In addition to mines, *Breese* also received one CGM (Miner's-Mate) from

West Loch for duty. *Gamble* transferred fourteen Mk 6 depth charges to the depot for temporary custody, and loaded eighty-five Mk 6 mines and two marker buoys. *Tracy* also obtained eighty-five mines.[2]

The minelayers stood out of Pearl Harbor in mid-afternoon, bound for Palmyra Island, a coral atoll lying 960 miles south of the Hawaiian Islands. Once clear of the harbor, the ships formed a line abreast (*Breese* and *Tracy* taking station 1,000 and 2,000 yards, respectively, on the starboard beam of *Gamble*, the flagship and formation guide), and set course 290°T, speed 15 knots. On the morning of 25 July, the group entered Palmyra Lagoon and moored alongside a fueling barge.[3]

Diagram 12-1

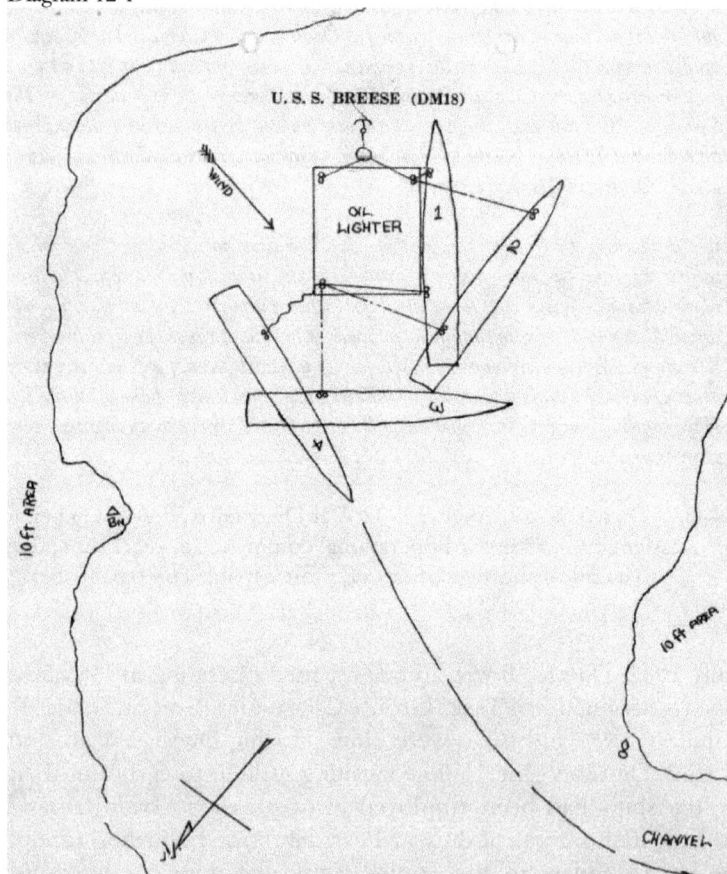

Use of mooring lines by USS *Breese* (DM-18) to depart Palmya Lagoon.
Breese War Diary, July 1942

A stiff, following sea with rollers across the reef made entry into the channel difficult, requiring *Breese*'s commanding officer to use right and left standard rudder to maintain course. While the ships refueled, a squadron of Gruman fighters came in to replace the Brewster fighters at the Naval Air Station. Due to a slight cross-wind, one crashed, nosing over on landing. Its propeller was bent, but there were no casualties. Following this brief stop for fuel, the ships set a southwest course for Suva, Fiji Islands.[4]

Breese had cleared the fuel barge, twisted round the oil lighter, by "winding" (see Diagram 12-1). This involved shifting the #4 line to the opposite corner of the lighter, and then taking in all but #2 and #4 lines. Number 2 line was then slacked, letting the wind ease the bow away from the lighter, before taking in the line, and allowing the wind to carry the bow around to position three. Kicking astern on the starboard engine slacked #4 line, and pointed the ship fair for the channel. After taking in the line, *Breese* headed outbound the restricted lagoon.[5]

On 29 July, the group passed Wailangilala Light to port, signifying they were then less than a half-day's travel from their destination. Crossing the International Dateline, the ships lost a day. Entering Suva Harbor the morning of 31 July, they moored at King's Wharf. Staying only long enough to fuel and provision, the ships stood out of the harbor, proceeded through Mbenga Channel and headed west, en route to Espiritu Santo Island, New Hebrides.[6]

MINELAYING OPERATIONS AT ESPIRITU SANTO

The ships passed through Selwyn Strait in early morning darkness on 2 August, and three hours later Bougainville Strait, making the run in to Espiritu Santo. Upon arriving at its destination at 0936 that morning, Task Group 2.7 commenced practice runs preliminary to laying mines at the entrance of Segond Channel. The purpose of these fields was to keep enemy submarines away from Allied ships in the new fleet anchorage at "Santo." At 1215, *Gamble* left the others to plant a buoy for the middle field, and rejoined at 1611. At completion of these runs, *Gamble* and *Tracy* stood out to sea. *Breese* closed the beach and landed a working party to paint white a prominent rock located midway between the two proposed eastern fields, for use as a navigation aid. That evening, a representative of Comairsopac (commander, Aircraft, South Pacific Force, Rear Adm. John S. McCain Sr., USN) came aboard *Gamble* for a conference with Commander Whitemarsh.[7]

The following day, the ships laid minefields at the eastern and western ends, and middle of the entrance to Segond Channel, expending a total of 171 mines. A "Q message" was sent to the fleet, warning of the danger areas when entering Espiritu Santo. Unfortunately, when radio circuits were crowded with higher precedence communications, ships requiring such information did not always receive them promptly. This was the case that night aboard the destroyer USS *Tucker* (DD-374) and the tanker SS *Nira Luckenbach*.[8]

Photo 12-1

Rear Adm. John S. McCain Sr. shaking hands with Vice Adm. Aubrey W. Fitch on the occasion of McCain's relief by Fitch as commander, Aircraft, South Pacific Force. Navy Photograph #80-G-43065, now in the collections of the National Archives.

LOSS OF THE DESTROYER USS *TUCKER* TO A MINE

As the unnamed patrol craft USS *YP-346* neared Espiritu Santo in the early evening on 3 August 1942, a destroyer challenged her via flashing light message. Following receipt of the correct response, the DD, also bound for Espiritu Santo, signaled "Good luck to you guys." The YP entered Segond Channel an hour past midnight and shortly after sighted a wrecked ship ahead that was breaking in two and folding up like a jackknife—the same one with which she had exchanged signals the previous night.[9]

After having bid them farewell, *Tucker* had unknowingly entered a defensive minefield laid earlier that day, and struck a mine at 2145. The destroyer was unaware of the danger, as she had received no radio warning regarding the existence of the new field. The explosion broke the *Tucker*'s back, killing three crewmen, with an additional three missing and presumed lost. In an effort to keep the DD from breaking in two, her captain ordered topside weight jettisoned, and sailors heaved depth charges, torpedoes, 20mm shells, and other portable gear overboard. Upon the arrival of *YP-346* at the scene, her skipper, Bos'n Joaquin Theodore, entered the minefield in an attempt to save the warship.[10]

Photo 12-2

Destroyer USS *Tucker* (DD-374) Jackknifed amidships and under tow by USS *YP-346* at Espiritu Santo, New Hebrides.
Naval History and Heritage Command photograph #NH 77030

The YP was able to tow the foundering *Tucker* clear of the field, fortunately without detonating any more mines, but despite her best efforts she was unable to beach the destroyer. The DD later grounded in the surf off the northwest coast of Malo Island, set there by strong wind and seas. In the interim, those aboard were either taken off or abandoned ship. On her arrival on scene around noon, the minelayer *Breese* found that most of the survivors had already been put ashore. The *YP-346* had taken about half the crew off and the *Nira Luckenbach* another dozen, while others had abandoned ship in the destroyer's whaleboat and life rafts. The minelayer took aboard the remaining thirty-eight men and three officers and, in the late afternoon, offered the services of three of its own officers to guide the *YP-346* and *Nira Luckenbach* through the field. The YP stood into port, while the merchantman chose to turn around, proceed out the channel, and make passage south of Malo Island. The beached destroyer later broke apart and sank after shifting offshore.[11]

(*YP-346* would likewise soon be lost. As Photograph 9-2 depicts, the former San Diego tuna clipper *Prospect* was destroyed by the Japanese light cruiser IJN *Sendai* off Guadalcanal on 8 September 1942. In later describing the incident, crewman Vincent Battaglia bitterly exclaimed, "Do you want to know where the Japanese navy was September 8, 1942? They were at Tulagi sinking a fishing boat.)[12]

The minefield claimed another unnecessary victim on 25 October 1942, when the Army transport *President Coolidge* was lost in the same location as had been the *Tucker*. Arriving at Espiritu Santo that morning, loaded with heavy weapons and 5,050 troops of the 172nd Infantry, 43rd Division, the ship's master had declined to wait for the pilot boat. As *Coolidge* proceeded inbound, in violation of regulations, a shore signal tower flashed an ominous message: STOP YOU ARE STANDING INTO MINES. Two explosions opened the transport's hull, and her master beached the ship a few hundred feet offshore at 0935. She hung there for about an hour, before wind and wave moved her off her perch, and she disappeared into the deep.[13]

GUADALCANAL-TULAGI LANDINGS

Two days after the minelaying operation at Espiritu Santo, *Tracy* took receipt of *Breese* and *Gamble*'s extra mines, and left that evening for Guadalcanal. Nearing her destination, *Tracy* sighted a vessel in a rain squall on 7 August, which identified herself as the seaplane tender *Mackinac* (AVP-13). The two ships then proceeded in company for the remaining short distance to Guadalcanal, and that afternoon, *Tracy* reconnoitered the channel between Maramasike and Malaita Islands.[14]

Japanese forces had occupied Tulagi, a small island nestled in a bay at Florida Island opposite Guadalcanal in the Solomons, on 3 May 1942. Finding it fit only for a seaplane base, the Japanese landed across the New Georgia Sound on the much larger island of Guadalcanal, on 5 July and began rapid construction of Lunga Point Airfield, from which the Empire's planes could menace the shipping lanes to Australia.[15]

In response to the Japanese establishing a toehold in the Solomons, U.S. naval forces had begun building a base on nearby Espiritu Santo Island on 28 July. In the escalating contest for control of the Solomons, 11,000 members of the 1st Marine Division landed at Guadalcanal on 7 August and captured the airstrip at Lunga Point the following day, as well as the principal Japanese encampment at Kukum located on the west side of Lunga Point. That same afternoon, Marines discharged at Tulagi took the Japanese-held island after fierce fighting, as well as the smaller islands of Gavutu and Tanambogo. *Tracy* and *Mackinac*, which arrived at Guadalcanal on the day of the initial Marine landings, were there to support these operations.[16]

Map 12-2

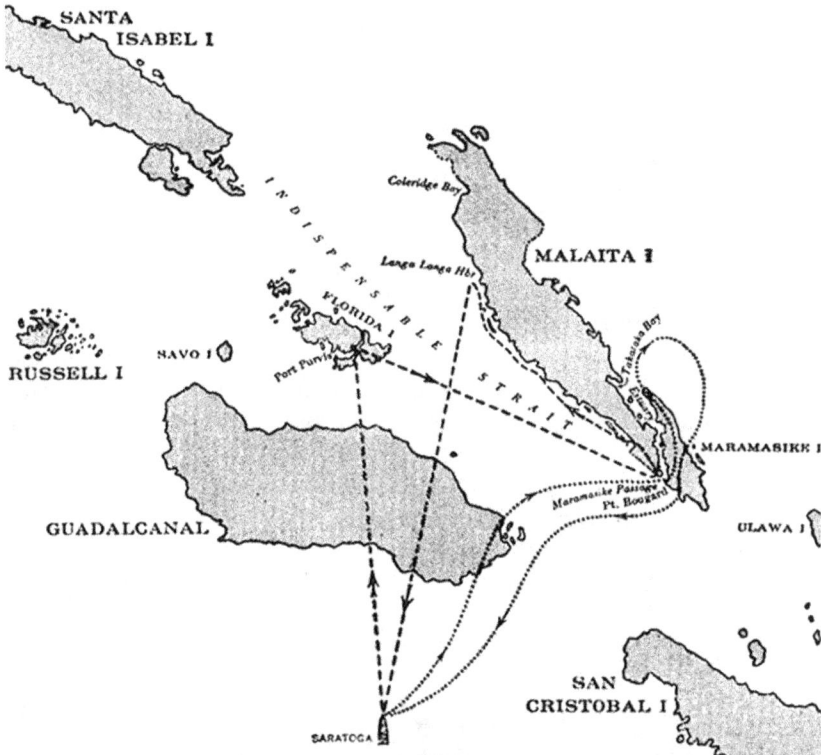

Air searches made by aircraft aboard the carrier *Saratoga* (CV-3), during the period in which the *Mackinac* was based in an estuary separating Malaita and Maramasike Islands to enable her patrol aircraft to search ocean waters to the northwest for enemy forces. *ONI Combat Narratives: Solomon Islands Campaign: I. The Landing in the Solomons, 7-8 August 1942* (Washington DC: 1943)

In support of the landings, *Mackinac* entered Maramasike Estuary at the south end of Malaita Island in the early afternoon on 7 August. One of the eight PBYs maintained aboard her flew ahead as she entered the passage between Malaita Island and Maramasike Island, dropping float lights upon coral heads to assist the ship in navigating the shoal water. From 7 to 9 August, her planes scouted for enemy forces in the waters to the northwest in the direction of the Japanese base at Rabaul, located 564 nautical miles from Guadalcanal on the northern tip of the island of New Britain.[17]

To help protect *Mackinac*, by denying Japanese entry into the estuary, *Tracy* sowed eighty-four mines in Maramasike Passage, along the east coast of Malaita Island on 8 August. Afterward, she maintained a patrol off the minefield until dawn when she anchored in the passage with *Mackinac*. That night, 9 August, an enemy task force of warships from Rabaul and Kavieng arrived off Guadalcanal. In the ensuing Battle of Savo Island, the Japanese scored a resounding victory, sinking the heavy cruisers USS *Astoria* (CA-34), USS *Quincy* (CA-39), USS *Vincennes* (CA-44), and HMAS *Canberra* (D33). "Navy brass" ceded the area temporarily to the Japanese the following day, and *Mackinac* and *Tracy* were ordered to retire to Espiritu Santo, due to the threat that enemy forces operating in the area posed to them.[18]

Month's end found *Tracy* back at Guadalcanal. Now attached to Rear Adm. Richmond K. Turner's Task Force 62 (Amphibious Force, South Pacific Force), she had been part of an escort for the transport USS *William Ward Burrows* (AP-6) and cargo ship *Kopara* (AK-62) to the Solomons. A 29 August entry in her war diary describes the minelayer coming under air attack, and provides an inkling of the conditions at Guadalcanal:

1100 Went to anti-air defense quarters. 1200 Eighteen silvery Japanese bombers at about 20,000 feet over Guadalcanal airfield dropped a stick of bombs and wheeled out over Lunga Roads where we were steaming preparatory to anchoring. A stick of about 5 bombs landed about 1,000 yards on the starboard bow of the TRACY. Dog fights over the TRACY drifted to the westward. About 6 miles westward a Japanese plane burst into flames and the pilot parachuted. We raced to the chute at flank speed. The pilot appeared to be loosening himself and hanging by his hands when he hit about a mile ahead. When we reached the chute there was no pilot. The harness and lines of the recovered chute were scorched and burned.

TRACY expended no ammunition because we have only small caliber A.A. guns and no plane came within range. We evaded attack by using full and flank speed and turning with easy rudder, using full rudder whenever planes were in position to drop bombs.

There were no casualties in spite of shrapnel hitting the decks.[19]

GAMBLE SINKS JAPANESE SUBMARINE *I-123*

On this pass through the area the First Lieutenant on the after deckhouse observed [an] air bubble six feet in diameter break the surface (7 minutes after the attack). Observed considerable submarine deck planking and other debris.... The deck planking was made up of sizes approximately 1" x 2" x 2 foot and 1" x 4" x 6 foot. Some of the planking still had 4" spikes or bolts in them and some showed that the bolts had pulled through leaving a 3/4" hole. A few planks had clean breaks across their length revealing a very white cross-sectional area indicating that the wood might be white oak. On other passes through this area small bubbles of air and oil were observed by men along the side of the ship.... A sounding on the fathometer showed the depth of the water to be 800 fathoms. All of the afore-stated facts lead to the conclusion that this submarine went to the bottom.

—Lt. Comdr. Stephen Noel Tackney, USN, commanding officer, USS *Gamble* (DM-15), describing in a report the aftermath of depth charge attacks made by his ship on a submarine. The victim of the minelayer proved to be the Japanese *I-123*.

Photo 12-3

Minelayer USS *Gamble* (DM-15) in December 1944
Courtesy of Richard J. Peterson, last commanding officer of *Gamble*

After *Gamble*, *Breese*, and *Tracy* laid the minefields at Espiritu Santo on 3 August, *Gamble* was assigned patrol and pilot vessel duties off the eastern entrance to Segond Channel. Her tasking entailed intercepting approaching groups of ships, delivering a pilot to each, and leading the ships through the protective minefields into the channel. The morning of 27 August, *Gamble* rendezvoused at sea with Task Unit 62.2.4, upon which Comdr. Ross P. Whitemarsh, USN (aboard *Gamble*) assumed duties as commander of the task unit.[20]

Once joined, the task unit formed up with the transport USS *William Ward Burrows* (AP-6) as guide, and cargo ship *Kopara* (AK-62) astern of her. The destroyer *Calhoun* (DD-85), minelayer *Tracy*, fast transport *Little* (APD-4), minelayer *Gamble*, and destroyer *Gregory* (DD-82) formed an anti-submarine screen—in that order from left to right— around the two large ships. On the morning of 29 August, a sighting was made at 0805 aboard *Gamble* of the conning tower (truncated in shape) of a large Japanese submarine, broad on her starboard beam at a distance of 9,000 yards. As the sub quickly submerged a minute later, Stephen N. Tackney, USN, the commanding officer of the minelayer, initiated a search for the enemy.[21]

Gamble came right to course 000°T, and proceeded at maximum speed to intercept the submarine, whose course was estimated to be 260° true. The minelayer slowed at 0822, and began a sonar search. She obtained a sound contact at 0844, bearing 295°T, range 900 yards. The range was too short for a good attack approach, but five depth charges were dropped, with depths set at 200 and 100 feet, alternately. Sonar contact was lost due to depth charge explosions. When regained, the sub had opened the range to 1,800 yards. At 0905, *Gamble* commenced a second attack on her wary adversary:

> Ranges were erratic indicating that submarine first headed towards the ship at six knots and then turned to the right and began running away at three knots. Propeller noises were heard at 900 yards range. Contact was lost at 280 yards. Ship swung left to establish lead angle and dropped five depth charges set for deep barrage (300-200 feet alternately).[22]

This attack was likewise unsuccessful, as were successive ones, it first appeared. Since *Gamble* had a limited supply of depth charges, Tackney decided not to drop on the fourth run, but to instead, analyze the submarine's evasive tactics. Would the enemy continue a high-speed run toward *Gamble* throughout the attack, or swing to either side when just short of a thousand yards from the ship? During the attack, one depth charge was dropped for good measure, even though it appeared to be too late. The short length of the submarine's sonar signature (due to her high relative speed) suggested that she had endeavored to "tuck in" under *Gamble*, and the small amplitude that she was running deep.[23]

Contact had been maintained on the fourth run, and a fifth run was begun at 1001. Because the contact was poor, no depth charges were dropped, and contact was lost. An extensive search was begun, during which evidence was found that the submarine had suffered damage from one or more of the previous attacks. After passing through the tail of an oil slick at 1126, *Gamble* followed the shiny path to a large pool of oil about a mile down the slick.[24]

The smell of diesel oil was now very pungent, and destruction of the minelayer's wounded quarry soon followed:

> It was quite evident now that the submarine had been stopped but upon hearing the destroyer [*Gamble*] approaching went ahead in an endeavor to get away. At 1140 contact was obtained, range 2,000 yards. Increased speed from 10 to 15 knots. Sound officer requested that the run be made at 10 knots. This was done. It was decided to drive home this attack. Range rate was very regular; bearing was constant; indicating submarine did not desire or was not able to use evasive tactics. The submarine's speed worked out to be two knots directly away. Contact was lost at 250 yards. At 1147 carried through the attack, dropping ten (10) depth charges with alternate settings of 300-200 feet.... The ship was slowed to regain contact. No contact could be obtained.[25]

At 1330, the search was discontinued on orders of commander, Task Unit 62.2.4, in order for *Gamble* to proceed to the rescue of four aviators from the *Saratoga* (CV-3) Air Group, stranded on Nura Island. A short time later, her motor whaleboat retrieved the pilot and crew of a downed TBF-1 Avenger torpedo bomber from the island.[26]

AFTERMATH

Within days of *Gamble* sinking the Japanese submarine *I-123*, enemy forces sent to the bottom three of the seven ships that had comprised Task Unit 62.2.4. On 30 August, Japanese bombers sank the destroyer *Calhoun* (DD-85) off Lunga Point, Guadalcanal, with the loss of 51 crewmembers. A week later, a Japanese surface force sank the destroyer *Gregory* (DD-82) and the fast transport *Little* (APD-4) off Lunga Point on 5 September 1942. *Gregory* and *Calhoun* were World War I vintage destroyers, as was *Little*, but modified to carry assault troops. Two boilers and two stacks along with all her torpedo tubes were removed during conversion to APD to provide space for landing craft, associated handling machinery, and troop quarters and stowage.[27]

PERSONAL AND UNIT AWARDS

Lt. Comdr. Stephen Noel Tackney, USN, was awarded the Navy Cross for his role in the destruction of the Japanese submarine; his medal citation follows.

> The President of the United States of America takes pleasure in presenting the Navy Cross to Lieutenant Commander Stephen Noel Tackney, United States Navy, for extraordinary heroism and distinguished service in the line of his profession as Commanding Officer of the Destroyer-Mine Layer U.S.S. *GAMBLE* (DM-15), during attacks on an enemy Japanese submarine in the Solomon Islands Area on 29 August 1942. While escorting ships supplying our newly-seized bases, Lieutenant Commander Tackney skillfully sighted and located by sound-search an enemy submarine lurking in the vicinity. For a period of four hours he made persistent and determined attacks against the Japanese craft. Although warned of an imminent hostile air assault, Lieutenant Commander Tackney and the men under his command dauntlessly continued their action until oil and wreckage on the surface mutely convinced them of the submarine's destruction. Lieutenant Commander Tackney's leadership, skill, and outstanding devotion to duty reflected great credit upon himself, his men, and the United States Naval Service.[28]

Five destroyer minelayers earned battle stars for operations in the area of Guadalcanal during the war, with *Gamble* and *Tracy* each garnering two. Chapter 1 describes the qualifying action for *Montgomery*, *Preble*, and *Tracy* between 31 January and 2 February 1943.

Anti-submarine Operations

Gamble (DM-15)	28 Aug 42	Lt. Comdr. Stephen Noel Tackney, USN

Guadalcanal-Tulagi Landings

Tracy (DM-19)	7-9 Aug 42	Comdr. John Leon Collis, USN

Capture and Defense of Guadalcanal

Gamble (DM-15)	30 Aug 42	Lt. Comdr. Stephen Noel Tackney, USN
Montgomery (DM-17)	31 Jan-2 Feb 43	Lt. Comdr. John Andrews Jr., USN
Preble (DM-20)	31 Jan-2 Feb 43	Lt. Comdr. Frederick Samuel Steinke, USN
Tracy (DM-19)	31 Jan-2 Feb 43	Comdr. John Leon Collis, USN

13

Consolidation of Southern Solomons

It now appears that we are unable to control the sea in the Guadalcanal area. Thus our supply of the positions will only be done at great expense to us. The situation is not hopeless, but it is certainly critical.

—Assessment by Adm. Chester W. Nimitz, commander in chief U.S. Pacific Fleet, on 15 October 1942 of the bleak situation American naval forces then faced in the Guadalcanal area. That same day, Nimitz chose Vice Adm. William F. Halsey (a man known throughout the Pacific for his fighting spirit) to replace Vice Adm. Robert L. Ghormley as commander, South Pacific Force.[1]

Attack – Repeat – Attack!

—Order sent by Vice Adm. William F. Halsey to Rear Admirals Thomas C. Kinkaid and George D. Murray, commander, Task Force 16 and 17, respectively, on 26 October 1942.[2]

During September and October 1942, there was an escalation in the fighting between Japanese and American forces on and around Guadalcanal. United States military planners were determined to keep the supply lines with Australia open while the Japanese were just as determined to cut them. In October, the Japanese decided to launch a major offensive to gain control of Henderson airfield on Guadalcanal, eliminate the 10,000 American troops on the island, and destroy all allied warships in the Solomons area. A massive naval force left Truk in the Caroline Islands, home base of the Japanese Combined Fleet, on 11 October to provide cover for the invasion forces with four aircraft carriers, four battleships, ten cruisers, and 30 destroyers. Truk lay 795 miles north-northwest of Guadalcanal.[3]

Two American carriers, *Enterprise* (CV-6) and *Hornet* (CV-8), the battleship *South Dakota* (BB-57), six cruisers, and fourteen destroyers were available to oppose this formidable armada. Vice Adm. William F.

Halsey, who had just taken command of the U.S. South Pacific Force, ordered his force to move north of the Santa Cruz Islands to intercept the Japanese Fleet and keep them from supporting the invasion force.

Photo 13-1

World War II poster featuring Adm. William F. Halsey, USN.
U.S. Naval History and Heritage Command

BATTLE OF THE SANTA CRUZ ISLANDS

The ensuing battle of the Santa Cruz Islands took place on 26 October without contact between surface ships of the opposing forces; all of the action was between aircraft. The *Enterprise*'s planes bombed the carrier *Zuiho*, knocking her out of action as far as the battle was concerned, and planes from the *Hornet* severely damaged the carrier *Shokaku* and the heavy cruiser *Chikuma*. The *Shokaku* would be out of the war for nine months. The Battle of the Santa Cruz Islands was the fourth carrier battle of the Pacific campaign and the fourth major naval engagement fought between United States and Japanese naval forces during the lengthy and strategically important Guadalcanal campaign.[4]

AMERICAN FORCES PREVAIL ON GUADALCANAL

Japanese forces made several attempts to retake Henderson Field between August and November 1942. The naval and land battles, and the smaller skirmishes and raids of the Guadalcanal Campaign culminated in the naval battle of Guadalcanal fought between 12 and 15 November. The battle was the last Japanese attempt to land enough troops to retake Henderson, but it was unsuccessful. The inability of the Japanese to capture the airfield doomed their effort on Guadalcanal, and they evacuated their remaining forces by 7 February 1943, conceding the island to the Allies. The importance of the Guadalcanal Campaign was summarized by Adm. William F. Halsey, USN, commander, South Pacific Force and South Pacific Area, "Before Guadalcanal the enemy advanced at his pleasure—after Guadalcanal he retreated at ours."[5]

CONSOLIDATION OF SOUTHERN SOLOMONS

KILL OR BE KILLED

—Declaration in the name of Vice Adm. William Halsey, USN, posted on a sign at Tulagi in 1943, reminding Navy servicemen of the realities of war.[6]

"Consolidation of Southern Solomons" is a phrase used to describe the period from 8 February to 20 June 1943, following the completion, of the enemy's evacuation of all its forces from Guadalcanal on 7 February 1943 (described in Chapter 1). Subsequently, the Japanese strengthened their positions in the upper Solomons, while still carrying out air raids and submarine attacks on shipping in the southern islands.[7]

Enemy resistance on Guadalcanal ended in early February 1943, in concert with the evacuation of Japanese troops on the island, but the Solomon Islands were far from being won. In the coming months, destroyer minelayers plied their trade all through the Solomons, as Halsey's forces pushed the Japanese farther and farther up the island chain. By May 1943, U.S. Navy offensive operations had moved a couple of hundred miles northwest of Guadalcanal to Kolombangara and New Georgia.[8]

Map 13-1

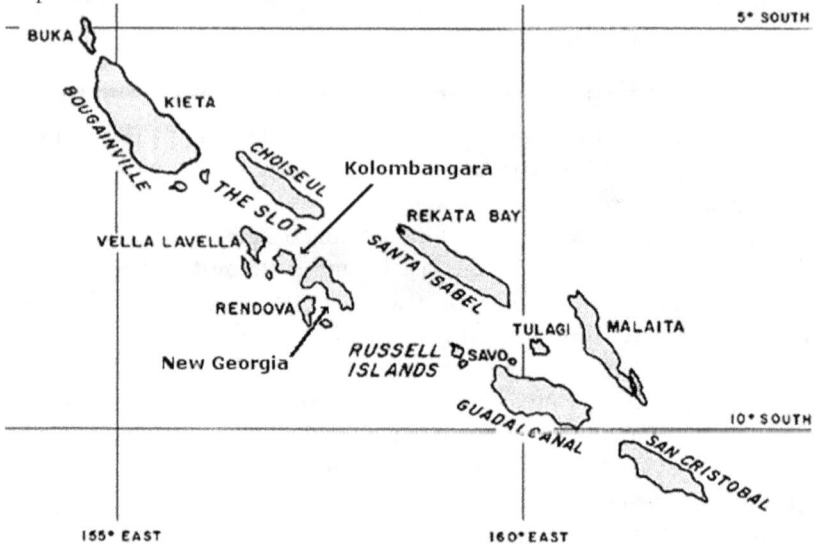

Southern Solomon Islands
http://www.ibiblio.org/hyperwar/USN/ACTC/img/actc-35.jpg

MINING OF BLACKETT STRAIT, KOLOMBANGARA

> *The Japanese advance base at Vila on Kolombangara Island is being supplied from their bases in the Shortlands area via Blackett Strait by convoys composed of small transports or barges. They may or may not be escorted by surface combatant vessels such as light cruisers or destroyers, and movement has been almost exclusively at night.*
>
> *This force will mine Blackett Strait near Ferguson passage during the night of 6-7 May, 1943, and will destroy any enemy forces encountered en route to or from that area.*

—Commander, Task Force 18 Operation Order No 6-34, 4 May 1943.

On 3 May 1943, *Gamble*, *Breese*, and *Preble* loaded eighty-five Mk 6 mines each at Espiritu Santo. *Breese* received hers while moored alongside the Liberty ship SS *James B. McPherson* in Berth 23, Segond Channel. The following afternoon, Task Group 35.6 sailed from Espiritu Santo for the Solomon Islands. After standing out, the destroyer *Radford* (DD-446), and minelayers *Preble, Gamble,* and *Breese* formed column open order, 700 yards spacing between ships, and increased speed to 17 knots. The ensuing transit was uneventful and in late morning on 6 May, the ships entered Tulagi Harbor to receive fuel. At completion in early afternoon, they proceeded northwestward up the Solomon Islands to carry out their orders.[9]

Map 13-2

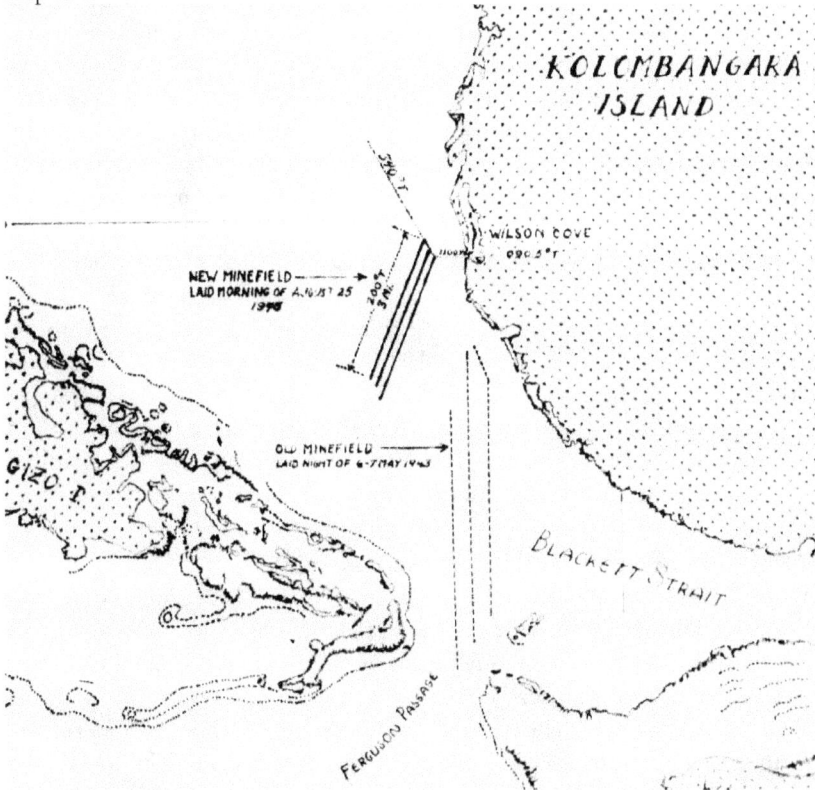

Locations of mines laid on 6-7 May 1943, and on 25 August 1945, in Blackett Strait. Commander, South Pacific, New Minefield, West Coast Kolombangara Island, 7 September 1943

Upon arrival off the southwest coast of Kolombangara, the sea was calm, the sky partially overcast, and the night dark. However, because the moon had not set, visibility was good. At five minutes after midnight, the three minelayers turned simultaneously. Though rain squalls had developed, they broke at times to reveal each ship to the other in perfect formation. Proceeding in a line abreast at 15 knots, each ship dropped a mine every 12 seconds, sowing over 250 mines in 17 minutes across Blackett Strait, the western entrance to Kula Gulf and part of the route of the "Tokyo Express."[10]

At completion of the mining, the task group sped north to join the protective screen of Rear Adm. Walden Lee Ainsworth's Task Force Eighteen, until detached to proceed independently to Tulagi to refuel. Task Group 35.6 arrived at Espiritu Santo in mid-afternoon on 8 May, fueled that day, and received mines from the SS *James B. McPherson* the day following. In late afternoon on the 10th, *Radford* and the three minelayers proceeded to sea, en route back to the still hotly contested Solomon Islands.[11]

Meanwhile, the mines laid in Blackett Strait produced rapid results. On the night of 7-8 May, four Japanese destroyers steamed through the mined waters. One, *Kurashio*, went down, two others *Oyashio* and *Kagero*, were badly damaged, bringing the fourth destroyer *Michishio* to the scene to provide assistance. Allied aircraft, alerted by a coastwatcher, Sub Lt. Arthur Reginald Evans, RANVR, intercepted the rescue operation, sinking the two destroyers and sending *Michishio*, now also badly damaged, limping back to port. (Evans would later become famous after witnessing on 2 August 1943 from atop Mount Veve on Kolombangara, the explosion of John F. Kennedy's *PT-109*. He did not realize, at the time, that it was an Allied craft, but later received and decoded a message that the PT boat was missing, and dispatched Solomon Islander scouts in dugout canoes to find the crew.)[12]

MINING OPERATIONS IN KULA GULF

Aboard *Breese*, in early morning darkness on 11 May, loose-play was discovered in the mines on both tracks. Unlike the previous such transit, this one could have quickly ended in disaster. General Quarters was sounded two minutes later at 0450, calling "all hands" to assist in securing the mines. The formation slowed to 10 knots to minimize pounding, and potential catastrophe. At 0543, formation speed was again increased to 17 knots after the mines were made fast.[13]

Task Group 36.5 entered Tulagi Harbor in late morning on 12 May, refueled, and stood out that afternoon. Upon joining Task Force 18 at 1559, the ships fell in astern, en route to New Georgia Island, Solomons. Midnight found *Radford, Preble, Gamble,* and *Breese* steering 270°T, speed 20 knots, formed in a column, preparatory to carrying out an order for a mining operation in Kula Gulf—which lay between the islands of Kolombangara to the west, and New Georgia to the south and east. After changing to the mining course, 110°T, the ships began their work. Concurrent with this action, Task Force 18 began bombardment of enemy positions on the adjacent islands. Upon completion of this field, the four ships joined the task force, which ceased bombardment at 0120, and retired eastward.[14]

As before, Task Group 36.5 was subsequently detached to proceed to Tulagi to fuel, and then make its way back to Espiritu Santo, where it arrived the afternoon of 14 May 1943.[15]

Photo 13-2

Light cruiser USS *Honolulu* (CL-48) conducting night bombardment on 13 May 1943, of Vila on Kolombangara Island, and Munda on New Georgia. Naval History and Heritage Command photograph #NH 76498

BATTLES OF KULA GULF AND KOLOMBANGARA

The consolidation of the Southern Solomons (8 February-20 June 1943) would be but one part of prolonged fighting, as Halsey continued his drive up the Solomons toward Japan. Almost immediately, Ainsworth's light cruiser force engaged Japanese surface forces in the Battle of Kula Gulf (6 July) and the Battle of Kolombangara (13 July). Around mid-afternoon on 5 July, Halsey's Third Fleet learned that the "Tokyo Express" was en route south from Bougainville with Japanese reinforcements for the central Solomons.   Ainsworth's Task Force 36.1—the light cruisers *Honolulu* (CL-48), *Helena* (CL-50), and *St. Louis* (CL-49) and four destroyers—intercepted the Japanese force in the Kula Gulf a few minutes past midnight.[16]

In a confusing action that lasted almost until dawn against the ten enemy destroyers (seven of which were being used as transports) under Rear Adm. Teruo Akiyama, Akiyama's flagship, the *Niizuki*, was sunk and *Nagatsuki* driven ashore, where she was destroyed by U.S. planes during the day.   However, amidst the remaining action, the destroyer-transports succeeded in unloading their troops on Kolombangara, and *Helena* was sunk by three torpedoes fired by the *Suzukaze* and *Tanikaze*.[17]

On his fifteenth patrol up the Slot, Ainsworth fought another night action with the "Tokyo Express" on 13 July.   The Japanese force under Rear Adm. Izaki Shunji was coming down the Solomons from Rabaul to land troops at Vila, on the southeastern shore of Kolombangara. Ainsworth had the light cruisers USS *Honolulu*, USS *St. Louis*, and HMNZS *Leander* (replacement for *Helena*) and 10 destroyers. The battle was joined past midnight and initially favored the Allies.   This changed as "long lance" torpedoes fired by Japanese destroyers began to find their targets. *Leander* was badly damaged by a torpedo, knocking her out of action, and the flagship *Honolulu*, cruiser *St. Louis* and the destroyer *Gwin* (DD-433) were also hit by torpedoes. Both cruisers were damaged but survived. *Gwin* exploded in a burning white heat—a terrible sight— and, unsalvageable, was later scuttled.[18]

UNIT AWARDS

The three minelayers received battle stars for the two mining operations. U.S. Navy ships could only receive a single star for any one operation, no matter how many engagements with the enemy.   The 20 March 1943 start of *Gamble*'s eligibility period reflects her being part of a convoy which came under a Japanese bombing attack on that date.

Consolidation of Southern Solomons		
Breese (DM-18)	6-13 May 43	Lt. Comdr. Alexander Bacon Coxe Jr., USN
Gamble (DM-15)	20 Mar-13 May 43	Lt. Warren Wilson Armstrong, USN
Preble (DM-20)	6-13 May 43	Lt. Comdr. Frederick Samuel Steinke, USN

14

New Georgia Campaign

Each of these ships carried 84 mines and it was not desired to unduly hazard them in a gunnery action. Unfortunately, they had to be stationed ahead to set the pace. Because of their age and their heavy mine loads there was some question as to their best sustained speed and, also, the possibility of a breakdown. Our suspicions were entirely unjustified. Not only did they maintain a constant speed of 26 knots en route to the Shortlands—which permitted the bombardment to start within two minutes of the time predicted—but they returned the entire distance of 300 miles at an average speed of 27 knots.

...This unit did a remarkably fine piece of work. The unit commander in PRINGLE, with her S.G. radar, did an excellent bit of navigation and the minelayers displayed superb seamanship in accurately maintaining their formation in rough weather during conditions of almost zero visibility and while making radical course and speed changes on clock time without signal.

—Observations by Rear Adm. Aaron Stanton ("Tip") Merrill, USN,
commander, Task Group 36.2, in his action report on minelaying
and bombardment in the Shortland-Faisi-Kolombangara areas
on the night of 29-30 June 1943.[1]

By early 1943, some Allied leaders, and most notably General MacArthur, wanted to focus on capturing the powerful enemy base at Rabaul, on the northeast coast of New Britain, but Japanese strength there and a lack of landing craft meant that such an operation was not practical in that year. Instead, on the initiative of the Joint Chiefs of Staff, Operation CARTWHEEL was developed, which proposed to envelop and cut off Rabaul without capturing it, by simultaneous offensives in New Guinea and northward through the Solomon Islands. The New Georgia Campaign, a series of battles that took place in the New Georgia group of islands, in the central Solomon Islands from 20 June-25 August 1943, was part of CARTWHEEL, the Allied grand strategy in the South Pacific.[2]

Map 14-1

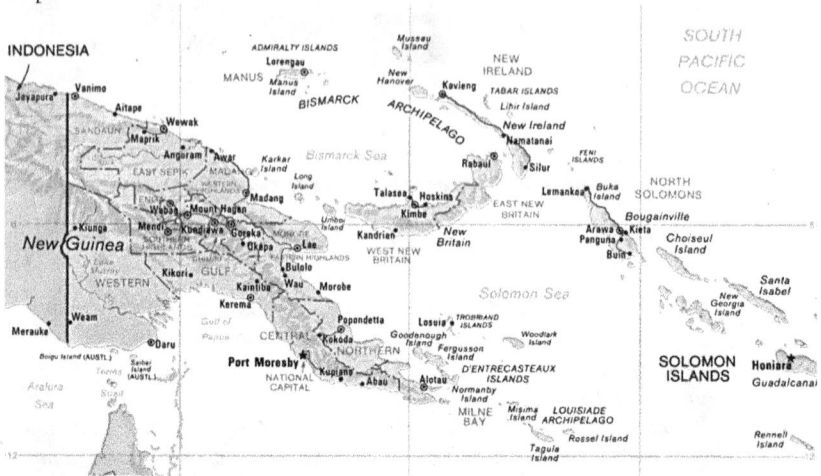

New Guinea and Solomon Island area. The strategy of Halsey's South Pacific forces and Gen. Douglas MacArthur's Southwest Pacific forces were to move northwest up through the Solomon Islands and up the New Guinea coast, respectively, then converge and break through the Japanese-held Bismarck Archipelago, upon which lay Rabaul.

Commander, Third Fleet, Admiral Halsey, provided the initial planning guidance for the occupation of New Georgia on 17 May 1943. Forces of the South Pacific Area were to seize and occupy positions in the southern part of the New Georgia Island Group preparatory to a full-scale offensive against Munda Point on New Georgia, and Vila on Kolombangara Island, also a part of the island group. From there, assigned forces would move to the northwest to attack Buin airfield on the south end of Bougainville Island, and to Faisi, a small island off Bougainville southeast of Buin, which hosted the headquarters of the Japanese 8th Fleet and a seaplane base. Bougainville—about 130 miles long and 30 miles wide—was the largest of the Solomon Islands, and mountainous, dominated by the Emperor and Crown Prince ranges. The specific tasks for the assigned forces were to seize, hold, and develop:

- A staging point for small craft in the Wickham Anchorage Area in the southeastern part of Vangunu Island and 50 miles from the Munda airstrip.
- Viru Harbor on New Georgia Island 30 miles southeast of the Munda airstrip.
- A fighter airstrip at Segi, New Georgia, 40 miles from the Munda airstrip.

- Rendova Island, whose northern harbor was just 10 short miles south of Munda, as a supply base, advanced PT Base, and an adequate support base to accommodate amphibians prior to their embarkation of troops for an assault on Munda and/or Vila.

Map 14-2

The Allies invaded the New Georgia Islands to capture Japanese airfields at Munda and Vila, and other enemy positions in the islands, as a part of Operation CARTWHEEL, a strategy designed to isolate and bypass Rabaul without capturing it. http://www.ibiblio.org/hyperwar/USN/ACTC/maps/actc-p486.jpg

More detailed planning for the operation was predicated upon the arrival in the South Pacific of an adequate number of LSTs (tank landing ships), LCI(L)s (large infantry landing craft) and LCTs (tank landing craft), some being built by commercial shipyards themselves newly constructed. Ultimately, Halsey's desire for an April or May D-Day was not realized due to constant slippage of the arrival dates of the necessary landing craft.[3]

PRIOR OCCUPATION OF THE RUSSELL ISLANDS

Immediately following the evacuation of all Japanese forces from Guadalcanal in February 1943, Admiral Halsey had set in motion plans for occupying the Russell Islands, so that they might be used as a staging point for the planned advance into New Georgia. About 9,000 infantrymen and Marines from New Caledonia landed unopposed at three different points in the Russells at about dawn on 21 February. By early June, the Seabees had completed building two air strips and the fuel tanks and other supporting facilities necessary for an invasion of New Georgia Islands.[4]

NORTH TO THE DANGEROUS UPPER SOLOMONS

In support of Operation TOENAILS, the planned invasion of New Georgia, Task Group 36.2 sailed from Tulagi in the early afternoon on 29 June, bound for the Shortland Islands. The Shortlands lay in the extreme north of the Solomons, just to the south of and within sight of Bougainville. Comprising the task group under Rear Admiral Merrill (commander, Cruiser Division Twelve) were two task units.

Task Group 36.2: Rear Adm. Aaron Stanton Merrill, USN

Task Unit 36.2.1	*Montpelier* (CL-57), *Denver* (CL-58), *Columbia* (CL-59), *Cleveland* (CL-55), *Philip* (DD-498), *Renshaw* (DD-499), *Waller* (DD-466), *Saufley* (DD-465)
Task Unit 36.2.2	*Pringle* (DD-477), *Preble* (DM-20), *Gamble* (DM-15), *Breese* (DM-18)[5]

The mission of Task Unit 36.2.2 was to lay mines in Bougainville Strait to temporarily disrupt Japanese surface raids out of Buin (on Bougainville Island), via the Shortland Island route, during the early phase of the New Georgia operations. Other commander, Third Fleet objectives for the mission by Merrill's task group included the bombardment of enemy targets ashore:

- Lengthen the Japanese surface routes to New Georgia
- Place a suitable surface force in position to cover the New Georgia landings
- Provide a diversion for the New Georgia operations
- Temporarily reduce enemy air strength and installations on Ballale in the Shortland Islands group
- "Soften up" enemy installations in the Shortland Island-Faisi Island areas of the Shortlands[6]

The cruisers and destroyers of Task Unit 36.2.1 were to provide protection to the mining detachment, and also conduct bombardment against Japanese-held islands in the vicinity. The two task units had proceeded separately from Havannah Harbor, Efate, New Hebrides, and joined up at Tulagi.[7]

North of the Russell Islands, the sky was mostly overcast with occasional local rain squalls. As the group drew nearer the Russells, the weather cleared considerably, making conditions less encouraging as the ships were then more vulnerable to detection by Japanese aircraft. Fighter protection from Allied land-based air could not be extended to the area of the forthcoming operation because of the distance involved, and due to requirements to support landings on Rendova and New Georgia.[8]

As the task unit continued on toward the Shortlands, *Pringle* was the guide, with the minelayers in column astern. The cruisers were formed in a line of section guides with the destroyers in an anti-submarine screen ahead of the formation. General Quarters stations were manned just prior to sunset, and destroyers *Renshaw* and *Waller* were sent ahead at 32 knots to bombard Vila on Kolombangara. Judging that it was likely that the group of twelve ships had been sighted by enemy planes, this action was meant as a diversion. The bombardment was visible from *Montpelier*'s bridge twenty-six miles to the north, as were flashes from enemy shore batteries returning fire—slow and deliberate from the northeast coast of Kolombangara.[9]

The remainder of the approach was uneventful. Radio reports of enemy sightings during the day led to the belief there were seven enemy destroyers, and perhaps two heavy cruisers and one light cruiser, in the Shortlands area. Prior to darkness, Merrill sent a message to *Pringle*, "Should enemy destroyers or other surface craft be encountered generally ahead during the night, mining group execute radical turn away. Avoid detection and clear the range for cruiser fire."[10]

A cheerful development was encountering increasingly foul weather, as the ships continued their transit up the Slot. Black clouds seemed to be within reach overhead, the occasional rain squalls merged into a solid downpour, and visibility decreased almost to zero with little improvement during lightning flashes. Somber news followed that a planned air raid by MacArthur's forces on Rabaul had been cancelled due to the bad weather. A similar report from Guadalcanal cancelled all planned strikes against the Japanese Kahili airfield, Ballale field, and the Faisi floatplane base. This was very bad news, as those strikes were expected to render enemy air less potent during the group's long retirement through Japanese waters, without Allied air coverage, following the pending operation.[11]

MINING OF BOUGAINVILLE STRAIT

Thirty minutes past midnight, *Pringle* changed course to 270°T and slowed to 18 knots. The final approaches to the mining course and the bombardment course, respectively, of the two task units was made at reduced speed to lessen the possibility of enemy aircraft detection of ships' wakes, and to permit time for the very accurate navigation required. At 0127, when Munia Island bore 024°T, distance 2.8 miles, and water depth was 100 fathoms, the ships came left to the mining course 250°T.[12]

Preble turned in the wake of *Pringle*, and the other two minelayers turned at the same time, forming a line of bearing from *Preble* of 115°T. In heavy rain with visibility less than 500 yards, the ships put down a standard three-row minefield in Bougainville Strait. Despite these conditions, a high level of phosphorescence in the water made the wakes of the ships distinctly discernable. *Pringle* laid a smokescreen which, aided by winds from the south, served as an effective cover for the minelayers. On completion of the mining operation, the task unit increased speed to 27 knots and turned to the retirement course, crossing the path of the bombardment group at 0147, before its commencement of naval gunfire against shore targets.[13]

As the mining operation had been undetected, bombardment was withheld until the mining detachment cleared the area. Gunfire by the cruisers was conducted in a downpour of rain, accompanied by lightning and strong southeasterly winds. Tracer fire could not be seen by gun crews much beyond the muzzles. It was improbable the enemy could see the gun flashes, and observers aboard the ships could not see fires believed caused by their gunfire, six to nine miles distant.[14]

RETIREMENT TO SAFETY

Retirement of the bombardment group was at 30 knots until the minelaying detachment was overtaken, when the speed of the task group was reduced to 27 knots. Dawn broke with a clear sky to the south and west, but passage close to land where heavy cloud cover still persisted, offered screening. At daylight, seven USAAF P38s appeared, cancellation of night strikes having made them available. They cruised back and forth at low altitude against the clear sky background, where they could see the ships and be seen by any threats to them. Just east of Rendova, many bogies appeared on radar, evidence of the many air battles taking place over the island.[15]

AFTERMATH

In his report on the operation carried out by Task Group 36.2, Rear Admiral Merrill expressed his belief that future mining operations should be conducted without attendant bombardment, as such action would give the enemy warning of the mining of waters:

> Except in the forward enemy held areas, where the Japanese are unable to maintain adequate sweeping services, it appears that simultaneous bombardment and minelaying operations by own forces are of decreasing value, save as the mining operation tends to delay pursuit by superior enemy forces. Three such combined operations have been undertaken recently by South Pacific surface forces of which the first laid mine field only is known definitely to have produced a considerable "bag" of enemy ships. It is now felt that each bombardment suffered by the enemy will alert him to the immediate need of extensive sweeping prior to initiating any ship movements in the area involved. For this reason, it is believed that mining and bombardment henceforth should, as a general rule, be undertaken separately.[16]

As the Japanese submarine *RO-100* transited Bougainville Strait on 25 November 1943, a mine exploded beneath her hull, portside amidships, two miles west of Oema Island. The explosion threw the commanding officer, Lt. Okane Hisao, and lookouts on the bridge into the sea. The engineering officer gave the order to abandon the sinking submarine. While trying to reach Buin, some of the survivors were killed by sharks. Of her fifty-man crew, all but twelve perished owing to the explosion, being trapped in the submarine, or perils of the sea.[17]

MINES LAID OFF KOLOMBANGARA IN AUGUST 1943

At midnight on 24 August 1943, Task Group 34.9—*Pringle*, *Preble*, *Montgomery*, and *Breese*, covered by Destroyer Division 41—arrived off Wilson Cove, on the western side of Kolombangara. The ships had made their way to the designated area by hugging the western shore of the island. From a starting point 1,100 yards to the west of Wilson Cove, the minelayers laid a three-line field, using radar ranges and bearings for navigation. The mining course and speed was 200°T, 15 knots.[18]

Upon completion of mining at 0056, the minelayers increased speed to 20 knots and turned to starboard, during which *Preble* and *Montgomery* collided. Both were heavily damaged, but able to return to Tulagi; *Preble* under her own power, and *Montgomery*, taken in tow that afternoon by the fleet tug *Pawnee* (AT-74). While the destroyers *Nicholas*

(DD-449), *O'Bannon* (DD-450), *Chevalier* (DD-451), and *Taylor* (DD-468) covered the mining operation in Vella Gulf, enemy aircraft dropped flares and bombed the formation. No damage was reported, and the planes were thus diverted from interrupting the minelaying.[19]

Additional air attacks were averted by a fortuitous rain squall as the task force retired across Kulu Gulf. Transit speed was limited to 9 knots, due to the damage to the minelayers, and the destroyers formed a protective screen around the mine craft.[20]

PERSONAL AND UNIT AWARDS

Comdr. Alexander Bacon Coxe Jr., USN, received the Bronze Star for his command of the minelaying operation on 29 June 1943; his citation reads in part:

> For meritorious service to the Government of the United States as Commanding Officer of a Destroyer-Minelayer operating as a unit of a Task Group during minelaying operations and bombardment of Japanese shore positions on Kolombangara, Shortland and Bougainville, British Solomon Islands, on the night of 29 June 1943. Operating in poorly charted waters under cover of inclement weather, Lieutenant Commander Coxe skillfully executed his assigned missions and brought his ship through without damage. By his courage and able seamanship, he contributed materially to the successful mining of waters used extensively by enemy vessels.

Four destroyer-minelayers earned a battle star for their support of the New Georgia campaign.

New Georgia-Rendova-Vangunu Occupation		
Breese (DM-18)	29 Jun-25 Aug 43	Lt. Comdr. Alexander Bacon Coxe Jr., USN
Gamble (DM-15)	29-30 Jun 43	Lt. Warren Wilson Armstrong, USN
Montgomery (DM-17)	24-25 Aug 43	Lt. Comdr. Dwight Lyman Moody, USN
Preble (DM-20)	29 Jun-25 Aug 43	Lt. Comdr. Frederick Samuel Steinke, USN

15

Occupation and Defense
of Cape Torokina

There were centipedes three fingers wide whose bite caused excruciating pain for a day, butterflies as big as little birds, thick and nearly impenetrable jungles, bottomless mangrove swamps, man-eating-crocodile-infested rivers, millions of insects, four types of rats larger than house cats, and heavy torrents of rain bringing enervating humidity. And sacred skull shrines, reminders of days of cannibalism and head-hunting.

—James Bradley and Ron Powers, describing in the book *Flags of Our Fathers* the conditions encountered by U.S. Marines on Bougainville in the Solomons

As a part of the plan to take Rabaul, the Marine 4th Raider Battalion, with Army reinforcements, made an amphibious landing at Segi Point on the southeast cape of New Georgia, Solomon Islands, on 21-22 June 1943 to begin the New Georgia offensive. By 25 August, this force had pushed through impenetrable jungles across a portion of the forty-five-mile-long island, swept the last Japanese defenders into the sea, and captured Munda airfield as well as Bairoko Harbor along the northwestern shore of New Georgia. The next objective was to seize an area on Bougainville, in order to build an airstrip from which bombers could fly airstrikes against Rabaul.[1]

BOUGAINVILLE

Bougainville, the largest island of the Solomons, lay near the northwestern end of the island chain, some 190 miles east of Rabaul. The mountainous island is dominated by the Emperor and Crown Prince Ranges, which include two active volcanoes, the largest of which, Mt. Balbi, rises to 10,171 feet. The dense jungle on the lower mountain slopes and coastal plains, and the swamp areas immediately inland from the beaches, are due to an annual average precipitation of 100 inches. These conditions made the island, as one Marine termed it, "a wet hell;" malaria and other tropical diseases were prevalent. There were approximately 54,000 natives, and Bougainville was one of the few places in the South Pacific with credible reports of headhunting still taking place. The Japanese had invaded the island, and the adjacent smaller Buka, in early 1942 when fewer than twenty Australian troops and a few naval coast watchers were based there. The soldiers withdrew into the jungle to observe the enemy until evacuated, while the coast watchers remained to report enemy air and sea movements. Their warnings of convoys and air raids coming down the Slot had helped the Marines to achieve victory on Guadalcanal.[2]

Map 15-1

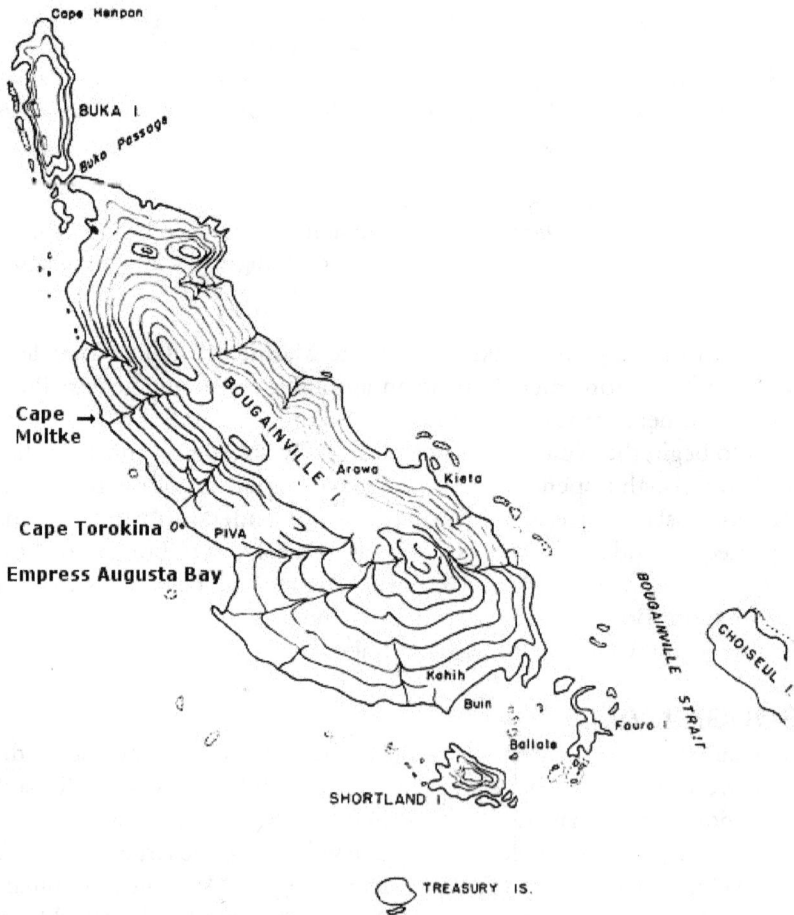

Bougainville, Solomon Islands
http://www.ibiblio.org/hyperwar/USN/Building_Bases/maps/bases2-p271.jpg

On 1 November 1943, the 3rd Marine Division (Reinforced) of the I Marine Amphibious Corps launched Operation CHERRY BLOSSOM with an amphibious assault landing at Cape Torokina, Bougainville. The division's orders were to seize an initial beachhead and to occupy and defend the area between the Laruma and Torokina Rivers to a distance of 2,250 yards back from the water's edge. The Marines were to prepare to continue the attack in coordination with the U.S. Army's 37th Infantry Division upon the latter's arrival, subsequent to D-Day, in order to extend the beachhead sufficiently to establish naval facilities and an airfield in the Torokina area.[3]

The Japanese opposed the landing with airstrikes from Rabaul. American forces countered with fighters launched from airfields at Munda, New Georgia, and Vella Lavella. Located west of New Georgia, the latter island was home base for Marine Fighter Squadron VMF-214 (the "Black Sheep" led by Maj. Gregory Boyington), which at the time was enjoying some rest and relaxation in Sydney, Australia, before returning to a second combat tour on 27 November 1943.[4]

MINELAYING OFF CAPE MOLTKE, BOUGAINVILLE

0010: Mine control ready to commence mining.

0030: Sighted enemy float plane close aboard, circling the formation. Did not fire on plane, as ships were within range of possible shore batteries. Plane flashed green light twice.

0149: Enemy plane dropped white parachute flare on port quarter. Two or more planes shadowing the formation.

0158: Plane dropped float flare on port bow.

0213: Enemy plane dropped flare ahead of [destroyer] RENSHAW.

0230: Radar contact on friendly cruiser force on starboard bow. Cruiser force passing to starboard at high speed on approximate northerly course. This force was covering unit for the mining operation.

0240: Intercepted TBS conversation by covering force indicating that they made surface contact with enemy forces.

0247: Covering force opened fire on enemy.

—USS *Sicard* (DM-21) 2 November 1943 deck log entries, describing conditions off Bougainville, as she, *Breese*, and *Gamble* laid mines at Cape Moltke and then retired with the destroyer *Renshaw* at high speed back down the Slot to the Russell Islands.[5]

Five days before the Marine landings at Cape Torokina on 1 November, *Breese, Sicard,* and *Gamble* left Espiritu Santo in the early afternoon on 27 October—bound for the Solomon Islands loaded with mines. Outside the port, a torpedo bomber appeared overhead to provide gun crews practice in anti-aircraft fire. As the aircraft made repeated runs towing a sleeve, each ship in turn had opportunity to fire 3"/50 and 20mm batteries. On the morning of 29 October, the ships entered Purvis Bay at Florida Island (across the Slot from Guadalcanal) to receive fuel from the oil barge *YON-146*.[6]

In early evening on 31 October, the minelayers cleared the harbor and proceeded northwestward. Task Unit 31.8.1 was headed up the Slot to sow an offensive minefield off Cape Moltke, Bougainville. The cape lay beyond Torokina, up the northwest coast of Bougainville, and the waters offshore in the path of "Tokyo Expresses" coming down from Rabaul. Ten minutes past midnight on 2 November, the ships (joined, less than two hours earlier by destroyer *Renshaw* DD-499) arrived off the cape and were preparing to commence operations, when an enemy float plane was sighted circling the formation. The ships were darkened, showing no lights, and the plane eventually moved away. Mining commenced at 0110, and two hundred fifty-five Mk 6 mines went into the water in sixteen minutes.[7]

Retirement down the Slot to the Russells was made without incident, except for searchlights and shore battery fire in the general vicinity of the Masamasa Island-Faisi Island-Buin, Bougainville area firing at an unknow target. The minelayers stood into Sunlight Channel, Russell Islands, in the late afternoon and moored alongside the *Erskine M. Phelps*. Built in 1898, this hulk of a former steel-hulled four-masted barque had been taken over by the Navy and designated YO-147 for use as a bunker oil depot ship.[8]

BATTLE OF EXPRESS AUGUSTA BAY

A Japanese task group sent from Rabaul to Empress Augusta Bay to attack U.S. Navy transport ships in order to break up their landing, found that Rear Admiral Merrill's Task Force 39 was blocking the bay entrance. The ensuing battle took place the night of 1-2 November 1943, forty-five miles west-northwest of Empress Augusta Bay, well off Cape Moltke. In it, Merrill's four light cruisers and eight destroyers defeated and put to flight the Japanese force.[9]

Task Force 39 included Destroyer Squadron 23, commanded by Capt. Arleigh Burke. After covering the initial landing at Bougainville, the "Little Beaver" squadron would participate in twenty-two separate engagements over the next four months, destroying one enemy cruiser, nine destroyers, one submarine, several smaller ships, and approximately thirty aircraft. During that period, the future Chief of Naval Operations received his nickname "31-knot Burke" for pushing his destroyers to just under boiler-bursting speed when in battle.[10]

ADDITIONAL MINELAYING MISSIONS

The mission of this task group is to reinforce a previously laid mine field and thus to destroy enemy shipping enroute to or from the BOUGAINVILLE-SHORTLAND area. This mission shall be accomplished without detection if practicable. Radio and visual silence will be effective from sunset each night until sunrise. Changes of course and speed will be made without signal.

— Comdr. William Julius Richter, USN, commander, Mine Division
One, in his Operation Plan No. 2-43, dated 6 November 1943.

Owing to the fact that many additional personnel were required for mines, certain modifications in the stationing of gun crews were made. Gun crews were stationed on 3 inch guns #2 and/ or 3, and #4, plus 20MM battery. Gun crew for 3 inch #1 gun, the majority of the forward repair party and all of the after repair party were stationed along the mine tracks to assist in the launching. No jams were encountered. S.C. radar, visual and radio silence were maintained in vicinity of plant.

— Lt. Comdr. Richard C. Williams Jr., USN, commanding officer
of the minelayer USS *Pruitt* (DM-22), in a report on mining
operations during the Bougainville Campaign.[11]

After loading at Mine Assembly Depot No. 2, Espiritu Santo, *Tracy* and *Pruitt* got under way in late afternoon on 28 October, bound for Purvis Bay. The next day, *Tracy* conducted some impromptu 20mm gunfire practice using an empty oil drum as a target. Following an uneventful transit, the two minelayers entered Purvis Bay in late afternoon on 30 October, and moored in berth 14.[12]

They left Purvis Bay around dawn on 1 November, accompanied by the destroyer *Eaton* (DD-510), to lay a minefield near Bougainville. Task Unit 31.8.2 was under Comdr. William J. Richter, USN, who was both the commanding officer of *Tracy* and commander, Mine Division One. In transit, the ships formed a column, with *Eaton* leading *Tracy* and *Pruitt*. Late that night, Richter set General Quarters at 2310, and at 2347 ordered mining stations manned. Shortly after midnight, as *Eaton* changed course to 195°T, *Tracy* turned outboard of her and took station

300 yards starboard of the guide, while *Pruitt* maintained station astern of *Eaton*. *Tracy* and *Pruitt* began mining at 0023, and completed operations at 0037, having sown 170 Mk 6 mines equipped with anti-sweep devices. A short time later, searchlights and gun muzzle flashes were sighted at 0058 in the direction of Masamasa, an island in Bougainville Strait. Retirement to the Russell Islands was made at 27 knots, via Gizo Strait and Blanche Channel.[13]

Tracy and *Pruitt* arrived back at Espiritu Santo in late morning on 4 November, moored at the Mine Assembly Depot wharf and began loading. That afternoon, they fueled from the tanker SS *Chester Sun*, in preparation for another mission off Bougainville.[14]

BACK TO BOUGAINVILLE

The following day, 5 November, *Tracy* and *Pruitt* stood out in late afternoon, joining *Breese*, *Gamble*, and *Sicard* to form Task Unit 31.8. The ships took station by division in two columns, 1,000 yards spacing between ships; 1st Division: *Tracy* (formation guide) and *Pruitt*, and 2nd Division: *Breese* (guide), *Sicard*, and *Gamble*. On the morning of 7 November, the task unit entered Leugo Channel and proceeded to Purvis Bay. After "topping off" from the oiler *YON-146*, the five minelayers departed. Joined by the destroyer *Baine* (DD-630), they formed a loose column; *Baine* leading, followed by *Tracy*, *Pruitt*, *Gamble*, *Breese*, *Sicard*.[15]

Proceeding up the Solomons, Richter made all possible speed while in the lee of the surrounding islands, with intent of slowing down if necessary when in the open seas. However, forecasts of bad weather proved unfounded, and a speed of 23 knots was held throughout the day. As the task unit approached Bougainville, the seas were smooth with scattered low clouds on the horizon and intermittent cumulus clouds overhead. Drawing nearer the mining area, a high three-quarters moon provided excellent visibility of the islands used for navigation, but the lunar light also illuminated the ships, which were trying to avoid detection by enemy forces. Fortunately, cloud movement obscured the moon before the ships' arrival in the area of the planned minefields.[16]

In early morning on 8 November, the ships slowed at 0145 to mining speed of 18 knots without signal, and adjusted stations. Twenty minutes later, a course change to 275°T preceded movement into their mining formation by ripple effect from ahead. *Breese*, *Sicard*, and *Gamble* commenced mining at 0212, and thirteen minutes later, they retired on a northerly course, having each laid eighty-five Mk 6 mines with anti-sweep devices. *Pruitt* and *Tracy* made their runs on course 286°T, and departed the area proceeding eastward.[17]

At 0250, while still headed east, *Pruitt* and *Tracy* observed many explosions close aboard to the south-southwest, in the general direction of the eastern minefield. These were likely bomb explosions, and not premature detonation of mines as was originally believed. *Sicard* had reported detection of an aircraft on radar, enemy float planes had been active in the area at night, and had frequently attacked shipping with bombs. Once joined, the two groups of minelayers retired at 27 knots.[18]

Their field was laid to close the northeastern entrance to the South Bougainville area, and cut off enemy supply of Japanese forces in that area. Three other existing fields assisted in fouling the eastern and southern entrances and a large part of the western entrance.[19]

ADDITIONAL U.S. TROOPS LAND AT TOROKINA

From 6 to 19 November 1943, the remaining regiment of the 3rd Marine Division and the Army 37th Infantry Division landed at Torokina to strengthen the force already ashore and expand the beachhead. There was no thought of pushing across the 250 square-mile island to eliminate the 25,000 Japanese in what would be a brutal, costly, and slow action. Instead, the Allies planned to take only a small piece of Bougainville, perhaps six square miles, including the deep port at Empress Augusta Bay. This would be the site of a major airfield from which American planes would range over the South Pacific to help provide security for the Allied convoys and task forces which would invade the Philippines in October 1944.[20]

AFTERMATH

Commander, Mine Division One highlighted in a post-action report, the limitations of employing antiquated World War I vintage destroyer minelayers and the need to stage mines closer to areas of operations to minimize wear and tear on these type ships:

> It is desired at this time to reiterate the previously made recommendation that consideration be given to the development of a modern type of fast minelayer. The present 1200 type have exceeded all expectations in the past, however, they do have definite shortcomings which are believed so serious as to limit their usefulness to defensive mining unless very strongly supported by better equipped and more powerfully armed vessels.

> The distance between the mine depot and the site of the mine field was almost a thousand miles. Six days elapsed between the laying of the first and second fields because of the travel required. At least four days would have been saved if it had been practicable to deliver

the prepared mines to GUADALCANAL. It is recommended that if cargo vessels or CM's are available for future operations, they be spotted significantly close to the mine fields to eliminate as much travel by DM's as practicable.[21]

PERSONAL AND UNIT AWARDS

Comdr. William Julius Richter, USN, commander, Mine Division One, received the Legion of Merit for directing the minelaying operations off Japanese-held Bougainville:

> The President of the United States of America takes pleasure in presenting the Legion of Merit to Commander William J. Richter, United States Navy, for exceptionally meritorious conduct in the performance of outstanding services to the Government of the United States while commanding a mine division in the Pacific area during World War II. Commander Richter operated at night without the use of visual or radio signals and conducted the laying of one defensive and three offensive mine fields quickly and accurately. The successful completion of these missions was largely responsible for the inability of Japanese garrisons to receive reinforcements and supplies.[22]

The officers and men of the five minelayers which laid the fields earned battle stars to affix to the Asiatic-Pacific Campaign ribbons on their uniform blouses.

Treasury-Bougainville Operation: Occupation and Defense of Cape Torokina

Breese (DM-18)	1-8 Nov 43	Lt. Comdr. Alexander Bacon Coxe Jr., USN
Gamble (DM-15)	1-8 Nov 43	Lt. Warren Wilson Armstrong, USN
Pruitt (DM-22)	7-8 Nov 43	Lt. Comdr. Richard Claggett Williams Jr., USN
Sicard (DM-21)	1-8 Nov 43	Lt. Comdr. John Vavasour Noel Jr., USN
Tracy (DM-19)	7-8 Nov 43	Comdr. William Julius Richter, USN

16

Central Pacific Campaign

The capture of Tarawa knocked down the front door to the Japanese defenses in the Central Pacific.

—Observation by Adm. Chester Nimitz regarding the acquisition of the Gilbert Islands, which came at a high cost in terms of lives lost.

Photo 16-1

Adm. Raymond A. Spruance, USN commander, Central Pacific Force, U.S. Pacific Fleet, on 23 April 1944, days before it was redesignated the 5th Fleet. Navy Photograph, now in the collections of the U.S. National Archives.

Allied advances in 1943 included the initiation of a drive by Vice Adm. Raymond Spruance's forces through the Central Pacific on the road to Tokyo, as well as those in progress by Halsey in the South Pacific and MacArthur in the Southwest Pacific. Adm. Chester Nimitz established the Central Pacific Force (Task Force 10) with Spruance in charge on 5 August 1943. Its title would be short-lived. After overseeing the battle of Tarawa in November 1943, Spruance would guide his force as it advanced through the Gilbert Islands, and assaulted Kwajalein and Majuro Atolls in the Marshall Islands. For these successes, Spruance was promoted to admiral in February 1944. Subsequently, the Central Pacific Force became the U.S. Fifth Fleet on 26 April 1944, which Spruance would command through war's end.[1]

Map 16-1

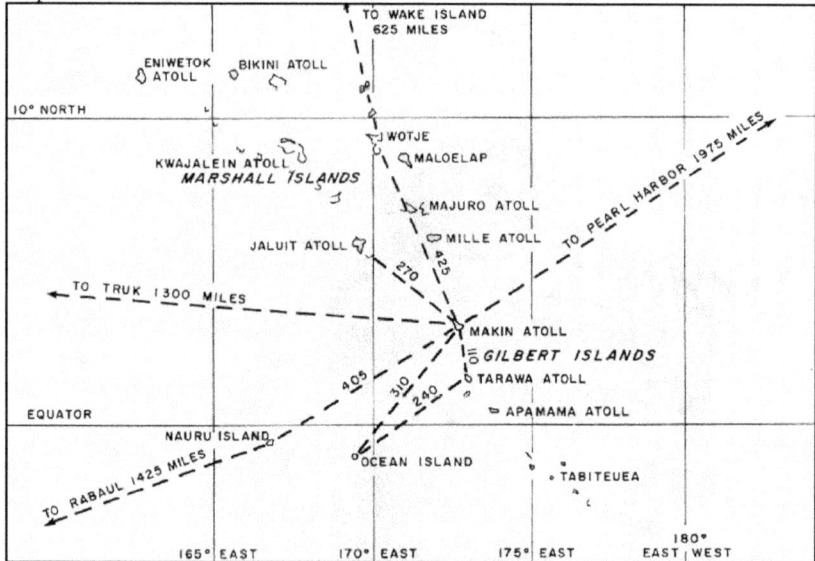

Gilbert and Marshall Islands in the Central Pacific
http://www.ibiblio.org/hyperwar/USN/ACTC/maps/actc-16-p612.jpg

The Gilbert Islands, which included Tarawa, Apamama, and Makin, were a group of coral atolls orientated in a roughly north-south line across the equator. The Japanese had invaded the British-held islands on the same day as the attack on Pearl Harbor and occupied them by 10 December 1941. The location of the islands immediately south and east of other important Japanese bases in the Carolines and Marshalls added to their strategic importance to the enemy. For the Allies, the Gilberts offered sites for airfields necessary for progress along the road through the Central Pacific toward Japan. The planned assault and capture of

Tarawa and Makin in November 1943 was a part of the overall American strategy of conducting an offensive through Micronesia—the Gilbert, Marshall, and Caroline Islands—at the same time as MacArthur's New Guinea-Mindanao approach to Japan.[2]

Map 16-2

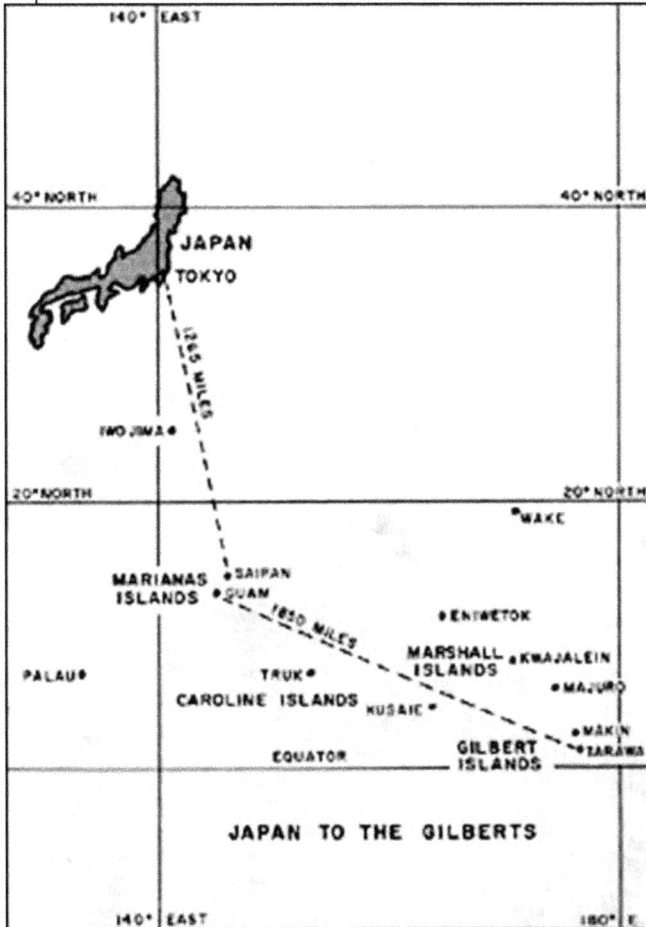

Principal route of the Central Pacific forces toward Japan
http://www.ibiblio.org/hyperwar/USN/ACTC/maps/actc-p735.jpg

In order to gain airbases capable of supporting operations across the Central Pacific to the Philippines and on to Japan, the U.S. needed to take the heavily defended Mariana Islands. However, use of land-based aircraft, to weaken enemy defenses and provide some measure of protection for the invasion forces, necessitated capturing the Marshall

Islands, northeast of Guadalcanal. An enemy garrison and air base on Betio, one of the islands of Tarawa Atoll in the Gilberts, guarded against invasion forces arriving from Hawaii. Thus, the starting point for the planned invasion of the Marianas lay far to the east, at Tarawa.[3]

MINELAYER *TERROR* JOINS THE PACIFIC FLEET

Photo 16-2

USS *Terror* (CM-5) loading Mk 6 mines at Yorktown, Virginia, circa October 1942. Naval History and Heritage Command Photo No. NR&L(M) 31592

Joining the Central Pacific Force in the Gilberts was USS *Terror* (CM-5), the Navy's only minelayer built specifically for minelaying. Commissioned on 15 July 1942, with Comdr. Howard Wesley Fitch in command, she could nominally carry eight hundred or so Mk 6 mines, which weighed about 1,400 pounds each. The actual number varied with the type of mine staged on the mine deck—which could accommodate 534 mines in six rows of 89 mines each. Another 268 mines could be stowed below in the holds. At the forward end of each mine track was a bogie car, operated remotely from the mine control space aft on the second deck on the waterline. These cars pushed mines aft and propelled them over the stern via openings in the transom. At completion, two large doors closed the mine breeches.[4]

Built at the Philadelphia Navy Yard, *Terror* joined the Atlantic Fleet following fitting out and shakedown. She arrived at New York on 30 October 1942 to prepare for her first large-scale operation, the Allied invasion of North Africa, Operation TORCH. In the months that followed after her return to the United States, *Terror* was based at the Naval Mine Depot, Yorktown, Virginia. She frequently participated in exercises in the Chesapeake Bay, occasionally stopping at Norfolk for repairs or overhaul, and served as "school ship" for students from the Mine Warfare Training Facility.[5]

In late September 1943, *Terror* began loading mines in preparation for her departure from the Atlantic coast. After leaving Norfolk on 2 October with Task Unit 29.2.6, she proceeded down the eastern seaboard, through the Caribbean and Panama Canal, and up the west coasts of Central America, Mexico, and California. On the morning of 19 October, *Terror* anchored in San Francisco Bay. The following day, she passed back under the Golden Gate bridge bound for Pearl Harbor. Arriving at Oahu in early afternoon on the 25th, *Terror* entered the harbor and moored port side to berth S-19. The next day, her crew began unloading cargo, received at Norfolk.[6]

On 28 October, Adm. Chester Nimitz and Vice Adm. William L. Calhoun, commander, Service Force, Pacific Fleet, came aboard in late morning for a visit. That afternoon, *Terror* shifted berths to Naval Mine Depot West Loch, Pearl Harbor. After loading cargo, she returned to the Naval Station. Departure finally came on 2 November, when *Terror* (Task Unit 16.15.6) stood out en route to Funafuti, Ellice Islands. She arrived there five days later, entering Te Ava Fuagea, a deep and narrow channel, in early morning and proceeding to anchorage. Upon securing from anchoring detail, crewmembers began unloading pontoon barges staged on the fantail.[7]

That same morning, 7 November, the seaplane tenders *Curtiss* (AV-4) and *Mackinac* (AVP-13) arrived at Funafuti at 0645 to support the planned U.S. invasion of Tarawa. *Curtiss* was the flagship of Task Force 57.   Embarked aboard her was Rear Adm. John H. Hoover, commander, Aircraft, Central Pacific Force.   During forthcoming operations, Hoover's headquarters would be aboard *Curtiss*, anchored in Funafuti lagoon with communication cables to shore and anti-submarine nets and defensive minefields protecting the lagoon. Hoover's tasking from Vice Adm. Raymond A. Spruance, was to:

- Support the Central Pacific Campaign by Air Operations
- Defend and develop positions captured
- Construct and activate airfields on Makin, Tarawa, and Apamama[8]

At a little past noon on 9 November, Capt. Herbert M. Scull, USN, came aboard *Terror* and set up his commander, Service Squadron 4 organization aboard the minelayer.   The squadron consisted of the destroyer tender *Cascade* (AD-16) and twenty-three other ships and craft ranging from the repair ships *Phaon* (ARB-3) and *Vestal* (AR-4), through tugs and patrol craft, to fuel-oil barges and 500-ton lighters.[8]

ENEMY BOMBING ATTACKS ON ELLICE ISLANDS

On the eve of the commencement of B24 bomber strikes on the atolls, Hoover spoke to his military commanders on 12 November about the importance of focusing all efforts on one objective—that of killing Japanese and destroying their installations.   The admiral explained the search objectives and emphasized that patrol planes must cover their assigned sectors accurately.   He also suggested that island commanders obtain all the fire extinguishing apparatus possible, and as best they could, set up anti-aircraft and fighter direction control.[9]

At a little past midnight, 0003 on 13 November, the radar operator aboard *Curtiss* reported several "Bogies" (unidentified planes).   This report was confirmed by radar ashore, which also indicated two groups of planes bearing 330 degrees true.   Condition Blue was set and General Quarters sounded on all ships and on Funafuti.   This alert status meant "Air attack probable, unidentified aircraft in vicinity.   All ships set Condition II."   All hands were at battle stations when the planes were definitely identified as enemy and upgraded condition Red was established.   Aircraft batteries on shore opened at 0045.   Three minutes later, the planes made their first run and dropped bombs.   The second run occurred at 0055 when the last bombs fell.   As the flight of aircraft had approached Funafuti, it separated into two groups.   One group of three planes made its run from the north. The other group of three went wide around the island to the east, then came in from the south. After the planes had disappeared from radar, condition White (All Clear) was reestablished. None of the ships had opened fire on the high-flying bombers, which were beyond the range of their anti-aircraft guns.[10]

Japanese aircraft (nine single planes and a three-plane formation) returned four days later and again attacked the base at Funafuti. The aircraft, believed to be twin-engine medium bombers, made individual runs from every direction dropping seventy to eighty large, medium and small bombs from 7,000 feet. The first of the aircraft, approaching singly from the northwest, was picked up on radar at 0337. The air raid warning sounded at 0400, enabling all hands to take cover before the first of ten bombing runs began at 0416. The island's Marine Corps anti-aircraft batteries—two 90mm batteries, three 40mm guns, and one .50-caliber machine gun—opened fire at 0420. Some ships in the lagoon were able to deliver 3-inch and 40mm fire as well. *Terror*'s 5"/38 battery opened at 0500 and ceased firing after expending fifteen rounds five minutes later. *Curtiss* fired twelve rounds of 40mm after a shore search light illuminated one of the planes, which appeared to be a "Sally." Dropped ordnance killed two Navy men, destroyed a B24 bomber and a C47 cargo plane, and damaged ten other aircraft.[11]

DEFENSIVE MINING AT FUNAFUTI COMPLETED

On 21 November, Capt. Robley Westland Clark, USN (commander, Mine Squadron 1, embarked aboard *Terror*) reported to Hoover that all mining operations had been completed for the Funafuti Atoll, Ellice Islands Group. A total of 151 Mk 6 mines were used in establishing fields to help protect Naval forces inside the lagoon. These minefields bore native names—Te Afua Sari, Te Ave Kum, and Te Buka Villi Villi—romantic sounding words belying the mines' deadly mission. It appears that the pontoon boats *Terror* brought from Hawaii laid these mines as she remained at anchor during her stay in the lagoon.[12]

Terror left Funafuti on 28 November for the return trek to Pearl Harbor. Arriving there in late morning on 4 December, she moored at Naval Mine Depot West Loch and detached Capt. Robley W. Clark and other officers and men aboard her for temporary duty in connection with the minelaying at Funafuti. *Terror* also offloaded to the depot a portion of her remaining mines. Captain Clark would later be killed at Okinawa on 1 May 1945, when a kamikaze hit his flagship USS *Terror*.[13]

CAPTURE OF TARAWA AND MAKIN

Two regiments from the U.S. 27th Infantry Division landed at Butaritari Island, Makin Atoll, Gilbert Islands, on 20 November and, following light Army losses (64 killed and 150 wounded), signaled "Makin taken," three days later. The assault on Tarawa that same day was bitterly contested by several thousand Japanese troops on Betio, the principal island of the atoll. Tarawa had been attacked repeatedly from the air for weeks preceding the assault, and the previous day by naval shore bombardment. Although these efforts killed approximately half the Japanese troops, the enemy was able to concentrate the remaining forces beside the only beach where a landing was possible and inflicted heavy casualties. The 2nd Marine Division suffered 871 killed, an additional 124 men who would succumb to their wounds, and 2,306 wounded or missing in action. The fighting lasted nearly four days, at the end of which time Betio Island was considered secure, although subjected to air raids and isolated sniper action.[14]

ESTABLISHMENT OF MINE DETAIL 19 AT TARAWA

After loading mines and gear at Pearl Harbor, *Terror* sailed for Tarawa on 9 December in company with the destroyer *Clarence K. Bronson* (DD-668). Aboard the minelayer were Captain Clark, three officers on his staff, and 27 men from NMD West Loch including Lt. R. W. Burgis, E-V (S), USNR, and two officers of Mine Detail Nineteen.[15]

Photo 16-3

Coastal transport USS *APc-108*, circa 1945-1946. A sister ship to *APc-109*, she evidences hard use during the war.
Naval History and Heritage Command photograph. #NH 81409

The minelayer anchored in Tarawa Lagoon on 14 December. As her crew unloaded gear for Mine Detail nineteen over the next couple of days, Captain Clark and a surveying party left for Makin Island in the 103-foot, wooden-hulled, coastal transport USS *APc-109*, to conduct a survey preparatory to future mining operations. On 19 December, *Terror* began unloading Mk 12-1 and Mk 13 mines into LCM landing craft for movement ashore.[16]

Mines Mk 12 Mods 0 and 3, were magnetic, ground mines designed to be planted from 21-inch submarine torpedo tubes. Mod 1 was the air-laid version, which included a parachute and could be fitted with soluble washers for an additional arming delay feature. These type ground mines were used in water depths of 16-125 feet against surface craft, and in depths up to 500 feet against submarines. Mk 13 was an aircraft-laid magnetic mine, delivered without a parachute, which could also be used as a bomb.[17]

Clark and the surveying party returned from Makin Island the following day, as the crew of *Terror* worked to establish a camp for Mine Detail Nineteen on Buota Island. On Christmas Day, *Terror* sailed from Tarawa for Espiritu Santo.[18]

AIRCRAFT MINING IN THE MARSHALLS

Between 30 December 1943 and 12 January 1944, Patrol Squadron VP-72, and Bombing Squadrons VB-108, VB-109, and VB-137 airdropped sixty-four Mk 12-1 mines and seventy-five Mk 13 mines in the waters of atolls in the Marshall Islands:

- Jaluit Atoll (Southeast Pass, Jabor Harbor)
- Maloelap Atoll (Enijun Channel, Tarao Anchorage, South Opening, Enibin Channel, Torappu Channel)
- Wotje Atoll (Schischmarov Strait, Wotje Anchorage, Legediak Strait, Meichen Channel)
- Mille Atoll (Tokowa Channel)[19]

The purpose of the operation was to make Japanese use of their bases in the Marshalls dangerous for a month or more, by mining anchorages and channels at Jaluit, Maloelap, Wotje, and Mille.

Squadron	Commander	Aircraft	Capability
VP-72	Comdr. S. J. Lawrence, USN	PBY-5	2 Mk 12-1 mines each
VB-108	Comdr. E. C. Renfro, USN	PB4Y-1	4 Mk 13 mines each
VB-109	Comdr. C. L. Miller, USN	PB4Y-1	4 Mk 13 mines each
VB-137	Comdr. E. C. Sanders, USN	PV-1	1 Mk 12-1 mine each[20]

Patrol Squadron 72 was based aboard the seaplane tender *Mackinac* at Tarawa. Bombing Squadrons VB-108 and VB-109 were at Apamama, and VB-137 on Bititu Island at Tarawa. At Tarawa, LCMs carried mines from Detail Nineteen to *Mackinac* and to Bititu Island. Mine delivery to Apamama was via the CenCats Seaplane Transport Service at Tarawa, and the tank landing craft USS *LCT-278*.[21]

Pilots used low approaches in carrying out their missions to avoid detection by enemy radar. When within 40 or 50 miles of the target, PBY-5 aircraft dropped down to between 100 and 300 feet altitude and 110-125 knots; gaining altitude to 700 feet (speed 80-100 knots) over the target. Mines were dropped at 5- to 8-second intervals. PV-1 aircraft approached at somewhat higher speeds and lower altitudes, climbing sharply to 700 feet to reduce speed below 150 knots and diving quickly after release to gain speed for retirement at low altitudes.[22]

Photo 16-4

Navy PB4Y patrol plane taking off from Carney Field, Guadalcanal.
National Archives photograph #80-G-89383

PERSONAL AND UNIT AWARDS

Capt. Howard W. Fitch, USN, commanding officer of USS *Terror*, was later awarded the Legion of Merit for exceptionally meritorious conduct during the war. *Terror* received a battle star for the period in which she was at Funafuti, supporting the Gilbert Islands Operations.

Gilbert Islands Operations		
Terror (CM-5)	13-28 Nov 43	Capt. Howard Wesley Fitch, USN

"Aussie Cat" Minelaying Operations

From these first long trips of over 2,000 miles, crews came back with fresh ideas on tactics. They found that shore defences were unwilling to fire and disclose their positions to aircraft which also had a reconnaissance role, unless some important nearby point was attacked.

Moonless nights were now regarded as the most suitable, for then night fighters were more easily avoided. There was a collision risk which could be reduced by sending fewer aircraft on any one night, those backing up later could use a glide approach to their targets.

—RAAF Command Headquarters - 7th Fleet reports
on RAAF minelaying operations 1943 – 1945.[1]

Photo 17-1

Royal Australian Air Force PBY Catalina flying boat, with a sea mine under each wing.
Australian War Memorial photograph NEA0347

AERIAL MINING OPERATIONS IN 1943

Operating from Cairns, Australia, between April and July 1943, number 11 Squadron RAAF Catalinas mined enemy harbors and dropped supplies to coast watchers in the Solomons, New Britain, and New Ireland. Catalinas of 20 Squadron, flying from Bowen in Queensland, were similarly engaged. Previously tasked with seaward reconnaissance throughout the New Guinea area, the Solomons, and New Caledonia, squadron crews increasingly performed night attacks on enemy bases and in mid-1943, began mining enemy harbors and dropping supplies to coast watchers as well.[2]

Between September and December 1943, 11 Squadron participated in combined attacks against Lakunai, Vunakanau, and Rapopo airfields on New Britain, which provided umbrella-type defenses for the powerful Japanese base at Rabaul on the island's northeast coast. The locations of other targets included Ambon, Netherlands East Indies; Kavieng, New Ireland; and Gasmatta, New Britain. Number 20 Squadron Catalinas also took part in attacks against New Britain in support of Allied plans to recapture the island.[3]

Number 43 Squadron was raised at Bowen, Queensland, on 1 May 1943, and then moved to its operational base at Kurumba, in the Gulf of Carpentaria, Queensland. Headquarters were established there on 19 August 1943. The first of the unit's aircraft arrived on 7 August, and the squadron flew its first strike on 8 September, against targets on Ambon. Subsequent bombing attacks were made against Babo, Jefman, Kaimana, and Langgoer in Papua and New Guinea; Ambon; and Taberfane in the Aru Islands southwest of Western New Guinea.[4]

PLANE CREWS

In an interview in 1989, Dr. John Charles Lane, a flight lieutenant medical officer with 20 Squadron during the war, described the forming of plane crews and subsequent operations. He was at RAAF Base Rathmines, in New South Wales on Australia's east coast, for about five months of training before being posted to 20 Squadron in Cairns, North Queensland (northeastern area of Australia). Lane described the operations the squadron flew out of Cairns and later from Darwin:

> When I first started, there was an anti-submarine patrol they used to run up in the Solomon Sea mostly, and that stopped pretty soon after I got there.... They used to do some bombing, ... and about that time they started the main program, which was mine laying. And, of course, they were called on for all sorts of other jobs from time to time, a bit of air-sea rescue, flying people in and out of the

Sepik River—you know, intelligence people—but I guess a certain amount of bombing and mine laying were the main operations.[5]

The flight crews of the Catalinas were close-knit groups. With the exceptions of replacements owing to personnel casualties or transfers, they remained intact. This practice, and a belief by the men that their missions were important, helped foster and maintain high unit morale:

> When the crews were being trained down at Rathmines, there was a process of so-called crewing up, so that the crew was, from there on, more or less fixed, apart from people who had to drop out for one reason or another, or perhaps got promoted or something, so that the crews … stayed together … for the entire tour of duty.

> … The morale of the squadrons was very high. They felt they were doing a … job which was really an effective one—that's the mine laying program, which was essentially an American navy program— and they were doing it well, they were doing it accurately…. I think there was a very strong sense of being a special kind of mini air force and doing a very useful … operational job.[6]

CHARACTERISTICS OF THE AIRCRAFT

> *Well, as an aeroplane to fly it wasn't particularly nice. Now I'm just talking about the pilots' feel of the controls. But as an aircraft engaged on interesting operations compared with the other aircraft in the RAAF, you couldn't beat it.*

—Clifford Dent Hull, a former flying officer with 11 and 42 Squadrons RAAF in World War II, describing the PBY Catalina seaplanes used to conduct minelaying.[7]

The PBY Catalinas employed by the RAAF for precision mining of Japanese harbors were slow and cumbersome, with a cruising speed of about 105 knots when laden with bombs and fuel. However, the planes did offer a measure of comfort that probably didn't exist in any other Australian military aircraft. Because of the long flights, it was necessary for some of the crew to be able to rest, so the Catalinas were equipped with four bunks and a very nice galley, which usually offered better food in the air than was available on the ground. Clifford Hull, a flying officer with 11 and 42 Squadrons during the war, described in an interview in 1989, the practice of crewmen relieving one another to allow for breaks:

Well, we'd change duties. The two pilots wouldn't sit up in the cockpit for eight or ten hours until one fell asleep. They would, in a sense, run watches like in a ship and one would go off and have a rest while the other continued the flying. Likewise, with the flight engineer. There were two flight engineers and they would take it in turns to man the engineer's panel. The poor old navigator had no-one to relieve him but, on the other hand, he didn't have to be doing something every minute.[8]

Photo 17-2

Crew of a PBY Catalina flying boat, of No. 43 Squadron RAAF.
Australian War Memorial photograph NEA0343

For self-protection, the planes had a bow turret, port and starboard blister turrets (each with a half-inch Browning machine gun), and a machine gun position in the after compartment which pointed downwards but wasn't a turret. The large blisters provided gunners an exceptional view of the exterior. Situated forward of the blisters was the bunk area, with the navigator's compartment forward of it just behind the cockpit in front.[9]

Photo 17-3

RAAF armourers hoist a magnetic mine into place under the wing of a PBY Catalina.
Australian War Memorial photograph NEA0348

The number, and weight of mines a Catalina carried on a mission was contingent on the distance the plane had to fly. Hull described the types of mines carried and the procedures used to load them onto the bomb racks of the aircraft:

Now, dependent on the distance we had to fly we would carry 2,000 pounds in mines or 4,000 pounds in mines. How many actual mines would that mean? ... Up to four.

In fact, it looked perhaps a little bit more like a bomb and there were several types of mines. There were acoustic mines and there were magnetic mines and there were combined acoustic and magnetic.... We had bomb scows which were flat-bottomed boats and a winch would be attached to the wing of the aircraft and the mines, or bombs, would be wound up to clip on to the bomb racks.[10]

MINELAYING MISSIONS

... When we started mine laying it was thought that the obvious time to do it was at full moon so the crews had maximum visibility. But after a few raids it was discovered that there was quite adequate visibility for the task when there wasn't any moon at all. So, we reversed our tactics where mining raids were carried out with no moon rather than full moon. You know, you could see an aircraft runway quite clearly from 6,000 feet as long as there wasn't cloud between you and the runway.

—Clifford Hull, a former flying officer with 11 and 42 Squadrons.[11]

In addition to anti-submarine patrols, and bombing, mining, and at-sea rescue missions, Catalinas also landed AIF (Australian Imperial Force) intelligence patrols behind Japanese lines and returned later to lift them out. Hull described waterborne landings being made on the Sepik River, the longest river in New Guinea, to do so, (which he did not participate in) and mining being the most interesting missions in which he was involved:

... There were quite a few landings made on the Sepik River which was a very large river and, oh, I think it's navigable for probably fifty miles or more from the coast. I think the mine-laying was the most unusual one [type mission] and as an American naval officer remarked after the war: "Mine-laying by Australian Catalinas was probably the ultimate use of the aircraft."[12]

Initially, "top military brass" had believed that "the concept of going a thousand miles in enemy territory in daylight with a Catalina and then bombing something would be just plain bloody silly." However, at Allied Headquarters in Brisbane was an Australian Mine Warfare officer with fleet air-arm experience, Lt. Comdr. Palgrave Ebden Carr, who was very keen on the idea, as was also the U.S. naval staff at Brisbane. Moreover, the RAAF had some Catalina pilots who had experience with the RAF in laying mines off the coasts of France and Norway. Hull recollected:

> So, there was a body of people of adequate rank who were very keen on the idea so, between the Yankee navy and our navy and ourselves, we started setting up a mine-laying campaign.[13]

Ensuing Catalina operations involved dropping mines in a channel or channels in a harbor and reinforcing these minefields from time to time. The goal might be to prevent Japanese cargo ships from leaving port with important supplies for troops somewhere; or from delivering raw materiels to Japan; or to blockade naval shipping. Carr, in consultation with his USN counterparts, would select a target and identify a "datum" point around the harbor as an aid to navigation for aircraft in making the mining run. Hull explained this process:

> It might, say, for argument's sake, be a jetty or a cape or a lighthouse or some feature that could be picked up in the dark, and the objective for a particular aircraft was to fly over that datum point at a certain height and a certain air speed on a certain course for so many seconds and then let go their mines and, if everything was working all right, they go plop into the water across a shipping channel. Now, there might be several aircraft raiding on the one night. They would not go there together; they would be separated some time apart and they would also fly over the target at different altitudes. The lowest we used to fly was about 200 feet. Anything lower was considered getting pretty dangerous particularly if there was a battleship [combatant ship] there, you know, and if you're a few feet out on your altimeter you could hit the mast or something like that. And the highest was probably about 900 feet because we reckoned anything higher than that was getting pretty inaccurate.[14]

When a Catalina flew over the datum point, the navigator would start his stopwatch, and as he counted, the two pilots would listen to his count knowing that, on a particular count, the first mine had to be released, and on a subsequent number, another one, and so on. Once the Japanese were aware that a harbor had been mined, minesweepers would do their work, necessitating a later mission to reseed it.[15]

VARIED DUTIES IN 1944

Much of 11 Squadron's efforts in 1944 were devoted to minelaying operations as well as to anti-submarine patrols, shipping searches, and seaward reconnaissance. Late in the year, the squadron participated in the mining of Manila Harbor on 14 December, prior to the invasion of Mindoro Island by Allied forces.[16]

In 1944, four aircraft of 20 Squadron were detached for basing at, and operations from Manus Island, in the Admiralties off northern New Guinea. The remainder of the squadron continued flying from Cairns, carrying out minelaying operations in the Caroline Islands; New Ireland; Surabaya, Java; and Kaimania, Manokwari, and Sorong in Papua. On 2 September, the squadron moved to Darwin in Australia's Northern Territory and began flying sorties over Bangka Strait, the Celebes, and Java. Later in the year the squadron also helped mine Manila Harbor.[17]

NEW SQUADRON JOINS THE MINING EFFORT

Number 42 Squadron RAAF was formed at Darwin on 1 June 1944, and relocated on 11 July to Melville Bay on Melville Island, located in the eastern Timor Sea off the coast of the Northern Territory. Aircraft and additional personnel arrived in August. Like the Catalinas of the other minelaying squadrons, its "Black Cats" were painted matt black and operated primarily at night. The squadron conducted its first operation on 27 August when three PBYs flew patrols searching for shipping. No. 42 Squadron commenced minelaying operations in September off the Celebes islands, and suffered its first loss on 23 September when a Catalina made a forced landing behind enemy lines. Its crew were rescued by another Catalina.[18]

The squadron laid mines off Makassar and Pare Pare Bay in Celebes during October, in which two additional aircraft were lost to enemy anti-aircraft fire. In November, a detachment of aircraft was deployed to Morotai in the Dutch East Indies, from which Catalinas operated against Brunei Bay, Tarakan, Sandakan and the Balabac Straits. Another detachment of No. 42 Squadron deployed to Leyte in the Philippines, and mined Manila Bay on the night of 14-15 December.[19]

An account of 42 Squadron's second aircraft loss, and rescue efforts by a 43 Squadron Catalina, follows.

EPIC SEA RESCUE OF A DOWNED PBY AND CREW

Catalina *A24-100* had completed a mining raid on Makassar Harbor, on the western side of the Celebes, on the night of 23-24 October 1944, when its starboard engine was damaged by Japanese anti-aircraft fire.

Unable to maintain height and with the second engine failing, Flight Officer Clifford Dent Hull made a forced landing in the open sea south of the southwestern Celebes Peninsula.[20]

Hull described, in an interview in 1989, his approach to Makassar and damage suffered due to Japanese anti-aircraft fire:

> ... As I recall it, to the west of Makassar, there's a low line of islands, or possibly high reefs, and we were going to pick our datum point through a gap in this chain of reefs or islands.... And we got through the reef and the navigator started counting out his seconds and, just before it was time for the first mine to be released, I remember the second pilot, Flight Lieutenant Terry McNally, saying to me in a quiet steady voice, "There's a ship over here, Cliff," and on went the counting and we let the first mine go and, WHAM, we got hit in the starboard engine and the navigator went on counting and we let the second mine go and then the flight engineer said again in a quiet voice, "We've been hit, Cliff, and there's petrol streaming down inside."[21]

As Hull broke away in a turn to port, he saw tracer coming over his head and ducked. Clearing the target area, it was quite obvious by the amount of petrol streaming into the aircraft that they weren't going to get home. Hull decided to make for what was known as an escape point, in this case, the island of Salajar off the southern Celebes. He climbed up high to about 8,000 feet to be clear of all other aircraft in the raid and, in case the plane caught on fire, to give the crew plenty of time to bail out. Everything heavy that could be removed was thrown out of the plane, except the dinghies, in an effort to remain airborne.[22]

Despite these actions, the plane couldn't maintain altitude on one engine, was gradually losing height, and at about 3,000-4,000 feet the good engine cut out. Hull glided down in the black night, not knowing whether the plane was over land or water:

> I didn't – wasn't to turn the landing lights on because that would give our position away to any Japanese forces that might observe the aircraft going down and it was a moonless night. So, I glided down hoping to make a landing judging the landing in the dark on the radio altimeter which was really the only guide I had. I couldn't look over the side and say, "Oh, there's the water down there" .... Anyway, we landed on the water with a fairly resounding crash and bounced once and we were down.[23]

Hull and his flight crew spent several hours on the water, ten to twenty miles offshore, uncomfortably close to four Japanese airfields in the southern Celebes. At dawn, the first engineer repaired shrapnel damage to the fuel system and was able to get the good engine started. Hull did not want to remain close to the coast, particularly since they had heard some ships in the night, and he began to taxi the plane further out to sea.[24]

This action proved fortuitous as a B24 bomber passing overhead sighted the wake of the seaplane far below:

> … About mid-morning an American Liberator flew over, he was doing a patrol down the Celebes and across the East Indies and back to a base near Darwin, and he picked us up by the wake of our taxiing. Because when we saw him he was so high up and he had unusual tail markings that we were a bit nervous. Anyway, he flew down low and flew around us for a while and dropped a survival pack and later on he was replaced by another Liberator who stayed with us until we were rescued.[25]

In early afternoon, a Catalina (*OX-U*) of 43 Squadron arrived to rescue the men. Captained by Flight Lieutenant Etienne, it landed and took the other crew aboard. After the disabled Catalina was sunk by machine gun fire, the rescue Catalina took off and returned to Darwin. The 1,800-mile round trip was mainly through Japanese held territory. At that time, it was the longest air-sea rescue in the Pacific War, and likely remains so today.[26]

Plane Crew of the Rescue PBY Catalina

FO Armand Andre Etienne (captain)	Sgt. Robert Richard Tingman
FO Ian McCallister Robson	Sgt. Albert Leslie Warton
Flt. Sgt. John Joseph Sweeney (navigator)	Sgt. Thomas Roy Elphick
Flt. Sgt. Raymond Victor Tumeth	Corp. James Francis Burgess Oliver
Flt. Sgt. Derek Fanshawe Robertson[27]	

1945 THROUGH WAR'S END

Number 76 Wing, to which the squadrons were assigned, focused on laying mines off Surabaya and the Laoet Straits in January and February 1945. Operations late in the war involved 42 Squadron Catalinas flying from Jinamoc Island in the Philippines staging through a refueling stop at Brunei Bay to lay mines off Sumatra. Others flew against Surabaya in the Netherlands East Indies. Aircraft also conducted harassment raids on Japanese air bases in the southwest Celebes, the Flores Sea and the Banda Sea.[28]

After the Japanese surrender was announced on 15 August 1945 (formally signed on 2 September), No. 42 Squadron assisted in the repatriation of Allied prisoners of war from Manila, and other Australian personnel from Labuan, an island off the northwest coast of Borneo. The squadron was disbanded at Melville Bay, Australia, on 30 November 1945. Number 11 Squadron dropped food and medical supplies to POW camps across southeast Asia and also flew survivors back to Australia. Its aircraft were withdrawn in early 1946, and the squadron was disbanded on 15 February.[29]

Number 20 Squadron missions in 1945 included laying mines off the coast of China, and dropping and extracting special agents behind enemy lines. The squadron was disbanded on 27 March 1946. After the war, 43 Squadron flew patrols over the Netherlands East Indies and the approaches to Darwin until notified on 10 November 1945, that it would be moving to Rathmines, New South Wales, at the end of the month. Relocation from Doctors Gully, Darwin, was completed by 3 December 1945. Soon, Headquarters Eastern Area informed the squadron that it would cease operations on 1 February 1946, and No. 20 Squadron was disbanded on 11 March 1946.[30]

SUMMARY

Royal Australian Air Force PBY-5 Catalinas performed a truly herculean effort, involving 1,130 sorties to lay 2,512 mines (1,851 American and 661 British) between April 1943 and March 1945. Australian "Cats" prowled the night skies, carrying mostly U.S. ground mines from Mine Assembly Depot No. 1 at Darwin. The mining was over a wide area from Kavieng, New Ireland to Wenchow, China, with most of the effort devoted to four general areas:

- Celebes   519 mines
- Borneo   489 mines
- China   436 mines
- Java   425 mines
- All other   646 mines[31]

The fabric-clad Catalina patrol planes were slow, cumbersome and poorly armed; requiring them to rely upon stealth by arriving over their targets at night in the dark and at a low altitude. Nonetheless, the aircraft carried mines into almost every Japanese-held harbor of importance. By laying their minuscule loads with great precision, they held enemy shipping in port, or forced it into deeper waters to become the prey of USN submarines and easier targets for USAAF bombers.[32]

Remaining undetected was paramount to success. Pilots flew low and slow in an effort to avoid discovery by enemy radar and fighters. Nevertheless, a considerable number of aircraft were lost (32 CATS and

330 crew) on dangerous long-range missions. Australia was credited after the war with sinking nine Japanese merchant and naval vessels with mines. These ships are identified in Appendix A.[33]

PERSONAL AWARDS FOR VALOR

Three individuals were awarded the Distinguished Service Order, and another forty the Distinguished Flying Cross. A summary of these and lesser awards received by personnel may be found in Appendix B.

Number 11 Squadron RAAF
Distinguished Service Order

Redmond Forrest Michael Green	David Vernon

Distinguished Flying Cross

Colin Stuart Brown	Denis Russell Lawrence
Reginald Bruce Burrage	Robert Michael Seymour
Francis Blomfield Chapman	James Farrell Spears
Arthur John Cleland	Vernon Eric Townsend
Terence Lawless Duigan	Athol Galway Hope Wearne
Brian Hartley Higgins	William George Searle White
Stuart Austin Ikin	

Number 20 Squadron RAAF
Distinguished Service Order

Athol Galway Hope Wearne

Distinguished Flying Cross

Harold Gordon MacMurray Brown	Gordon Neville Read
Alexander Ronald Emslie	Norman Valentine Robertson
Robert Maxwell Hirst	Henry Corbett Scott
Victor Allan Hodgkinson	Thomas Vincent Stokes
Leslie Harold Hokanson	Gilbert Robert Thurstun
James Lionel Mutch	

Number 42 Squadron RAAF
Distinguished Flying Medal

Harold Richard Longworth

Number 43 Squadron RAAF
Distinguished Flying Cross

Anthony Noel Lee Atkinson	Reginald Joseph Marr
Noel Charles Edward Barr	Patrick John McMahon
Robert Trevor Clark	Ronald Nevill Damian Miller
Armand Andre Etienne	Brian Ortlepp
Robin Henry Gray	Charles Frederick Thompson
Lindley Maxwell Hurt	John William Thompson
John Kenneth Longmuir	Benjamin Alfred Titshall[34]

18

Capture of the Marianas and South Palau Islands

This force will capture, occupy and defend SAIPAN, TINIAN, and GUAM, will develop airfields in these islands and will gain control of the remaining MARIANAS, in order to operate long range aircraft against JAPAN, secure control of the Central Pacific and isolate and neutralize the central CAROLINES.

—Excerpts from Commander, Fifth Fleet Operation Plan 10-44

Hell is on us.

—Mamoru Shigemitsu, Japan's Minister of Foreign Affairs, remarking on the invasion of Saipan, June 1944

In March and April 1944, Allied landings in the South and Southwest Pacific added the Admiralties, Emirau in the Bismarcks, and Hollandia and Aitape, New Guinea, to the rapidly growing numbers of breaches in the Japanese defensive perimeter. Shore-based Army, Navy and Marine air squadrons neutralized bypassed enemy strong points, while planes of Task Force 58 repeatedly struck Truk in the Caroline Islands, the operating base for the Combined Fleet, akin to the U.S. Navy's Pearl Harbor. Japanese leaders were well aware that the most decisive actions of the war were at hand. Adm. Soemu Toyoda, commander in chief, Combined Fleet, in a message to all of his commanding officers on 4 May 1944 warned:

> The war is drawing close to the lines vital to our national defense. The issue of our national existence is unprecedentedly serious; an unprecedented opportunity exists for deciding who shall be victorious and who defeated.[1]

OPERATION FORAGER

In the short span of the two months between 15 June and 12 August 1944, the Japanese lost their last chance for victory in the Pacific after one U.S. Army division and two Marine Corps divisions captured Saipan, Guam and Tinian.   Termed Operation FORAGER, the successful offensive had been undertaken to obtain a site from which U.S. Army Air Force B29s could conduct a strategic bombing campaign of Japan. The outlook for the Japanese had already been bleak. As a part of Nimitz's island leapfrogging strategy, amphibious forces had seized Kwajalein, Majuro and Eniwetok atolls in the Marshalls in February, tightening the noose around the Japanese homeland's outer zone of defense.[2]

FORAGER had involved a giant step forward, 1,800 miles, by U.S. forces across the Central Pacific from Majuro Atoll, where the Fifth Fleet was based, to the Mariana Islands of Saipan, Guam, and Tinian, and to the southern Palau Islands in the Carolines. By spring 1944, Allied forces had established naval bases and airfields in the Marshalls, Admiralties, and north central New Guinea, which threatened enemy positions in the Caroline Islands, eastern Netherlands East Indies and southern Philippines. Enemy lines of communication had been cut eastward of Hollandia, New Guinea, and to enemy bases in the Marshalls. Lines of communication to the Carolines had been weakened and enemy air strength there was deteriorated because of attacks by land-based and carrier aircraft.   As a result, the Japanese were strengthening their positions in the Mariana Islands as a barrier against continued Allied advancement in the western Pacific.[3]

The Japanese still had, in the Carolines, three airfields at Truk and one field each at Satawan, Puluwat, Woleai, Palau and Yap. On 1 April 1944, two additional fields in Palau were under construction, and two fields at Ponape were operational but normally had no aircraft. In the Mariana Islands, there were three enemy airfields on Saipan and Tinian, two on Guam, and one each on Pagan and Rota. A major part of the Japanese Fleet was based in the southern Philippines-Singapore area. Palau served as a Japanese submarine base, Truk as an advanced base, and submarines occasionally stopped at Saipan. The enemy was using subs for reconnaissance and supply in the eastern Marshalls and for attacks on Allied surface forces throughout the Central Pacific.[4]

The fifteen Mariana Islands are orientated in an arc stretching from Farallon de Pajaros south to Guam, with the four largest islands— Saipan, Tinian, Rota, and Guam—at the southernmost end. The Allied invasion of Saipan began on 15 June, with landings by the 2nd and 4th Marine Divisions opposed by the Japanese Army's 43rd Division. The ensuing fierce fighting led to the decision to commit at once the Allies reserve force, the U.S. Army 27th Infantry Division. By the time Saipan was secured three weeks later on 6 July, the American casualties were the highest suffered to date in the Pacific War—2,949 killed and 10,464 wounded. Because of the Bushido Code, the Japanese had been determined to fight to the death; enemy losses totaled 21,000 killed in action, 8,000 suicides and 921 prisoners.[5]

Map 18-1

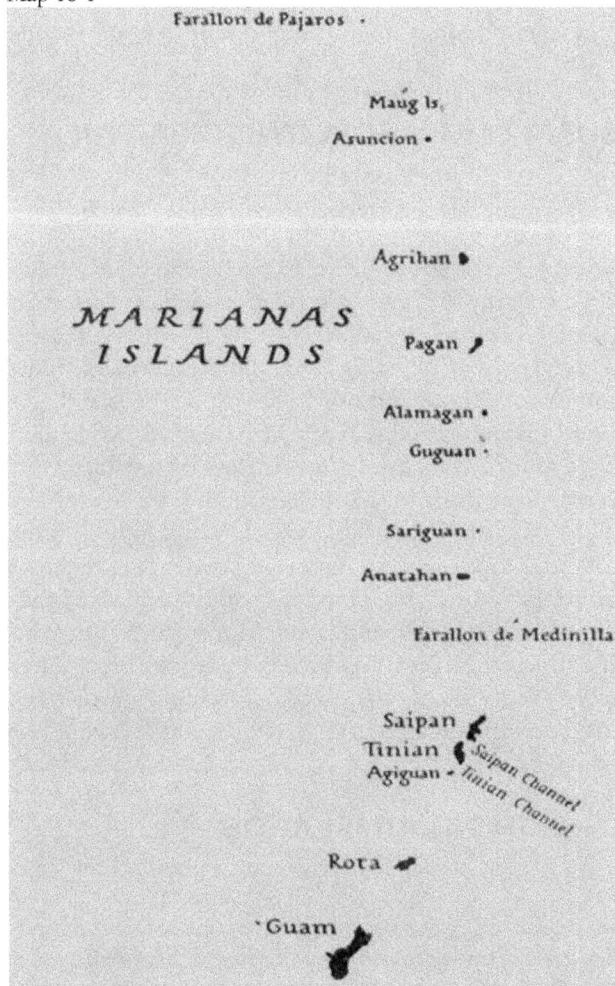

Farallon de Pajaros ·

Maug Is.

Asuncion ·

Agrihan ▶

MARIANAS
ISLANDS   Pagan ♪

Alamagan ·
Guguan ·

Sariguan ·

Anatahan ▬

Farallon de Medinilla

Saipan
Tinian   *Saipan Channel*
Agiguan ·   *Tinian Channel*

Rota

·Guam

Guam had fallen to Japanese forces at war's commencement, on 10 December 1941, because America's defense of the Philippines was negligible without ownership of nearby Saipan and Tinian which Japan had acquired in 1921 along with the remainder of the Marianas, except for Guam.

BATTLE OF THE PHILIPPINE SEA

While these assault forces and the enemy ashore fought the Battle of Saipan between 15 June and 9 July 1944, the Japanese Combined Fleet sortied to engage the U.S. Navy Fleet supporting the landings. In the ensuing 19-21 June Battle of the Philippine Sea, the American battle force defeated the Japanese in what would be the greatest carrier battle of the war. Following the loss of two carriers, the *Shokaku* and *Taiho*, 346 planes, and large numbers of pilots, Japanese naval air was unable for the duration of the war to engage Allied forces with parity. This reality resulted in Japanese leadership turning to the use of suicide planes.[6]

TINIAN AND GUAM TAKEN FOLLOWING SAIPAN

With Saipan taken, U.S. forces moved down the island chain. On the morning of 21 July, the 3rd Marine Division landed near Agana, about midway up the west coast of Guam, and the 1st Provisional Marine Brigade landed near Agat to the south. The 305th Regimental Combat Team of the 77th U.S. Army Infantry Division landed at Agat that afternoon. The remainder of the division came ashore during the next three days. After several days of fierce fighting, the Japanese defenders withdrew into mountainous central and northern part of the island. As fighting continued on Guam, the 2nd and 4th Marine Divisions landed on Tinian on 24 July and took the island after six days of combat.[7]

Guam was the last of the three islands to be secured. Following the unexpected toughness of the Japanese defense on Saipan, three divisions were committed to the assault instead of two. This required bringing the 77th Infantry from Oahu, Hawaii. Additionally, daily pre-assault naval bombardment and air bombing began on 8 July and continued through D-Day, the 21st. Organized Japanese resistance ended on 10 August, and Guam was declared secure. American casualties (1,435 killed or missing in action, 5,646 wounded), were less than half those on Saipan.[8]

ASSAULT OF SOUTH PALAU ISLANDS

We soon found ourselves hotly firing at the infernal machines [mines] and very close to the reef. Considerable maneuvering was necessary to stay in place due to an inshore set. Rifle fire from the beach in the vicinity of the point was experienced and the spurts of flame from the rifles plainly seen. MONTGOMERY claimed six mines destroyed and PREBLE destroyed three leaving one floating very close to the surf where it drifted down toward the tip of Peleliu and was destroyed by

PREBLE *later in the day. All caliber guns were fired from 3" down to .30 caliber. The remainder of the day was spent in mine detonation and recovery of our and* MONTGOMERY*'s Option buoys after they had served their purpose.*

—USS *Preble* (DM-20) War Diary entry for 12 September 1944.

The assault and occupation of Palau—located to the southwest of Guam and north of New Guinea—commenced in mid-September. In order to secure the flank, and also gain airfields for American forces preparing to invade the Philippines, U.S. Marine and Army forces landed on Peleliu and Angaur in the Palau Islands on 15 and 17 September, respectively. The fighting was protracted on both islands, particularly on bloody Peleliu where, after heavy and intense combat, the island was finally secured on 27 November 1944. At Peleliu, the 1st Marine Division and the 81st Army Infantry Division collectively suffered 1,794 killed and 8,010 wounded or missing—the highest casualty rate for U.S. military personnel of any battle in the Pacific War.[9]

MONTGOMERY AND PREBLE SENT TO PALAU

In support of the assault of the south Palau Islands, *Preble* and *Montgomery* had left the anchorage at Purvis Bay, Solomon Islands, on the morning of 6 September for movement there with Task Group 32.5. Well out to sea, the ships formed a circular cruising disposition that afternoon for the transit; which would be characterized by smooth sea, clear sky, and sudden rain squalls that were over as quickly as they began. On the afternoon and night of 11 September, the task group split into three smaller groups.[10]

Preble and *Montgomery* remained with the Eastern Peleliu Fire Support Group until 0330, when the destroyer minesweepers USS *Hovey* (DMS-11), *Southard* (DMS-10), *Long* (DMS-12), and *Hamilton* (DMS-18), and two minelayers broke off, bound for a point southeast of Peleliu Island. Here the sweepers were to begin clearing Japanese mines in preparation for the landings. Two hours later, Comdr. Wayne R. Loud (commander, Mine Squadron Two) directed *Preble* at 0530 to proceed to the southeast edge of a shoal about midway between Angaur and Peleliu and, with siren sounding, drop twenty Mk 6 depth charges along the 100-fathom curve at 200-yard intervals.[11]

Aviators, in an earlier air strike on the islands, had sighted what they believed were floating "antenna masts" over the shoal and, flying low over them, witnessed disturbances on the water apparently caused by mines detonating (presumably acoustic ones). *Preble*'s tasking was to try to detonate the remaining mines with noise, using her siren first and then depth charges. No mines were found, however, as cited in an entry that day in her war diary:

> The run was made at 26 knots without incident but with considerable suspense and noise since both whistle and siren were blown. No antennae were seen, no mines exploded, and no mines were subsequently swept over this shoal which leads to the belief that our aviator friends could not tell the difference between a mine explosion and the splash of an A.A. shell, undoubtedly fired at them from one of the two Jap islands.[12]

Map 18-2

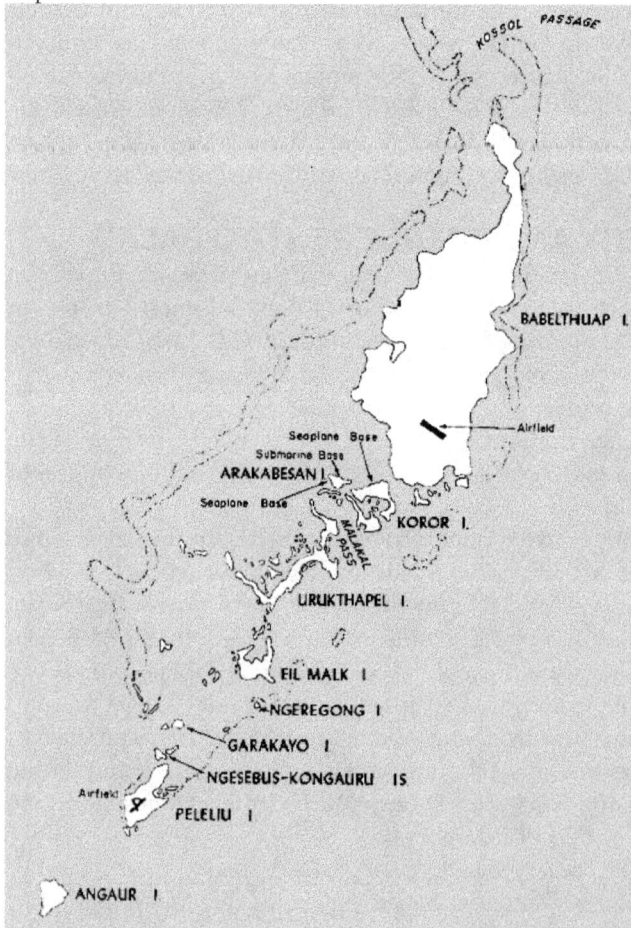

South Palau Islands
http://www.ibiblio.org/hyperwar/USMC/IV/maps/USMC-IV-2.jpg

At 0610 that morning, 12 September, the destroyer minesweepers formed a starboard "protective" echelon disposition and proceeded into the passage between Angaur and Peleliu Islands, with gear streamed. *Montgomery* followed close astern, for duties as a mine destruction vessel, and when *Preble* rejoined the formation she took station 50 yards inboard of the last minesweeper float. The plan was for the minelayers to drop yellow painted Mk 6 drill mine cases fitted with Option flags at the outer limits of the swept area. On the first pass, no mines were swept, and *Preble* and *Montgomery* each dropped three mine markers along the eastern and western limits of the swept areas to be used as reference points for commencing turns. During the second pass, closer to the Peleliu shore, *Hamilton* and *Long* swept (cut loose) several moored mines which *Montgomery* destroyed with gunfire.[13]

Diagram 18-3

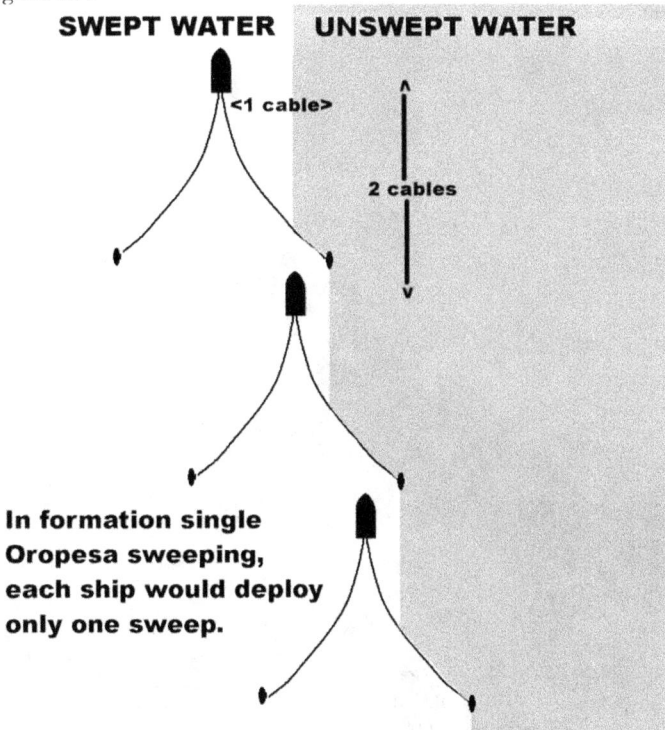

SWEPT WATER   UNSWEPT WATER

<1 cable>

2 cables

In formation single Oropesa sweeping, each ship would deploy only one sweep.

In this type of formation, the minesweepers proceed in echelon, each overlapping the sweep wire of the one ahead, with the lead ship supposedly in clear water (no danger from mines). When performed by the Royal Navy, as shown in this diagram, the approximate area covered by each ship's sweep was one cable (220 yards) and the approximate distance of one ship behind another was two cables.
Courtesy of Rob Hoole

LOSS OF MINESWEEPER *PERRY* TO A MINE

*In the opinion of the Commanding Officer USS PERRY, Lieutenant
Commander BALDRIDGE, USN, Commanding officer, USS PREBLE
exhibited extraordinary skill in bringing his ship in close to the scene of the casualty
in an endeavor to get a line aboard USS PERRY. Because USS PERRY was
swinging away, the gun line would not carry, but later a boat from USS
PREBLE carried the line across. His actions brought USS PREBLE's boats
to the scene much sooner than would have been otherwise possible. By his careful
and skillful ship handling in an area known to be mined, he avoided damage to
his own vessel while rendering the maximum possible assistance, undoubtedly
preventing additional loss of life.*

—Lt. William N. Lindsay Jr., USNR, commanding officer USS *Perry*[14]

Photo 18-1

Destroyer minesweeper USS *Perry* (DMS-17) abandoned and sinking, off Angaur Island,
after striking a mine, 13 September 1944.
Naval History and Heritage Command photograph #NH 92989

Minesweeping and mine demolition continued the remainder of the first
day, and after dark the minelayers joined other ships, well out to sea
beyond the range of Japanese shore batteries, for night retirement. The

following day, 13 September, *Preble* was assigned to Sweep Unit 1 Baker (DMSs *Southard* and *Perry*). At 1418 that afternoon, USS *Perry* struck a mine off the southeast coast of Angaur and subsequently sank. She was at the time in column astern *Southard* (having just completed a sweep off the northeastern tip of Angaur) and maneuvering to set up for an exploratory sweep off the Green landing beaches on the island. *Preble* rescued ninety-six members of *Perry*'s crew, and tried unsuccessfully to salvage the ship.[15]

The mine detonation lifted *Perry* bodily out of the water, and opened a large hole in her starboard side between the turn of the bilge and the keel. Water rushed in, flooding both firerooms and all power was lost at once. *Perry* initially took a 30-degree list to port which continued to increase, and abandon ship was ordered at about 1420. Riding to the sunken paravanes and other sweep gear made fast to her, she then drifted toward the southeast coast of the island through unswept waters. Heeled over with her portside main deck awash, she rolled over and began to sink by the bow. *Perry* broke in two at 1605, and sank two minutes later. Casualties were 6 missing and presumed dead, 1 dead of injuries, and 11 others injured.[16]

Perry had been off the starboard quarter of *Southard* in a starboard echelon with minesweep gear streamed to both port and starboard sides. Although this book only touches on minesweeping, an explanation of the type method used to counter moored-mines at Palau (and elsewhere) is probably in order. The word "mine" usually conjures up a mental picture of a spherical, spiked object anchored on a cable and riding beneath the surface of the sea. For moored, mechanical mines, which require physical contact to counter, this image would be correct.

For this type of sea-emplaced ordnance, U.S. Navy minesweepers employed moored gear essentially consisting of two cables (known as wires) streamed off the stern, diverted by "otters" to the port and starboard sides of the ship at a certain depth below the surface and held down by a "depressor." Mechanical or explosive cutters were attached to the wires at regular intervals. To keep the gear from "bottoming out," a float, or "pig," with a jackstaff and flag on top as a visual reference, was attached to the port and starboard wires. If the wires, streaming downward and outward astern of the ship, caught a mine cable, they guided it to the cutters, which severed it, causing the mine to surface so that it could be destroyed by gunfire.

Diagram 18-4

DEPRESSOR
WIRE

STERN ROLLER
CHOCK

SPAN PENDANT

SPAN PENDANT

DEPRESSOR

PORT SWEEP WIRE

STARBOARD
SWEEP WIRE

INTERMEDIATE
CUTTERS
BOTH SIDES

INTERMEDIATE
CUTTERS
BOTH SIDES

END CUTTER MK 9,
MODIFIED

STARBOARD
OTTER

END CUTTER MK 9,
MODIFIED

FLOAT PENDANT

PORT
OTTER

SIZE 1 FLOAT

MNV70215

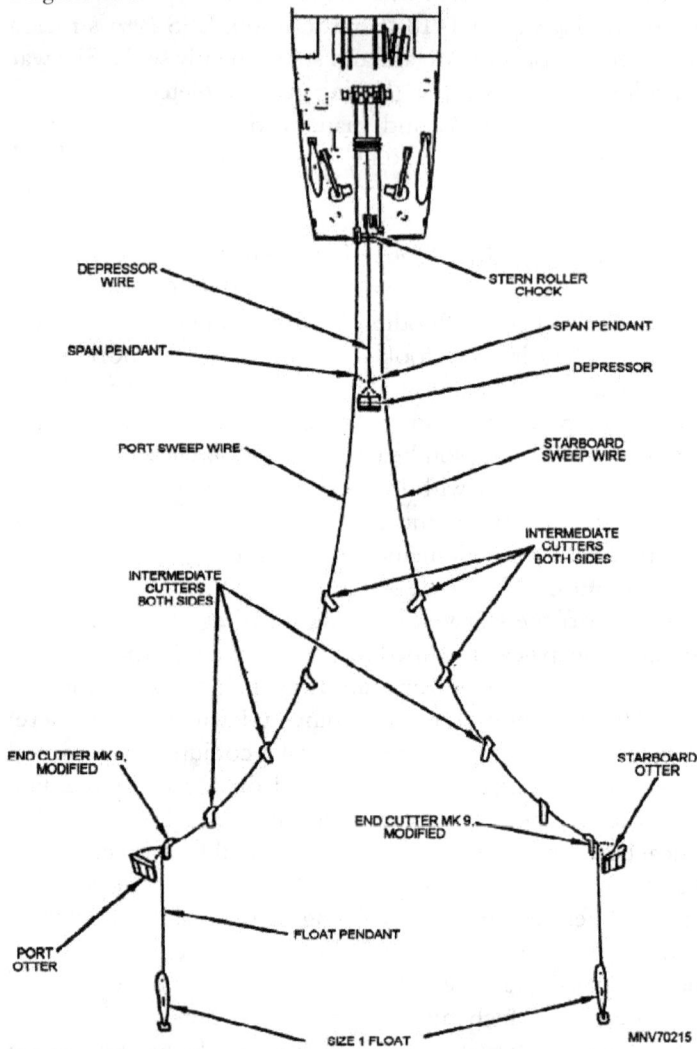

U.S. Navy Double-O moored minesweep

COMPOSITION OF TASK GROUP 32.9

Southard, (the now lost *Perry*), *Montgomery* and *Preble* were only part of the mine forces present at Palau to clear small shallow-laid minefields hampering landing operations along the coasts of Peleliu and Angaur, and a large shallow minefield in the waters of Kossol Passage. The sheltered waters of this passage were to serve as a temporary fleet anchorage and seaplane base. The identities of the mine warfare vessels comprising the task group under Comdr. Wayne R. Loud, USN, follows:

Task Group 32.9 (Minesweeping Group)

Sweep Unit	Ships
1A	*Hovey* (DMS-11), *Hamilton* (DMS-11), *Long* (DMS-12), *Montgomery* (DM-17)
1B	*Southard* (DMS-10), *Perry* (DMS-17), *Preble* (DM-20)
2	*YMS-140*, *YMS-177*, *YMS-180*
3	*YMS-1*, *YMS-33*, *YMS-324*
4	*YMS-2*, *YMS-275*, *YMS-325*, *YMS-238* (Hydrographic Unit)
5	*Zeal* (AM-131), *Token* (AM-126), *Tumult* (AM-127)
6	*Triumph* (AM-323), *Competent* (AM-316), *Velocity* (AM-128)[17]

The members of Sweep Units 5 and 6 were *Auk*-class steel-hulled minesweepers, six of ninety-five such rapidly constructed for war duty. Spanning 221 feet, the 890-ton ships boasted one 3-inch gun, two 40mm and eight 20mm guns, and had a top speed of 18.1 knots. In addition to these six ships and the five destroyer minesweepers and two destroyer minelayers present, there were also then ten YMS (Yard minesweepers) at Palau. Others arrived later.[18]

Photo 18-2

Minesweeper USS *Auk* (AM-57) off the Norfolk Navy Yard, circa May 1942.
Naval History and Heritage Command photograph #NH 84027

The largest production run of any World War II warship was not, as one might imagine, a particular class of destroyer, frigate, or submarine, but instead 561 scrappy little 136-foot wooden-hulled vessels characterized by Arnold Lott in *Most Dangerous Sea* as belligerent-looking yachts wearing grey paint. Because of their small crew size, cramped quarters, and limited storage for food and fuel, the Navy concluded YMSs would only be able to operate from a naval base or yard for several days before having to return for support. Accordingly, the motor minesweepers were designated District craft and known as "Yard" minesweepers. However, the necessities of war would result in them operating far from home waters and they plied the oceans of the world from the Aleutian Islands to Cape Horn. A second minesweeper casualty at Palau involved the loss of one of these type ships.[19]

YMS-19 SUNK BY A MINE

The explosion was like a bomb hitting a lumber yard.

—Lt. John K. Mahaffrey Jr., USNR, commanding officer USS *YMS-19*,
describing a mine detonation which blasted his ship into pieces;
sinking it in about two minutes with the loss of nine lives.[20]

Prior to D-Day, 15 September, when the 1st Marine Division stormed ashore at Peleliu, the minesweeping group had been able to clear the eastern and western coastlines of the island of mines to the extent that they were suitable for landing. Beach Red on Angaur had proved not mined; all the other beaches there were mined and awaited the availability of sweepers to clear them. The presence of minefields along the coast of Angaur restricted movements of landing craft and impeded the unloading of shipping. Commander, Task Force 32 (Rear Adm. George H. Fort, USN) directed Loud to sweep these fields at the earliest possible date, but not to delay the clearance of Kossol Passage.[21]

On 22 and 23 September, *YMS-2, 19, 140, 275, 293,* and *319* (under Lt. W. J. Bell, USNR) arrived at Angaur and swept close inshore for moored mines. The following day, the western coastline of Angaur plus Blue beach, at the northeast tip of the island, and Scarlet beach, at the southeast tip of Peleliu Island, were swept. One mine was cut in the first area, eleven in the second, and three in the last. Sweeping efforts then moved to the north and east coasts of Angaur on 24 September, in which twenty mines were cut and destroyed. During this operation, *YMS-19* struck a mine and sank with a loss of nine personnel.[22]

Photo 18-3

USS *YMS-12*, sister ship to *YMS-19* lost at Angaur, Palau, on 24 September 1944. Naval History and Heritage Command photograph #NH 86722

FINAL TALLY AND ASSESSEMENT OF OPERATIONS

A total of seventy-five mines were swept in the Angaur-Peleliu area; and an astounding 323 mines in Kossol Passage, which was reported cleared on 23 September. That same day, Angaur (the two-mile-long island south of Peleliu, wanted for a bomber strip) was captured by the U.S. Army 81st Infantry Division. Commander, Mine Division Two included in his after-action report on the operations at Palau, a brief assessment of the overall performance of the minesweepers, by ship class, and a recommendation that tenders be obtained to provide needed support for the 136-foot, wooden-hulled YMSs:

> The DMS and AM class vessels were all well-handled and efficient sweepers. The ships' offices and crews were veterans of several previous operations and performed like professionals. The YMS type vessels varied from smart, well trained ships to those with little knowledge of what they were trying to do or how to do it. Apparently, in several areas, YMSs are still confining their activities to convoy escort and patrol work and are receiving little training in their specialty – minesweeping. The DM class were invaluable in mine destruction, buoy laying and hydrographical survey operations.

> As usual, the YMS type vessels had difficulties with radio, radar and sonar equipment and with water and fuel supply. A few had engineering breakdowns and all experienced a lowest ebb in provisions. It is recommended that a ship be designed and

equipped as a YMS tender and accompany the attack force to carry spares, effect repairs, render technical service, and be a source of fuel, water and provision for YMSs at the objective.[23]

AFTERMATH

The capture of the Marianas and south Palau Islands enabled continued movement forward in the Central Pacific. In the Southwest Pacific, MacArthur was in control of Biak—a small island near the north coast of Papua on which a strategically sited Japanese Army airfield had served as a base for operations in the Pacific theater—as well as western New Guinea, and was poised to cross the Celebes Sea to fulfill his promise to return and liberate the Philippines.[24]

Before progressing to the landings on Leyte Island, which launched the Allied invasion of the Philippines, we will first visit two other mine warfare-related areas. Chapter 19 is devoted to the Navy's conversion of auxiliary minelayers (former Army mine planters) for service as YMS tenders, and Chapter 20 to Royal Navy and Royal Netherlands Navy submarine-minelaying operations in the Pacific.

UNIT AWARDS

Terror received a battle star for support she provided during the capture of Guam, and *Montgomery* and *Preble* for their contributions to the capture of the south Palau Islands.

Marianas Operation: Capture and Occupation of Guam

Terror (CM-5)	4 Aug 44	Comdr. Horace William Blakeslee, USN

Western Caroline Islands: Capture and Occupation of south Palau Islands

Montgomery (DM-17)	6 Sep-14 Oct 44	Lt. Dan Thomas Drain, USNR
Preble (DM-20)	6 Sep-14 Oct 44	Lt. Edward Francis Baldridge, USN

USN Auxiliary Minelayers
Pressed into New Duties

Photo 19-1

Auxiliary minelayer USS *Chimo* (ACM-1) passing movies to USS *Obstructer* (ACM-7), via light line, while en route to Saipan, September 1945.
Naval History and Heritage Command photograph 1978

The small U.S. Navy was propelled into war by the Japanese attack on Pearl Harbor, with wholly too few ships. As America's shipyards increased production to full capacity to meet requirements, the Navy brought World War I ships laid up in "ghost fleets" out of "mothballs," and procured commercial and privately-owned vessels as stopgap measures until purpose-built vessels could join the Fleet. The YPs (ex-San Diego tuna boats) which supported the 1st Marines on Guadalcanal were one such example. The Navy, as it is apt to do when having too many of one type ship and not enough of another, modified some of the former for new duties or, in some cases, simply reclassified them.

As the war continued, and scores of "bright and shiny" new ships joined the Fleet, this practice lessened but did not disappear altogether. Such was the case with the Navy's auxiliary minelayers. As discussed in Chapter 2, The U.S. Army's Coast Artillery Corps was responsible for maintaining minefields to protect American harbors and coastlines from hostile ships. By 1944, when it was clear that there was no longer any serious threat to American harbors, the Army began releasing its 188-foot twin-screw, steel-hulled mine planters for other uses. The Navy received three of the 1,300-ton ships in 1944 and five more in 1945.

COMMISSIONING, FITTING OUT OF INITIAL SHIPS

Planter (ACM-2) was placed in commission at St Helena Navy Yard, Norfolk, Virginia, on 4 April 1944 by Rear Adm. Felix Gygax, USN, and attached to Service Squadron 4, Atlantic Fleet. She remained moored in the yard for the next three-and-a-half weeks, undergoing conversion to serve as an auxiliary minecraft in the capacity of buoy layer, mother ship for minesweepers, and as a minelayer.[1]

Planter got under way in the Chesapeake Bay on 30 April to calibrate degaussing equipment, swing ship to compensate compasses, and test fire her 20mm and 40mm guns. At completion, she anchored for the night off Cape Charles City, Virginia. The following day was devoted to calibration of the ship's Radio Direction Finder, and 2-4 May she spent at Naval Operating Base, Norfolk, loading and restowing operational materiels and training personnel. On 5 May, *Planter* got under way in company with *Chimo* (ACM-1) and *Barricade* (ACM-3) for anti-aircraft machine gun practice. The next few days were spent in port, loading stores, followed by time at sea with the other ACMs practicing towing, fueling, buoy laying and gunnery; anchoring at night east of the Wolf Trap Magnetic Range.[2]

Departure day for Europe arrived on 13 May 1944, when at 0712 *Planter* got under way from the Naval Operating Base to join Convoy UGS-42 in the lower Chesapeake Bay for transit from Norfolk to the Mediterranean area. Task Group 63.4—*Chimo*, *Barricade*, and *Planter* as senior ship—were assigned rear positions in the convoy; detailed to pick up survivors, should it be attacked. The convoy was made up of 90 cargo vessels and tankers, three tank landing ships and eleven infantry landing craft. One DD (destroyer) and twelve smaller DEs (destroyer escorts), accompanied by an oiler, comprised the Naval escort.[3]

DUTY IN EUROPE AND THE MEDITERRANEAN

Discussion about the service of the auxiliary minelayers in Europe is not within the scope of this book. *Chimo* earned a battle star, along with the minelayer *Miantonomah*, for the invasion of Normandy, and *Barricade* and *Planter* ones for Southern France.

Invasion of Normandy

Chimo (ACM-1)	6-25 Jun 44	Lt. Comdr. John Winston Gross, USNR
Miantonomah (CM-10)	6-25 Jun 44	Comdr. Austin Edward Rowe, USN

Invasion of Southern France

Barricade (ACM-3)	15 Aug- 25 Sep 44	Lt. Charles Percy Haber, USN
Planter (ACM-2)	15 Aug- 25 Sep 44	Lt. Theodore Thomas Scudder Jr., USNR; Lt. Richard Albert Knapp, USN

ESTABLISHMENT OF MINECRAFT, PACIFIC FLEET

On 15 October 1944, the Navy established a new command, Minecraft, Pacific Fleet, to oversee the large numbers of widely dispersed mine forces in the Pacific. Service Squadron 6, to which the mine forces were previously assigned, was overgrown and could not accommodate their continuously expanding numbers. The first commander of the new Pacific Force was Rear Adm. Alexander Sharp Jr., USN, who had voluntarily stepped down from his vice admiral position as commander, Service Force, Atlantic Fleet, for the Pacific command.[4]

Following the end of the war, Rear Adm. Arthur D. Struble, USN, commanded the Pacific Fleet's mine force as it began the long process of clearing mines from the former combat zone. This book is also not able to devote adequate space to these operations, but reference is made to them in the next few pages to describe the service of the ACMs.

TRANSFER OF ACMS TO THE PACIFIC FLEET

The U.S. Navy needed lots of minesweepers and other minecraft in the Pacific and recognized that following Japan's eventual defeat, it would require even more to clear thousands of U.S.-laid mines. Accordingly, it began transferring mine warfare ships from the Atlantic to the Pacific even before the surrender of Germany on 7 May 1945. This continued for a while after it. Among the latter ships were the auxiliary minelayers.

Photo 19-2

Rear Adm. Arthur D. Struble, USN, commander, Visayan Attack Force, immediately prior to the launching of the Mindoro, Philippines, invasion, circa 12 December 1944. Naval History and Heritage Command photograph NH 97324

In preparation for Pacific duty, another former Army mine planter was put into naval service on 24 March 1945, as *Barbican* (ACM-5), commissioned at the Charleston Navy Yard, Charleston, South Carolina. Her primary mission was to be the operational headquarters of a squadron commander under whom a squadron of YMSs would operate. Her secondary duty was to act as a tender for the Yard minesweepers. *Barbican*'s preparation for service was similar to that of other auxiliary minelayers, with the addition of a full power run supervised by a Navy captain from the Board of Inspection and Survey. At month's end, she reported to commander, Minecraft Shakedown Group, at Little Creek, Virginia, for three weeks of various drills, ship-handling exercises, and battle problems.[5]

Following this training and evaluation, *Barbican* underwent an availability period at the Portsmouth Navy Yard for correction of material deficiencies. She then shifted to the Naval Operating Base at Norfolk to fuel and load the ship's allowance of 40mm and 20mm ammunition, and to prepare for her shakedown final inspection. On 23 April following an "all hands" personnel inspection; inspection of the ship for material readiness; and successful completion of a final battle problem in the Chesapeake, *Barbican* moored at Little Creek. The next day, she reported for duty to commander, Service Force, Atlantic, and returned to NOB Norfolk. The remainder of the month was devoted to loading ship's spares, commissary stores, and YMS spare parts, which were brought aboard and stowed ready for sea.[6]

May and early June 1945 was much the same as *Barbican* remained at NOB Norfolk, loading YMS spare parts while also awaiting the full allowance of spares to arrive. This period was also used for crew firefighting, lookout watches, and seamanship training. June 8th-10th brought final preparations for sea. In early afternoon on 11 June, *Barbican* joined sister ships *Chimo*, *Obstructer*, and *Trapper* for transit to the Panama Canal Zone—en route to duty with the Pacific Fleet.[7]

The four auxiliary minelayers reached Cristobal on the 19th, and remained there until departure on 22 June for passage through the Panama Canal. Upon entering the Pacific, operational command of the ships changed from Service Force, Atlantic Fleet, to Administrative Command, Minecraft, Pacific Fleet. *Obstructer* sailed independently for San Diego, stopping at Manzanillo, Mexico, to receive fuel en route, as did *Chimo*, arriving at the scenic California coastal city on 3 July. Lt. Comdr. F. K. Zinn, USNR, commander, Mine Squadron 101, and his staff embarked aboard *Barbican* at San Diego on 16 August, and *Barbican* and *Chimo* sailed for Pearl Harbor the following day. They arrived there on 27 August, joined by *Obstructer* a few hours before entering harbor.[8]

Trapper (commissioned on 15 March at Charleston Navy Yard) had proceeded from the Canal Zone to San Diego with the *Barbican*. She'd taken departure for Pearl Harbor on 10 August, steaming singularly, and arrived there on the 20th. The extended period the ACMs spent in San Diego had been well utilized for voyage repairs and engine overhaul, shakedown training and exercises, and loading lubricating oil for YMSs. On 23 August, Comdr. John D. Riner, USNR, reported on board *Trapper* as commander, Mine Squadron 102. He read his orders to the assembled crew, placing the squadron in commission, upon which his Broad Command Pennant was hoisted at the mainmast.[9]

SEVEN ACMS SEE DUTY IN THE WESTERN PACIFIC

Seven of the eight *Chimo*-class auxiliary minelayers subsequently served in Asian waters.   The remaining ship, *Barricade*, after transiting the Panama Canal on 2 September 1945 and reaching San Diego, operated along the coast of California through the end of her service. She was decommissioned on 28 June 1946 and transferred to the Coast Guard.[10]

Summary information about the eight minelayers follow: (1) ship commissioning dates, (2) dates reported for duty to the Pacific Fleet, (3) identities of mine squadrons to which ACMs were assigned flagship duties, and (4) identities of commanding officers:

Chimo-class Auxiliary Minelayers (former Army Mine Planters)

Ship	Comm.	PacFlt	Commanding Officer
Chimo (ACM-1)	7 Apr 44	23 Jun 45	Lt. Comdr. John Winston Gross, USNR
Planter (ACM-2) (Flagship for CoMinRon 105)	4 Apr 44	17 Apr 45	Lt. Theodore Thomas Scudder Jr., USNR; Lt. Richard A. Knapp, USN; Lt. Charles D. Bier, USN
Barricade (ACM-3)	7 Apr 44	2 Sep 45	Lt. Charles Percy Haber, USN
Barbican (ACM-5) (Flagship for CoMinRon 101)	24 Mar 45	23 Jun 45	Lt. Comdr. Alexander Anderson Jr., USN; Lt. Norman R. Burnett, USN
Bastion (ACM-6) (Flagship for CoMinRon 107)	9 Apr 45	Jul 45	Lt. Earl David Fatkin, USNR; Lt. James Paul Albertson, USN; Lt. George D. Harrelson, USN
Obstructer (ACM-7) (Flagship for CoMinRon 106)	1 Apr 45	23 Jun 45	Lt. Sammie Smith, USN
Picket (ACM-8) (Flagship for CoMinRon 104)	6 Mar 45	7 May 45	Lt. Ralph C. Wilson Jr., USNR; Lt. Robert F. Harwood, USNR
Trapper (ACM-9) (Flagship for CoMinRon 102)	15 Mar 45	23 Jun 45	Lt. Richard E. Lewis, USNR; Lt. Warren B. Shearon, USN

UNIT AWARDS

Chimo and *Obstructer* each earned a battle star for dangerous post-war clearance of U.S. mines in Japanese, Korean, and Chinese waters. *Chimo* received hers for clearing minefields in the East China Sea-Ryukyus Area; during her participation in Operation SKAGWAY, she was a part of Task Group 52.3 operating from Sasebo.

Obstructer received a battle star while serving as flagship for Lt. (jg) E. B. Byhre, USNR, commander, Task Unit 70.5.10; comprised of *Obstructer*, YMSs *4*, *336*, *363* and *392*, and *LCS(L)-30* and *-50*. For this operation, which involved sweeping the approaches to Swatow, and Amoy Harbor and its approaches, Byhre was commanding officer of *LCS (L)-30*, and also temporary commander of LCS(L) Group One.[11]

Photo 19-3

USS *LCS(L)-50* under way off Portland, Oregon, 19 September 1944.
Naval History and Heritage Command photograph #NH 81530

The LCS(L), an amphibious craft, was designed to close enemy shores during landing operations, blasting gun positions and personnel. They were along to provide protection for the YMSs. During this period, there was great tension between Chinese Nationalists under Gen. Tu Li-ming and Generalissimo Chiang Kai-shek's Chinese Communists. The U.S. government preferred to consider the communists agrarian reformers and sent Gen. George C. Marshall to China during the last week of December 1945. Marshall flew into the Shanghai airport.[12]

Map 19-1

Eastern China in 1941

In addition to *Chimo* and *Obstructer*, six destroyer minelayers earned battle stars for these operations as well.

Minesweeping Operations Pacific: "Skagway"
(East China Sea-Ryukyus Area)

Chimo (ACM-1)	7-9 Nov 45	Lt. Comdr. John Winston Gross, USNR
Breese (DM-18)	14-24 Aug 45	Lt. David James Pikkaart, USNR
Gwin (DM-33)	27 Oct 45	Comdr. Elmer Cecil Long, USN
Henry A. Wiley (DM-29)	14-24 Aug 45	Comdr. Paul Henrik Bjarnason, USN
Henry F. Bauer (DM-26)	20 Aug- 9 Nov 45	Comdr. Richard Claggett Williams Jr., USN
Robert H. Smith (DM-23)	14-24 Aug 45	Comdr. Wilber Haines Cheney Jr., USN
Shannon (DM-25)	14-24 Aug 45	Comdr. William Thomas Ingram II, USN

Special Operations: Minesweeping Operations Pacific

Obstructer (ACM-7)	10-18 Dec 45	Lt. Sammie Smith, USN

20

Royal Navy Submarines/Dutch
O-19 lay Mines in Far East Waters

At the beginning of 1944, Vice Adm. Sir James Somerville, RN, commander in chief Eastern Fleet, had under his command at Trincomalee, a deep-water harbor at Ceylon (today Sri Lanka), the 4th Submarine Flotilla. It consisted of the depot ship HMS *Adamant* and five T-class submarines: *Taurus, Trespasser, Tally Ho, Templar* and *Tactician.* T-class and also S-class submarines continued to arrive from the Mediterranean and British Home Waters throughout the year. When a second depot ship, HMS *Maidstone*, arrived in March, the S-class submarines were organized into the 8th Flotilla with her.[1]

A third depot ship, HMS *Wolfe*, joined in August and formed the 2nd Flotilla with six S-boats; upon which the *Maidstone* left for Australia, joined there by a mixed flotilla of both type submarines. Earlier in June, the minelaying submarine *Porpoise* had joined the 4th Flotilla, as would her sister ship, the *Rorqual*, in December 1944.[2]

With a substantial submarine force available, it was hoped that casualties to enemy shipping in the patrol area west of Malaya would correspondingly rise, but there proved to be a general dearth of targets in these waters. This was due in part to Japan's increasing shortage of shipping, and also to aircraft-minelaying which caused the enemy to resort mainly to small craft for movement of cargos by water in the area. Accordingly, in early 1944, it was decided to lay mines from submarines while on patrol, in an effort to force traffic into deeper waters where it could be more readily attacked.[3]

SCOPE AND DURATION OF MINING EFFORT

A majority of the resultant twenty-one minefields in 1944, were laid on inshore routes off the west coasts of Malaya and Thailand. Penang also received some attention, as did the Sumatran coast; and one field was laid close to the Burmese port of Mergui, already a target for regular air mining. The minelaying submarines were usually tasked with additional duties while on patrol, including destroying small craft carrying supplies for the enemy, taking part in special operations, bombarding shore targets, and serving as air/sea rescue vessels.[4]

Over a fourteen-month period between 14 March 1944 and 10 May 1945, eighteen British submarines laid a combined total of 557 mines, and the Dutch submarine HNMS *O-19* eighty mines.

Mines Laid by RN Submarines

Date	Submarine	No. Mines	Location
14 Mar 44	*Trespasser*	12 M Mk 2	Outer Mati Bank, Malacca Strait
19 Mar 44	*Taurus*	12 M Mk 2	Aroa Islands, Malacca Strait
18 Apr 44	*Taurus*	12 M Mk 2	South of Penang
13 May 44	*Surf*	8 M Mk 2	Palau Terusan
14 May 44	*Tally-Ho*	12 M Mk 2	Bunja Shoal, Malacca Strait
16 May 44	*Tactician*	12 M Mk 2	Langkawi Sound, Malaya
18 May 44	*Sea Rover*	8 M Mk 2	Sembilan Islands
2 Jun 44	*Tantalus*	12 M Mk 2	Dindings, Malaya
3 Jun 44	*Stoic*	8 M Mk 2	North of Penang
4 Jun 44	*Templar*	12 M Mk 2	Dindings, Malaya
7 Jun 44	*Tantivy*	12 M Mk 2	Sembilan Islands, Malacca Strait
14 Jun 44	*Surf*	8 M Mk 2	Palau Terusan
24 Jun 44	*Truculent*	12 M Mk 2	South of Klang Strait, Malaya
6 Jul 44	*Porpoise*	30 Mk 16	Belawan Deli
8 Jul 44	*Porpoise*	16 M16/ 10 M Mk 2	Balawan Deli
16 Sep 44	*Trenchant*	12 M Mk 2	Sembilang Channel, Sumatra
24 Sep 44	*Tudor*	10 M Mk 2	Off Pulo Lantar, Malaya
30 Oct 44	*Tradewind*	12 M Mk 2	Mergui Islands, Burma
19 Nov 44	*Thorough*	12 M Mk 2	Outer Mati Bank, Sumatra
9 Dec 44	*Porpoise*	35 Mk 16 & 15 Mk 16	Penang
16 Dec 44	*Thule*	12 M Mk 2	Off Palau Terusan, Malaya
23 Dec 44	*Thorough*	12 M Mk 2	Off Palau Terusan, Malaya
3 Jan 45	*Rorqual*	50 Mk 16	Phuket
3-4 Jan 45	*Rorqual*	12 M Mk 2	Papra Strait
Jan 45	*Porpoise*	12 M Mk 2	Penang
Jan 45	*Porpoise*	49 Mk 16	Penange (sub lost, date unknown)
22 Jan 45	*Rorqual*	50 Mk 16	Ritchies Archipelago
23 Jan 45	*Rorqual*	12 M Mk 2	Ritchies Archipelago
10 May 45	*Rorqual*	12 M Mk 2	Between Two Brothers and the Sumatran coast
		44 Mk 16	South of the Thousand Islands[5]

Mines Laid by Dutch Submarine HNLMS *O-19*

3 Jan45	40 T3	Channel between Bali Island and the Java coast
13 Apr 45	40 T3	Northern entrance to the Banka Strait[6]

Photo 20-1

Adm. James Sommerville, RN, with Capt. John H. Cassady, aboard USS *Saratoga* (CV-3) at Trincomalee, Ceylon, in April 1944.
U.S. National Archives photograph #80-G-K-15604

SUBMARINE-LAID MINES

The T- and S-class submarines employed M Mk 2 magnetic ground mines launched from their torpedo tubes. Because this type of mine was not sufficiently sensitive in water depths greater than 10 fathoms, much skill was required of commanding officers when operating in shallow, constricted waters, particularly when submerged. However, with considerable determination, and a little luck, these type boats of the 4th and 8th Flotillas completed nineteen minefields in 1944.[7]

Porpoise and *Rorqual* (*Porpoise*-class submarines) carried both M Mk 2 magnetic ground mines and Mk 16 moored contact mines, and in much greater numbers. The Mk 16 was designed for use in 100 fathoms or less, had Hertz horns and carried a 320lb explosive charge. Two of the M Mk 2 mines occupied the space of one torpedo. Since they rested on the sea floor, vice being suspended in the water column, they offered a much larger 1,000lb explosive charge.[8]

FIRST MINING OPERATION

The first minelay was made by HMS *Trespasser*, a relatively new T-class diesel-electric submarine commissioned on 11 September 1942 with a ship's complement of fifty-nine. She spanned 275 feet, displaced 1,090 tons, and was fitted with one 4-inch deck gun, three machine guns for anti-air defense, and a maximum of sixteen 21-inch torpedoes or twelve M Mk 2 ground mines. The location of her ten torpedo tubes (6 bow, 2 external bow, and 2 external amidships firing forward) enabled her to fire a formidable forward torpedo salvo. External tubes were necessary to avoid compromising structural integrity of the pressure hull with too many openings. They could not be reloaded from within the submarine, and it was not possible to conduct maintenance on or withdraw a torpedo once it was loaded into an external tube. T-class submarines could carry six reloads in their torpedo compartment, but since two mines occupied the space of one torpedo, these six torpedoes were left behind for mining missions. *Trespasser* would have had ten torpedoes, if all tubes were loaded, and a full complement of twelve mines when departing on such an operation.[9]

Photo 20-2

HMS *Trespasser* under tow off Barrow, England.
Imperial War Museums photograph FL 3790

Trespasser left Trincomalee on 7 March 1944 on her fourth and last patrol in Far Eastern waters before returning to the UK for a refit. As she was required to depart for Britain by 5 April, her planned patrol was short, limited to a minelay off the Sumatran coast, followed by a special operation on the Malayan side of the straits. In early morning on 14

March, *Trespasser* deposited her twelve Mk 2 mines in groups of four, inside the Outer Mati Bank. Minimum water depth was 10-fathoms.[10]

On completion, she withdrew to the northwest, and later watched an enemy sub-chaser on a southeasterly course, pass over the position of her first batch of mines, before they were due to become active. Somewhat disappointed that there were no resultant mine detonations, but hopeful that she had covered a well-used route, *Trespasser* proceeded for the area of the special operation. It was aborted, and she returned to Trincomalee on 24 March. Her minefield later claimed the 1,274-ton *Kasumi Maru*, sunk on 12 May 1944. (The cargo ship was the former British salvage vessel SS *Kathleen Moller*, which had been seized by the Japanese at Hong Kong in December 1941.)[11]

A British Naval Staff History study of 1973 cites *Kasumi Maru* being sunk on 12 May by one of *Trespasser*'s mines, whereas a Joint Army-Navy Assessment Committee study of 1947 (the major source for such information in Appendix A) concluded that the cargo ship was lost on 1 June to a U.S. Army airdropped mine. These type inconsistencies are not uncommon and are often resolved only after the wreckage of a ship or submarine, of sufficient importance to warrant examination by experts, has been located on the seafloor.

SUMMARY OF SUBMARINE-MINELAYING IN 1944

The majority of the minelaying operations carried out by Royal Navy submarines in 1944, involved tube-laid M Mk 2 ground mines. HMS *Porpoise* arrived on station in June, bringing with her a capability to lay large Mk 16 moored mines as well as the Mk 2 types. By year's end, a total of 416 mines had been laid, without loss of any RN submarines; claiming four vessels sunk, and three damaged including the Japanese submarine *I-37* on 27 April 1944.

I-37 had departed Penang that morning at 0500 for Singapore, escorted by her E14Y1 "Glen" floatplane. Around 0800, when she was some twenty miles south of Penang, an explosion occurred about 110 yards off her port bow, badly shaking the submarine. The lights went out, apparently caused by shorts in the electrical switchboard. After settling to the bottom at shallow depth to effect repairs, *I-37* returned to Penang the following day. A pierside inspection found the only other damage suffered was to the valves of two ballast tanks on her port side.[12]

MINELAYING IN EARLY 1945

Submarine minelaying continued on a much-reduced scale in 1945, being undertaken only by *Porpoise* and *Rorqual*. The latter submarine, also a *Porpoise*-class (commanded by Lt. John P. H. Oakley, RN), had arrived at Trincomalee on 14 December 1944. While on her first patrol of the year, *Porpoise* was lost in the vicinity of Penang, through unknown cause, but was suspected to have been the victim of bombing attacks by Japanese Naval aircraft.[13]

HMS *PORPOISE*'S DUTY WITH THE 4TH FLOTILLA

HMS *Porpoise*, under the command of Lt. Comdr. Hubert A. L. Marsham, RN, had arrived at Trincomalee on 12 June 1944 and joined the 4th Submarine Flotilla. Her ship's complement was 6 officers and 53 men. The 288-foot minelaying submarine boasted a 4-inch deck gun, two machine guns, six 21-inch bow torpedo tubes (with twelve reloads), and fifty mines. The mines were stored on a rail running nearly the entire length of the top of the pressure hull, enclosed by a non-watertight casing. When laying a minefield, large stern doors at the rear were opened and an endless chain (similar to a conveyor belt) inside the extra deep casing discharged mines into the sea. To accommodate the casing along the centerline of the upper hull, her conning tower and periscope were slightly offset to the side.[14]

Photo 20-3

British Submarine HMS *Porpoise* moored alongside the heavy cruiser HMAS *Australia*, with HMS *Salmon* (an S-class submarine) outboard of her. Australian War Memorial photograph P00604.012

Acting Lt. Comdr. Hugh B. Turner, relieved Marsham as her commanding officer on 21 November. *Porpoise* left Trincomalee on 2 December, on her 22nd war patrol (3rd in the Far East) to lay a minefield off Penang and operate off the Nicobar Islands. Turner conducted a periscope reconnaissance of the area to be mined during daylight on the 8th, and made a submerged approach from the south the following morning, complicated by the presence of many sampans and by fishing stakes to the northeast. After laying fifty Mk 16 mines in two rows as close as possible to the 10-fathom curve seaward of the island, *Porpoise* remained in the area until the following night, in which she sank an enemy sailing junk in the Strait of Malacca, then set a course for the Nicobar Islands, in the eastern Indian Ocean. The submarine's arrival at Trincomalee on 18 December marked the end of the patrol.[15]

Photo 20-4

Mk 16 mines were a slightly larger version of the Mk 6 mine. The mine case contained 635 lbs of HBX-1, over twice as much explosive charge as the Mk 6 mine; otherwise the two mines were very similar.
Courtesy of Ron Swart

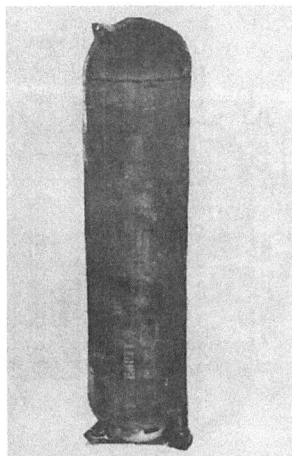

British M Mk 2 ground, magnetic or acoustic mines, laid by submarines or surface craft, carried 1,000 lbs of Amatol or 1,060 lbs Minol with Tetryl booster explosive charges.
Courtesy of Ron Swart

PORPOISE LOST ON MINING MISSION

On 2 January 1945, *Porpoise* sailed from Trincomalee to make two mine lays to the west of the southern end of Penang, in the general vicinity of a field she had laid in December. Her first operation was to involve twelve Mk 2 ground mines, and the other forty-nine Mk 16 moored mines. The plan was to lay both fields at low water on or about 9 January; when the height of the tide would be three to four feet above datum, and the moored mines set at a depth of six to seven feet below chart datum.[16]

Nothing was heard from *Porpoise* after her departure from port. At 0101 on 13 January, HMS *Stygian* received a signal from Capt. Hugo M. C. Ionides, RN, commander, 4th Submarine Flotilla, that *Porpoise* was in trouble seventeen miles northwest of Palau Perak. This information came from an Ultra decrypt indicating that at 1000 on 11 January, a Nakajima B6N2 "Tenzan" bomber (of the 331st Naval Air Group) had attacked a submarine, dropping two 132lb bombs. One missed its port bow by two meters and the other was a direct hit. At 1145, a Tenzan carried out a second attack, seventeen miles off Perak Island, dropping two bombs on a submerged submarine leaking oil. That night, a third attack was made in the same area with six bombs and the attack was continued at 1000 the following morning.[17]

Photo 20-5

Japanese torpedo bomber Nakajima B6N2 "Tenzan" with American markings. After capture, restored by Technical Air Intelligence Unit SWPA and flown by Allied pilots for assessment purposes.
Australian War Memorial photograph AC0195

Stygian was in the vicinity and carried out a surface search of the area and tried to contact *Porpoise* by radio, but nothing was seen or heard and no traces of wreckage found. The attacks by the Tenzan aircraft were consistent with the route *Porpoise* would likely have taken retiring after her minelaying operation. Divers have looked for the wreck without success, and the definitive cause of her loss remains unknown.[18]

Liberation of the Philippines

Should we lose in the Philippines operations, even though the fleet should be left, the shipping lane to the south would be completely cut off so that the fleet, if it should come back to Japanese waters, could not obtain its fuel supply. If it should remain in southern waters, it could not receive supplies of ammunition and arms. There would be no sense in saving the fleet at the expense of the loss of the Philippines.

—Adm. Soemu Toyoda, Imperial Japanese Navy, discussing Vice Adm. Takeo Kurita's mission to destroy completely the transports in Leyte Bay following the American invasion of the Philippines, and why there were no restrictions as to the damage that his force might take.[1]

The U.S. Sixth Army went ashore at Leyte Island on 20 October, 1944, two months and two years after the first landings were conducted in the Guadalcanal area of the Solomon Islands. Amphibious landings in the Marianas in June and July breached Japan's strategic inner defense ring and gave the Americans a base from which B29 bombers could attack the Japanese home islands. Following the invasion of Saipan, the Japanese counterattacked in the Battle of the Philippine Sea, fought between the First Mobile Fleet and American Fifth Fleet from 19 to 21 June. In what one American aviator termed "The Great Marianas Turkey Shoot," the U.S. Navy destroyed three enemy aircraft carriers— the *Hijo*, *Shokaku*, and *Taiho*—some 480 planes, and nearly as many aviators. The devastating loss left the Japanese with virtually no carrier-based aircraft or experienced pilots for the forthcoming Battle of Leyte Gulf.[2]

The naval force (comprised of units of the American Third and Seventh Fleets) assembled for the invasion of Leyte was not quite as large as the one that had taken part in June in the invasion of Normandy, but it had more striking power. Embarked aboard the assault vessels were the Sixth Army's Tenth and Fourteenth Corps. The weather at the entrance to Leyte Gulf at daybreak on 20 October was cloudy with altostratus and partial swelling cumulus, a visibility to seaward of twelve miles, and light winds from the southeast. Planners had been concerned that a typhoon might pass through the area and cause retirement or diversion of the forces en route from New Guinea. However, the conditions on "A-Day" (Assault Day) were perfect as described by commander, Third Amphibious Force:

> The assault proceeded on schedule following the preliminary bombardment by ships' gunfire and aircraft, a slight onshore tendency of the almost imperceptible wind conveniently drifting the smoke and dust of the bombardment off the beaches and into the interior. The airborne beach observer had made his required report earlier, but the report was unnecessary in this case due to the almost complete absence of surf.[3]

The landings at Tacloban, located in northeast Leyte on an inlet of the Leyte Gulf, and at Dulag, twenty-five miles to the southward, were made against little opposition. Naval historian Samuel Eliot Morison noted about the operation: "The Leyte landings were easy, compared with most amphibious operations in World War II—perfect weather, no surf, no mines or underwater obstacles, slight enemy resistance, mostly mortar fire." With this beginning, the liberation of the Philippines was off to a good start.[4]

MOVEMENT TO LEYTE BY USN MINE FORCES

Ten days earlier, a sizeable force of mine warfare ships under Comdr. Wayne Loud, USN, had left Seedler Harbor, Manus, Admiralty Islands, the morning of 10 October, bound for the Philippines. Comprising it were 7 DMSs, 10 AMs, 22 YMSs, 2 DMs, and an APD (a converted destroyer, termed a high-speed transport) carrying four LCPR landing craft fitted with small boat Oropesa minesweeping gear. En route to Leyte, the ships fueled from tankers off Palau on 15 October, and that night the weather turned bad. Driving rains reduced visibility, and strong winds and heavy seas pounded the minesweepers. The YMSs were making flank speed in order to maintain their stations in the cruising disposition. These 136-foot wooden-hulled vessels suffered mightily—many sustained radio, radar, gyro, and engine casualties—but managed to stay afloat, reaching the entrance to the Leyte Gulf the night of the 16th, battered and leaking.[5]

Sweeping began on schedule, at dawn the following day, but under near-typhoon conditions. By mid-afternoon, *YMS-70* had reached her limit. With water coming in through seams, opened by heavy seas, faster than the pumps could handle, Lt (jg) Orson L. Slackett, ordered his crew at 1730 to abandon ship. *YMS-341* stood by and as a result of skillful maneuvering, was able to rescue eighteen men. The remaining officers and men, totaling sixteen, were tossed around all night long in life rafts until rescued the next morning by YMSs *39* and *73*.[6]

Photo 21-1

Balikpapan Operation, Borneo, June-July 1945. An underwater demolition team's LCPR towing a rubber boat employed for swimmer discharge and pickups. National Archives photograph #80-G-274700

OPERATIONS IN SURIGAO STRAIT/LEYTE GULF

Between 16 and 23 October, the minesweeping force swept 306 mines of the moored, chemical horn variety. Protective echelon formations were employed by all sweepers, with standard distance between ships in formation, 1,000 yards for DMSs and 800 yards for AMs and YMSs. The minelayers *Preble* and *Breese*, and high-speed transport *Sands* (APD-13), buoyed the extremities of the minefields, marking the approaches to the landing beaches, and served as mine destruction vessels.[7]

The LCPRs, rigged as small boat sweepers and carried to Leyte in the *Sands*, were used for the first time in this role. They were charged with conducting sweeps among shoals, inaccessible to the deeper draft ships, and with sweeping the approaches to the landing beaches. They were able to determine that the approaches were clear up to the 3-fathom curve (18-foot water depth) or 500 yards off shore. Driven off once by enemy shore battery fire, the LCPRs returned, after fire support ships had silenced the battery, and completed their mission. The only damage sustained was one rudder shot away.[8]

BREESE SHOOTS DOWN FIRST JAPANESE PLANE

> *One plane, the first to be downed by surface anti-aircraft fire in Leyte Gulf, was definitely destroyed by this unit, and five others were taken under fire, one of which was probably damaged.*

—Lt. Comdr. David B. Cohen, USN, commanding officer of the destroyer minelayer USS *Breese* (DM-18), describing shooting down a Japanese torpedo-bomber on 21 October 1944.

In addition to aircraft attacks against the entire minesweeping force, Sweep Units One and Seven, made up of DMSs, and YMSs, respectively, and the *Preble* were targets of individual dive bombing and strafing attacks by two groups of four enemy planes. The composition of TG 77.5 of which these two Sweep Units were a part is shown below, as well as that of the Hydrographic Unit. The latter unit consisted of an Australian frigate and harbor defense motor launch, and two American YMS-class ships.[9]

Task Group 77.5: Comdr. Wayne R. Loud, USN
Commander, Mine Division Two

Task Unit 77.5.1 (Minesweepers)
Comdr. Wayne R. Loud, USN

Sweep Unit	Ships/Craft	Commander
One	*Hovey, Southard, Hamilton, Long, Howard, Palmer, Chandler, Preble, Breese, Sands,* fleet tug *Chickasaw* (AT-83)	Comdr. Loud
Two	*Zeal, Token, Tumult, Velocity, Scout*	Lt. Comdr. E. Woodhouse
Three	*Requisite, Pursuit, Revenge, Sage, Sentry*	Lt. Comdr. H. R. Pierce Jr.
Four	YMSs *1, 81, 140, 219, 319*	Lt. Stahli
Five	YMSs *238, 243, 286, 293, 335, 398*	Lt. W. A. Latta, USNR
Six	YMSs *6, 39, 49, 52, 340, 342*	Lt. Schminke
Seven	YMSs *70* (sunk), *71, 73, 314, 341*	Lt. Burns
Eight	Four LCPRs	Lt. (jg) Daly, USNR (CoMinRon 2 staff)

Task Unit 77.5.2 (Hydrographic Unit)
Lt. Comdr. Robert B. A. Atlee Hunt, RAN

Frigate HMAS *Gascoyne* (K354), *YMS-316, YMS-393,* HMAS *HDML-1074*[10]

On 21 October, the General Quarters alarm sounding aboard *Breese* at 0545, called all hands to their battle stations. A minute later, a low flying torpedo-bomber was sighted off the port quarter of the ship, a mere 2,000 yards distant, and coming in 50 feet off the water. Two of *Breese*'s port 20mm guns and one starboard twenty, and the 40mm gun opened fire at 0557. One minute later (operated by director control and after firing sixteen rounds), the 40mm hit the plane which crashed and sank. The 20mm gunners reported hits in the after fuselage and tail. This was the first enemy plane shot down inside the Leyte Gulf by surface ship anti-aircraft fire.[11]

Three days earlier, on the morning of 18 October, *Preble* had been attacked without warning by two enemy dive bombers, which released four medium bombs, bracketing the ship and spraying her sides with shrapnel, but causing no casualties. *Preble* took the planes under fire with her main battery—but scored no hits. Two days later, she opened fire with her 3"/50 battery on a formation of six Mitsubishi G4M "Betty" bombers attacking the force, but again brought none down.[12]

DEPARTURE OF A MAJORITY OF THE MINE FORCE

On 23 October 1944, the DMSs and DMs left the Leyte area in their wakes, and four days later the YMSs departed for Manus. The AMs remained at Leyte, including the *Requisite*, and kept busy with sweeping, anti-submarine patrol, and rescue missions before proceeding to support other landings in the Philippines.[13]

COMBAT DUTY OF THE MINESWEEPER *REQUISITE*

Photo 21-2

Minesweeper USS *Requisite* (AM-109) in a harbor, circa 1944.
Naval History and Heritage Command photograph #NH 89206

Near war's end, U.S. Navy ships were directed to submit their war history, essentially a summation of information from decklogs and diary entries. These documents were often compiled by someone without first-hand knowledge of many of the events in which their ship had participated. In the case of *Requisite* (AM-109), the drafter of her history noted about Leyte, "the ship's log and the WAR DIARY reflect none of the tension, the sleeplessness, and the hazard of those days." As an introduction for a narrative from an unnamed crewmember, he highlighted that the men tell it more vividly.

Requisite's duty was particularly laudable in that she was awarded the Navy Unit Commendation for the period 10 October 1944 to 1 March 1945 for service in Leyte Gulf, Ormoc Bay, Lingayen Gulf, and Subic Bay during the invasion of the Philippines. This award is second only to the Presidential Unit Citation for degree of heroism by a Navy unit. *Requisite* was the only mine warfare ship thus honored, and the account by the crewmember provides an overview of highlights. A summary of periods for which *Requisite* earned battle stars is provided to aid readers with the timeline. Because a ship could only receive one battle star for any single operation, otherwise qualifying actions at Leyte and Luzon were not recognized.

Battle Stars earned by USS *Requisite* (AM-109) in the Philippines

Date	Star	Action
10 Oct-29 Nov 44	yes	Leyte operation: Leyte landings
7-8 Dec 44	no	Leyte operation: Ormoc Bay landings
12-18 Dec 44	yes	Luzon operation: Mindoro landings
4-18 Jan 45	no	Luzon operation: Lingayen Gulf landing
29-31 Jan 45	yes	Manila Bay-Bicol operations: Zambales-Subic Bay

The sailor's account, with subject headings inserted, also introduces the remainder of the chapter:

LEYTE LANDINGS

It was off Samar. We had a couple of Carriers sunk down there [light carrier *Princeton* and escort carrier *Gambier Bay*]. We saw rafts floating and a lot of debris. We never learned whether these men supposed to be floating in this area had survived; but did learn later that natives thereabout had rescued and hidden American Sailors from time to time. We hoped these men had gotten ashore and escaped both the enemy and the sharks in this way.

PRELUDE TO THE BATTLE OF SURIGAO STRAIT

October 24th. That was the day we finished sweeping and went to anchor in San Pedro Bay. As we entered we saw some battle wagons and destroyers and D.E.'s [destroyer escorts] and tankers steaming out. The escorts were laying a thick smoke screen. Suddenly the air was thick with Jap planes. They hit some cargo ships. Anti-aircraft [fire] was dense.

November was pretty dull. Routine air raids daily. Routine sweeping. Usually a few mines swept each day, sometimes as many as a dozen. Work all day and stand G.Q. [man battle stations] at night.

LANDINGS AT ORMOC BAY

The 7th of December we were sent into Ormoc Bay to sweep ahead of the assault group of LST'S, APD'S, DESTROYERS, AND LCI'S. A group of destroyers had gone in a few days previously and two of them had been sunk. One of these was the *COOPER* [DD-695]. We got in at daybreak. We (the *REQUISITE* and the other AM'S) zig-zagged in close to the beach, attempting to draw fire, while the destroyers formed up in battle formation behind and at length commenced their bombardment of the beach. At about 0800 our forces began landing troops and the Jap planes started coming in. A KAMIKAZE missed the *REQUISITE* by 10 feet landing burning in the water. A Japanese convoy of 13 ships attempted to enter the bay. American planes left all 13 sinking. On that day we went to G.Q. at daybreak and secured long after sunset. But that wasn't different from any other day.

LANDINGS AT LUZON ISLAND IN LINGAYEN GULF

We streamed gear at 3 o'clock in the morning off Mindoro, with little dim lights on our pigs (floats). It was pitch black. We passed a Japanese ship on our way, about a quarter of a mile off. We didn't fire because we didn't know what caliber her guns might be (our main battery was a 3 inch). We notified the battle force by voice radio, however. A half hour later a battle started on the horizon that kept up for a dozen salvos. The Japanese ship caught fire and burned all day, off in the distance. There were KAMIKAZES round all day, attacking LST'S but they stayed too far away for us to fire at them.

LUZON CAMPAIGN/SWEEPING IN SUBIC BAY

So, passed December and January, with sweeping in the bloody Lingayen Gulf from the 2nd to the 19th of January and moving on to Subic Bay on January 25th.

The island in the middle of the entrance seemed covered with guns and pillboxes. We kept away from it the first day, but after that swept in the whole bay without paying any attention to the shore batteries. No guns were fired at us. One mine only was swept during the entire sweep. It was too bad for the crew that it couldn't know this operation was going to be a rest cure. Nice weather too. But we couldn't know until we left that we'd have no opposition.

Total score in the Philippines, two Jap planes shot down 43 mines swept, 110 men and officers ready for a let-up from G.Q.'S.[14]

LINGAYEN GULF LANDINGS

> *The enemy's large-scale employment of suicide planes gives rise to highly concentrated and desperate anti-aircraft fire.... About forty-three planes reached their targets [Allied ships], many after flying through the heaviest concentration of gun fire. A plane diving at speed in excess of four hundred fifty knots becomes in effect a low velocity projectile; its momentum is so great that it may not be deflected by volume of hits alone. The loss of a tail or the total disintegration of a wing generally, but not always, causes the plane to swerve from its general course. Killing the pilot may or may not deflect the plane.*

—Vice Adm. Thomas C. Kinkaid, commander, Luzon Attack Force, highlighting the grave danger that kamikaze aircraft presented to Allied shipping during Lingayen Gulf Landings in January 1945.[15]

At 0930 on 9 January 1945, assault forces of the U.S. Sixth Army under Lt. Gen. Walter Krueger, commenced landing on Lingayen Gulf beachheads in the Lingayen-Dagupan-San Fabian-Rabon area of Luzon. The amphibious operation, code named MIKE ONE, was a part of MacArthur's Operation MUSKETEER, a four-phased plan to liberate the Philippine Islands. The objectives of the Lingayen Gulf landings were the prompt seizure of central Luzon, the reoccupation of the Manila-Central Plains area, the establishment of bases for the support of operations to the north of the Philippines, and the complete occupation of Luzon Island. The weather conditions that day were favorable; sea and surf conditions were slight, there was no precipitation, and visibility was limited only by bombardment smoke carried offshore by land breezes that morning. As the Lingayen Attack Force came ashore, the enemy offered little opposition to landing craft approaching the beach or soldiers disembarking, the only resistance coming on the northeastern flank. This reflected new tactics adopted by the Japanese to withdraw from established beach defenses and to instead fight inland in the jungle. The landing of the San Fabian Attack Force, a few miles to the east, was unopposed. Over the next two days, ship-to-shore movement of follow-on assault troops and equipment continued smoothly and on 11 January, Krueger assumed command of Sixth Army forces ashore. That afternoon MacArthur and his staff left the light cruiser *Boise* (CL-47), for his headquarters at Dagupan near the Lingayen Gulf.[16]

LUZON ATTACK FORCE

Vice Adm. Thomas C. Kinkaid commanded the Luzon Attack Force (Task Force 77) comprised of Task Force 78 (San Fabian Attack Force) under Vice Adm. Daniel E. Barbey and Task Force 79 (Lingayen Attack Force) under Vice Adm. Theodore S. Wilkinson Jr. In the final days leading up to the invasion of Luzon, there had been 138 ships in the Lingayen area engaged in minesweeping and bombarding the coastline. On the morning of the invasion, 9 January, 344 additional ships arrived to take part in the amphibious landings at Lingayen and nearby San Fabian.[17]

TASK GROUP 77.6—LEYTE TO LINGAYEN GULF

Task Group 77.6 (hydrographic and minesweeping group) under Comdr. Wayne R. Loud, sailed from Leyte on 2 January for the MIKE ONE operation. In the group were ten destroyer minesweepers, ten minesweepers, forty-three Yard minesweepers, two destroyer minelayers, one high-speed transport (APD), the old minelayer *Monadnock* (CM-9), the Australian frigate HMAS *Gascoyne* (K354) and sloop *Warrego* (U73). The minesweepers were to sweep Lingayen Gulf for moored and acoustic mines prior to landings by joint expeditionary forces.

Task Force 77 Task Groups	Task Group Commander
TG 77.1 Fleet Flagship Group	Capt. A. M. Granum
TG 77.2 Bombardment and Fire Support Group	Vice Adm. Jesse B. Oldendorf
TG 77.3 Close Covering Group	Rear Adm. Russell S. Berkey
TG 77.4 Escort Carrier Group	Rear Adm. C. T. Durgin
TG 77.5 Hunter-Killer Group	Capt. J. C. Cronin
TG 77.6 Minesweeping and Hydrographic Group	Comdr. Wayne R. Loud
TG 77.7 Screening Group	Capt. John B. McLean
TG 77.8 Salvage and Rescue Group	Comdr. B. S. Huie
TG 77.9 Reinforcement Group	Rear Adm. Richard L. Conolly
TG 77.10 Service Group	Rear Adm. R. O. Glover[18]

MINESWEEPING FORCE SUFFERS THREE SHIPS LOST, OTHERS DAMAGED, BY AIRCRAFT ATTACKS

The Japs have, at last, apparently decided to break up Mine Squadron Two. Time and again, they passed up larger, more fruitful, targets and smaller, less well defended, targets to pick a DMS for bombing, strafing, torpedo and suicide dive attacks. More and continuous, day and night, air cover is required to protect these units operating away from the main body.

—Comdr. Wayne R. Loud, USN, commander, Mine Squadron Two and Task Group 77.6, highlighting the danger his Minesweeping Group faced from Japanese air attacks in Lingayen Gulf, Philippines.[19]

Photo 21-3

Japanese "Zeke" type kamikaze aircraft diving on light cruiser USS *Columbia* (CL-56) during the Lingayen Gulf operation, 6 January 1945.
Naval History and Heritage Command photograph #NH 79448

The minesweepers were fated for plenty of combat at Lingayen Gulf, not from mines, but from kamikazes, which literally rained out of the skies to damage six ships in the first six days of the operation. A minesweeper was the first ship sunk at Lingayen, and a minesweeper was the last ship sunk at Lingayen, and the only other ship sunk in the interim was a minesweeper. *Hovey* was sunk (120 survivors), *Long* sunk (85 survivors), *Palmer* sunk (120 survivors), *Brooks* damaged (153 survivors), *Southard* damaged, no casualties.[20]

Date	Ship	Commanding Officer	Disposition
6 Jan 45	*Long* (DMS-12)	Lt. Stanley Caplan, USNR	Sunk by kamikaze
6 Jan 45	*Brooks* (APD-10)	Lt. S. C. Rasmussen Jr., USNR	Damaged by kamikaze, 3 killed, 11 wounded
6 Jan 45	*Southard* (DMS-10)	Lt. Comdr. John Edward Brennan, USNR	Damaged by kamikaze, 6 wounded
7 Jan 45	*Hovey* (DMS-11)	Lt. Benjamin Nooe Cole, USNR	Sunk by kamikaze, 24 killed in addition to 24 survivors from *Long* and *Brooks*
7 Jan 45	*Palmer* (DMS-5)	Lt. William Edward McGuirk Jr., USN	Sunk by Japanese twin-engine bomber, 2 killed, 38 wounded, and 26 missing in action[21]

Aircraft attacks on Loud's Task Group 77.6 began during the transit to Lingayen Gulf. The numbers and diversity of enemy aircraft that would attack scores of Navy and merchant ships, indicated Japan's deep commitment to its defense of the Philippines. Moreover, among their many carrier-based and land-based fighters and bombers carrying out such attacks, were kamikazes—planes piloted by zealots willing to sacrifice their lives by crash diving into Allied ships.[22]

The first attack came just after sunset on 2 January, as the sweepers exited Surigao Strait and rounded Panaon Island into the Mindanao Sea, when three bombers appeared overhead, dropped their ordnance, and disappeared. Dawn on 3 January brought enemy reconnaissance planes, which were driven off. At 0729, six "Sallys" (Mitsubishi Ki-21 type 97 heavy bomber) and "Vals" (Aichi D3A Navy type 99 bombers) streaked out from Negros Island to starboard of the convoy and began an attack. Three made kamikaze dives, but missed their targets. One crashed close aboard *YMS-53*, but she emerged from a cloud of smoke and spray still intact. The wreckage of an exploding plane killed two sailors aboard the fleet oiler *Cowanesque* (AO-79) and one of the aircraft, whose pilot apparently deciding not to sacrifice himself, got away.[23]

In early morning on 4 January, the minelayer *Monadnock* (CM-9) received orders from Loud to proceed independently into Mindoro Harbor, and to await further orders. The 292-foot former cargo ship, under the command of Lt. Comdr. John E. Cole, USNR, stood into the harbor and anchored off the beach. At 0755, Army authorities ashore issued a Condition "Red" air raid warning, and all hands manned their battle stations. Enemy aircraft began attacking the seaplane anchorage at Mangarin Bay at 0820, and one bomb was observed dropping into the bay northeast of Ilin Island.[24]

Photo 21-4

Coastal minelayer USS *Monadnock* (CMc-4), circa late 1941, prior to her redesignation as a minelayer (CM-9).
Naval History and Heritage Command photograph #NH 104257

Shortly after this, a Val came in from the southeast in a 30-degree dive on the Liberty ship SS *Lewis L. Dyche*, anchored 1,580 yards from *Monadnock*. About 1,500 yards from the *Dyche*, the attacking plane leveled off about twenty feet above the water, before completing its suicide attack on the cargo ship amidships. Loaded with bombs, she disintegrated, killing all hands, including the 28-man Armed Guard. Debris from the exploding ship showered the *Monadnock*, killing one, injuring twelve, and causing minor damage to the minelayer. *Monadnock*'s first casualty of the war, Seaman 2nd Raymond Edsel Freeze, USNR, was buried in the Army Air Force cemetery at San Jose, Mindoro.[25]

ATTACKS AGAINST TASK GROUP 77.6 CONTINUE

Meanwhile, enemy attacks continued as the task group proceeded north from Mindoro to the Lingayen Gulf. In mid-afternoon on 5 January, sixty miles west of the entrance to Manila Bay, a kamikaze gave Lt. (jg)

F. W. Ketner's, USNR, *YMS-53* a close shave, and another just missed *Scrimmage* (AM-297). Later that day, six "Zekes" (zero fighters) with bombs strapped under their wings came in from all sides. In rapid succession, they crashed alongside *Sentry* (AM-299), *YMS-53*, *Requisite* (AM-109), the seaplane tender *Orca* (AVP-49), and fleet tug *Apache* (ATF-67). The minesweepers were lucky, but *Orca* and *Apache* were damaged. A sixth plane sliced the mast off the infantry landing craft *LCI(G)-70* and killed two men; but didn't stop her.[26]

That night, the moon was bright, the sky full of planes, and the convoy came under nine separate attacks. All were beaten off with *YMS-327* the only near casualty. Two bombs straddled her, opening some seams, but her crew kept her afloat and in operation.[27]

SWEEP OPERATIONS IN LINGAYEN BAY

At sunrise on 6 January, TG 77.6 dissolved and individual groups of sweepers moved to their assigned areas. The plan for Lingayen was to sweep the entire gulf—some 500 square miles—for moored mines, generally by the DMSs and AMs, and an influence sweep of all waters shallower than 20 fathoms by the YMSs. The DMSs—*Hopkins*, *Southard*, *Chandler*, *Hovey*, *Long*, *Hamilton*, *Howard*, *Palmer*, *Hogan*, and *Dorsey*—streamed their gear at 0930 to clear a channel into the gulf for battleships and cruisers of the bombardment group. Amidst kamikaze attacks, the sweepers carried out their work, and by mid-afternoon the BBs and CGs poured into the gulf. Their presence provided attacking kamikazes a greater number of and larger targets, at least temporarily.[28]

Before the day was over, two battleships, three cruisers, and five destroyers had been damaged by the "divine wind." Over the next two days, minesweepers kept sweeping for mines that were not there, and fighting off kamikaze attacks. The most determined one, on 7 January, came following the completion of work, when several planes streaked out of the setting sun. One twin-engine bomber dropped two bombs on *Palmer*'s portside, then splashed in to starboard. Six minutes after the bombs hit, the destroyer minelayer sank, stern first, with twenty-six men missing. *Hopkins* and *Breese* together picked up 123 survivors.[29]

Breese (accompanied by *Brooks*) had laid buoys the morning of 6 January, marking the entrance into the gulf, and had then taken up mine destruction vessel duties. *Preble*, with whom she normally operated, would spend almost all of January lying at anchor in San Pedro Bay, Leyte, owing to mechanical breakdown.[30]

Two days after *Palmer* was sunk, Army forces went ashore. With the capture of the beachheads at Lingayen, most of the minesweepers returned to Leyte. A few YMSs remained behind in case the Japanese conducted aerial mining. The score for the sweepers at Lingayen strongly favored the enemy. Three sunk, three seriously damaged, and several others moderately damaged for only three mines swept.[31]

MYSTERY SOLVED

It was later learned why the expected minefields off Lingayen beaches had failed to materialize. Lt. Col. Russel W. Volckmann, USA, who earlier in the war had escaped from Bataan, and organized Filipinos into possibly the largest guerrilla force in the world, had overseen their disposal. When the landings were made at Leyte, Volckmann knew that Lingayen would be next, so his guerrillas began eliminating the mines.[32]

Photo 21-5

Lt. Col. Russel Volckmann, USA
U.S. Army photograph

William Dwight Whitney, *The Century Dictionary and Cyclopedia: An encyclopedic Lexicon of the English Language* (New York, NY: The Century Co., 1889)

The Filipinos dragged for the mines using a manila line looped between two bancas sailing abreast one another. When the line snagged a moored mine, a diver went down, and unshackled the mine case from its mooring. The mine was towed to the beach, disarmed, disassembled and disposed of. In less than two months, over 350 Japanese mines were rendered harmless; the explosive charges were used against the Japanese and the mine cases as wash tubs.[33]

RENDER MINE SAFE OPERATIONS

Additional sweeping by AMs and YMSs ahead of assault landings at Zambales-Grande Island, and in Manila Bay, followed in late January through 12 April, when Seventh Fleet minesweepers cut their last mine in the bay. During one week in March, the sweepers had worked over a field of fifty U.S. mines dropped the night of 14 December 1944, by twenty RAAF Catalinas. Some of these aircraft had come from Darwin, and others from Rathmines, New South Wales. These mines had all been fitted with sterilizers timed so that they would become inactive before the date set for landings at Manila, and not one of the mines was still alive when the sweepers arrived on scene.[34]

Efforts by the U.S. Navy Philippine Sea Frontier (as minesweepers became available), and mine clearance by the U.S. Mobile Explosive Investigation Unit (MEIU) at Manila, continued until war's end. One of the individuals involved with bomb and mine disposal was Lt. Harold Leon Billman, Royal Australian Navy Volunteer Reserve. After nearly two-and-a-half years in the RAAF, he joined the Navy as a sub-lieutenant in late 1942. Billman completed a controlled mine course and render mine safe course at Flinders Naval Depot, Williamstown, Victoria, and was then sent to Army Bomb Disposal School. Qualifying Diver II in HMAS *Penguin* and promoted to lieutenant, his first duty was with HMAS *Basilisk* at Port Moresby, involving bomb and mine disposal in New Guinea forward areas.[35]

Photo 21-6

Distinguished Service Cross

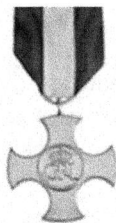

Mention in Despatches

Lt. Harold Leon Billman, RANVR, was awarded the Distinguished Service Cross for his service in the Philippines in 1944-45. He had previously earned a Mention in Despatches for his work in New Guinea.
Australian War Memorial photograph 057550

Landing with the Sixth Army at Leyte on 20 October 1944, Billman began a new assignment. Working with MEIU One, in the Philippines, he defused bombs, torpedoes and mines. Two of his most unusual assignments involved examining Japanese Shinyo Special Attack (Suicide) boats found on Corregidor Island, and a new type of Japanese mine. Although the aircraft-laid moored mine (subsequently designated

"Camote") was of an unusual shape, he found it had a standard safety switch and horns, and taking all precautions, rendered it safe. The 640Kg mine, with an explosive charge of 80 Kg was shipped by air to Washington, DC, for study by the Bureau of Ordnance.[36]

Photo 21-7

Japanese Shinyo explosive motorboat on the beach at Lingayen Gulf.
Naval History and Heritage Command photograph #NH 44316

Four other Australians, all RANVR, would also serve with MEIU One: Lieutenants John Hunter and Gavin Anderson, and Lieutenant Commanders Leon Goldsworthy and Geoffrey Cliff. The latter two individuals are revisited in the book's postscript.[37]

UNIT AWARDS

Three minelayers earned battle stars for support of Allied amphibious landings on Leyte and Luzon Islands in the Philippines.

Leyte Operation: Leyte Landings		
Breese (DM-18)	12-20 Oct 44	Lt. Comdr. David Barney Cohen, USN
Preble (DM-20)	12-20 Oct 44	Lt. Edward Francis Baldridge, USN

Luzon Operation: Lingayen Gulf Landing		
Monadnock (CM-9)	5-18 Jan 45	Lt. Comdr. John. E. Cole, USNR
Breese (DM-18)	4-18 Jan 45	Lt. George W. McKnight, USNR
Preble (DM-20)	4-18 Jan 45	Lt. Edward Francis Baldridge, USN

22

Assault and Occupation of Iwo Jima

At great cost, you'd take a hill to find then the same enemy suddenly on your flank or rear. The Japanese were not on Iwo Jima. They were in it! I'd known combat in the Solomons with its sly ambushes and jungle firefights, but Iwo was another kind of war. On Iwo by the 8th day, only two officers of my second battalion (26th Marines, 5th Marine Division) were standing.... We had one prisoner— unconscious, his clothes blown off.

—Col. Thomas M. Fields, USMC (Ret.)

The Battles of Iwo Jima and Okinawa were fought by Allied forces between February and August 1945 to secure island bases for a final B29 bomber assault on Japan. The Iwo Jima operation was conducted first because it was expected to be easier than an assault on Okinawa. Because of the enemy's prolonged and bitter defense of Leyte and Luzon, the planned dates for both actions had slipped. Thus, the Pacific Fleet had to cover and support both invasions, while the Seventh Fleet and its amphibious forces were concurrently engaged in liberating the Southern Philippines.[1]

On 19 February as, naval gunfire pounded the island, more than 450 ships massed off Iwo Jima. Marines of the 4th and 5th Divisions hit the four assault beaches shortly after 0900, initially finding little enemy resistance. Coarse volcanic sand hampered their movement as they struggled to move up the beach from the surf zone. As the protective naval gunfire subsided to allow for advancement, the Japanese emerged from fortified underground positions to begin a heavy barrage of fire against the invading force. The 4th Marines continued to push forward against heavy opposition to take the Quarry, a Japanese strong point, while the 5th Marine Division's 28th Marines isolated Mount Suribachi that same day. The 3rd Marine Division joined the fighting on the fifth day, charged with securing the center sector of the island. The fortified enemy defenses linked miles of interlocking caves, concrete blockhouses and pillboxes, which required frontal assaults to gain nearly every inch of ground. Maj. Gen. Harry

Schmidt, commanding the Fifth Amphibious Corps—of which the 3rd, 4th, and 5th Marines were a part—declared Iwo Jima secured on 16 March. Ground fighting, however, continued between then and the official completion of the operation on 26 March 1945.[2]

Photo 22-1

A few of the hundreds of landing craft staging for assault landings on Iwo Jima. Naval History and Heritage Command photograph #NH 104203

PREASSAULT SWEEPING

When small minesweepers are employed for close inshore sweeping, it is necessary that fire support DD type ships be detailed for counter battery fire if the sweeping is to be completed as scheduled. These close support ships should move in with the sweepers. The YMS's were turned back twice by heavy fire from the beach that could not well be countered by heavy ships lying off 6,000 yards. It was necessary for this command to detail two (2) old type DM's as close fire support. These ships have only 3 inch batteries.

—Rear Adm. Alexander Sharp Jr., USN, commander, Minecraft, U.S. Pacific Fleet, emphasizing the importance of Force Protection in his report on operations at Iwo Jima[3]

Photo 22-2

Shells explode ashore during the bombardment of Iwo Jima on D-Day-minus-two, 17 February 1945. The view looks east toward Mount Suribachi's west side. The ship in the right foreground appears to be a YMS engaged in pre-invasion minesweeping. U.S. Navy Photograph #NH 104142

GAMBLE (DM-15) DAMAGED WITH LOSS OF LIFE

One of the old type DMs, *Gamble*, had left Saipan the evening of 14 February for Iwo Jima, rendezvoused with other Naval forces at sea, and arrived in the battle area before dawn on the morning of the 17th. She spent the entire day supporting various sweep units by firing at targets ashore. The battleships, cruisers, and destroyers, had already commenced shore bombardment. *Gamble*'s mission was to lay as close to the beach as possible, and fire at any installations capable of hindering sweep operations. She was also responsible for destroying any mines as might be brought to the surface by the sweepers. To assist in this work, a mine disposal officer was aboard the minelayer for temporary duty.[4]

The minelayer's commanding officer, Lt. Comdr. Donald Noble Clay, USN, decided to use 3" 50 caliber AA common ammunition against shore installations because of its value as an anti-personnel weapon, and because of the quantity aboard. The ship's automatic weapons were to be used mainly against air attacks and ammunition saved until needed. During this period, sporadic fire from enemy gun installations occurred, and several other ships were hit. Enemy aircraft were reported in the vicinity from time to time by other ships, but no sightings were made aboard the minelayer. At sunset, *Gamble* retired (moved out to sea for the night) in accordance with guidance from commander, Task Group 52.3. Rear Admiral Sharp, embarked aboard

Terror (CM-5), was commander, Minecraft, U.S. Pacific Fleet, and for this operation commander, TG 52.3 and commander, Mine Group.[5]

The next morning, *Gamble* took up screening duties and patrolled her assigned area. That evening, she joined Task Force 54 for night retirement. At 2125, the minelayer was struck by two 250lb bombs from a twin-engine enemy plane, believed to be a "Nick" (Kawasaki Ki-45 Toryu Army type 2 two-seat fighter), flying at a very low altitude. The bombs struck in the vicinity of the bulkhead between the forward and after firerooms, just above the waterline, rupturing the hull and destroying all three boilers. Water rushed into the firerooms, and the ship lay dead in the water. As a result of this action, four men and one officer were killed, another five men and three officers were wounded, and one man missing, presumed dead.[6]

Photo 22-3

Six images of a "Nick" Kawasaki Ki-45 Toryu Army heavy fighter with U.S. markings. Japanese Operational Aircraft "Know Your Enemy!" CinCPOA Bulletin 105-45

Commander, Task Group 52.3 had earlier issued a Condition "Red" warning, upon which *Gamble*'s crew went to battle stations in order to be ready to repel an enemy air attack. The formation, of which *Gamble* was a part, consisted of two battleships, three cruisers, and thirteen escort vessels. As she began maneuvering on various courses to elude possible attack, several single unidentified aircraft appeared on her radar screen, but none closed the formation and soon disappeared. The threat condition was then downgraded to "White," and *Gamble* secured from General Quarters, leaving one-half the battery manned and all guns ready for immediate use. At 2125, while on station in a circular screen around the battleship *Nevada* (BB-36), the bombs struck.[7]

The plane came straight in on her starboard beam, flying very fast and only about fifteen feet off the water. The junior officer of the deck sighted it about 1,000 yards out, and warned "plane to starboard," but there was no time to bring the guns to bear. Following the twin explosions, GQ was reset, and all hands turned to putting out fires, jettisoning topside weight, and shoring endangered bulkheads. Eight Mark 6 depth charges were set on safe and jettisoned. At 2225, the *Hamilton* (DMS-18) came alongside to remove casualties. Twenty minutes later, the remainder of the casualties were transferred to her under the care of the medical officer.[8]

Dorsey (DMS-1) took *Gamble* under tow at 0150 on 19 February, moving her to a point about ten miles northeast of Iwo Jima. The short remaining night was spent jettisoning additional topside weight and shoring bulkheads. Dawn brought favorable weather and a calm sea, and the sub-chaser *PC-800* came alongside with hot coffee. She was followed by the rescue and salvage ship *Clamp* (ARS-33), which took off all personnel and their gear, except for the commanding officer, six officers and thirty-nine crewmen. The remainder of the day was used to pump engine room bilges, and (to further improve stability, and survivability of the ship) throwing overboard practically everything still removable topside, including:

- The mast
- 358 rounds of 3"50 caliber armor piercing
- 61 rounds of 3"50 caliber illumination (star shells)
- 6 3"50 caliber dislodging charges
- 16,530 rounds of 20mm
- 25,000 rounds of .50 caliber
- 3,000 cartridges of .30 caliber
- 200 cartridges of .45 caliber
- 600 primers and 63 detonators

- #1 stack
- Two 20mm guns from the galley deck house
- Ready lockers on the forecastle and galley deck house[9]

That evening a convoy was formed of miscellaneous LCIs (infantry landing craft), LSM (medium landing ship) and AMs (minesweepers) which had departed the battle area. The resultant transit to Saipan with *Gamble* under tow by *LSM-126* was uneventful. The minelayer was ultimately judged unsalvageable and decommissioned on 1 June. On 16 July she was towed outside Apra Harbor, Guam, and sunk.[10]

MINESWEEPING AT IWO PROVES UNNECESSARY

The assault on Iwo Jima was the first operation participated in by the newly constituted Pacific Minecraft Command. Rear Adm. Alexander Sharp, who put forty-three minecraft into the operation, watched from his flagship, USS *Terror* (CM-5). The sweepers worked hard, some under fire, while first clearing a path for fire support vessels shoreward from the 100-fathom curve and subsequently waters for assault force landings. On the morning of the first day, *Chandler* (DMS-9) sighted one lone mine, some ten miles southeast of Mt. Suribachi. Old and rusty, it was evidently a floater from some other island, and was sunk by *PC-800*. Although no one yet knew it, it would constitute the only mine at Iwo Jima.[11]

CALLS FOR FIRE FROM SHORE

Photo 22-4

Minelayer USS *Thomas E. Fraser* (DM-24) in the Atlantic on 27 September 1944.
National Archives photo 80-G-282662

Iwo Jima also marked the first time that *Shannon* (DM-25), *Gwin* (DM-33), *Lindsey* (DM-32), and *Thomas E. Fraser* (DM-24)—newly built destroyers converted to minelayers—made a front-line appearance with a fire support group. These ships were much more heavily armed than were World War I vintage minelayers like *Gamble*. *Thomas E. Fraser*, for example, was fitted with three twin 5"/38 dual purpose gun mounts, six twin 40mm gun mounts, eleven 20mm guns, two .50 cal. machine guns, and depth charges, rolled over her stern from tracks or launched off her sides by port and starboard projectors.[12]

Fraser's duties at Iwo included firing from a position a few hundred yards off the beach, point-blank into the mouths of fortified caves on Suribachi and blasting pillboxes about its base. Spotters on shore, communicating by "walkie-talkie" with the minelayer, directed the gunfire with telling accuracy. The movement of Marines could be seen clearly as they advanced or were driven back by Japanese fire from the caves and the pillboxes, and men aboard *Fraser*, witnessing the Marines fight for their lives, bent to their guns with a will.[13]

From dust to dawn, the ship's 5-inch guns fired star shells at short intervals to provide the Marines with illumination for their advances. During the night, this illumination revealed a detachment of Japanese soldiers attempting to swim around a point of land for a flank attack on the Marines. Gunfire from *Fraser* quickly eliminated this threat.[14]

Iwo Jima was an easy slough for minesweepers, but a tough battle for the Marines who made "uncommon valor a common virtue" on the bloody slopes of Suribachi.[15]

Photo 22-5

Stars and Stripes flying over Iwo Jima; U.S. Marines raised the American flag over Mount Suribachi, Iwo Jima on 23 February 1945.
Naval History and Heritage Command photograph #NH 104279

BATTLE STARS AWARDED MINELAYERS

Iwo Jima Operation: Bombardment of Iwo Jima

Gwin (DM-33)	24 Jan 45	Comdr. Frederick Samuel Steinke, USN

Iwo Jima Operation: Assault and Occupation of Iwo Jima

Terror (CM-5)	16-19 Feb 45	Comdr. Horace William Blakeslee, USN
Breese (DM-18)	16 Feb- 7 Mar 45	Lt. George W. McKnight, USNR
Gamble (DM-15)	16-19 Feb 45	Lt. Comdr. Donald Noble Clay, USN
Henry A. Wiley (DM-29)	16 Feb- 9 Mar 45	Comdr. Robert Emmett Gadrow, USN
Henry F. Bauer (DM-26)	19 Feb- 6 Mar 45	Comdr. Richard Claggett Williams Jr., USN
Lindsey (DM-32)	16-19 Feb 45	Comdr. Thomas Edward Chambers, USN
Robert H. Smith (DM-23)	19 Feb- 9 Mar 45	Comdr. Henry Farrow, USN
Shannon (DM-25)	19 Feb- 16 Mar 45	Comdr. Edward Lee Foster, USN
Thomas E. Fraser (DM-24)	19 Feb- 8 Mar 45	Comdr. Ronald Joseph Woodman, USN
Tracy (DM-19)	16 Feb- 7 Mar 45	Lt. Comdr. Richard Edward Carpenter, USNR

Assault and Occupation of Okinawa

The mission of TG 52.2 was to make the waters of OKINAWA GUNTO safe to surface forces from the danger of mines, by clearance sweeps in depths less than 100 fathoms, and by means of exploratory sweeps in other waters under 500 fathoms.

As vessels could be spared from minesweeping they were turned over to the Screen Commander (CTG 51.5) for use in the inner and outer screens and as radar pickets, except for logistic services, repairs and administration.

Enemy forces encountered were limited to shore batteries, submarines, planes, baka bombs, mines and suicide boats, rafts and swimmers. Other that the above, the minesweepers had few worries, except for reefs, shortage of fresh water and a deplorable and serious shortage of diesel spares, which was more often than not a lack rather than a shortage.

—Rear Adm. Alexander Sharp Jr., commander, Minecraft, U.S. Pacific Fleet and commander, Mine Flotilla, remarking on the duties of his forces and conditions experienced during the capture of Okinawa.[1]

The eighty-two-day-long Battle of Okinawa was fought on Okinawa in the Ryukyu Islands, and was the largest amphibious assault in the Pacific. The Ryukyus, a chain of Japanese islands stretching southwest from Kyushu to Formosa (now Taiwan), includes the Osumi, Tokara, Amami, Okinawa, and Sakishima Islands. The larger islands are mostly volcanic, the biggest being Okinawa, and the smaller ones are mostly coral. Four divisions of the U.S. Tenth Army (the 7th, 27th, 77th, and 96th) and two Marine divisions (the 1st and 6th) fought on Okinawa, supported by naval, amphibious, and air forces. The purpose of the operation, which lasted from 1 April through mid-June 1945, was to capture the island for use as a base for air operations during the planned invasion of Japan.[2]

Some Japanese accounts of the battle refer to it as *tetsu no ame* or *kou no kaze*, "iron rain" or "steel wind," respectively, due to the ferocity of the fighting, the intensity of kamikaze attacks from the Japanese defenders, and the sheer numbers of ships and armored vehicles that assaulted the island. The battle resulted in the highest number of casualties in the Pacific Theater during World War II. The Tenth Army suffered 7,613 killed or missing in action, and 31,800 wounded, while Marine Corps casualties overall—ground, air, and ships' detachments—exceeded 19,500. Navy losses were 34 vessels and craft sunk and 368 damaged, over 4,900 sailors killed or missing in action, and over 4,800 wounded. Japan lost over 100,000 soldiers, either killed, captured or committed suicide.[3]

Photo 23-1

Battleship USS *New Mexico* (BB-40) hit by a kamikaze on 12 May 1945, off Okinawa. Navy Photograph #80-G-328653, now in the collections of the National Archives

SWEEPING COMMENCES

In pre-dawn darkness on 24 March 1945, as Admiral Spruance's massive Fifth Fleet neared Okinawa, Sweep Unit Two (a handful of ships of Rear Adm. Alexander Sharp's Mine Flotilla) was the first to sight the ominous island. At 0615, three destroyer minesweepers—*Macomb* (DMS-23), *Forrest* (DMS-24), and *Hobson* (DMS-26)— streamed gear and began their work in area V-2. The job of the remaining member, the minelayer *Gwin* (DM-33), was to serve as a support ship for the minesweepers. From 24-28 March, the DMSs conducted combat sweep operations by day, and at night retired well out to sea to a safer area.

The weather was generally good, with the most noticeable thing being the brightness of the moon during the entire period.[4]

Map 23-1

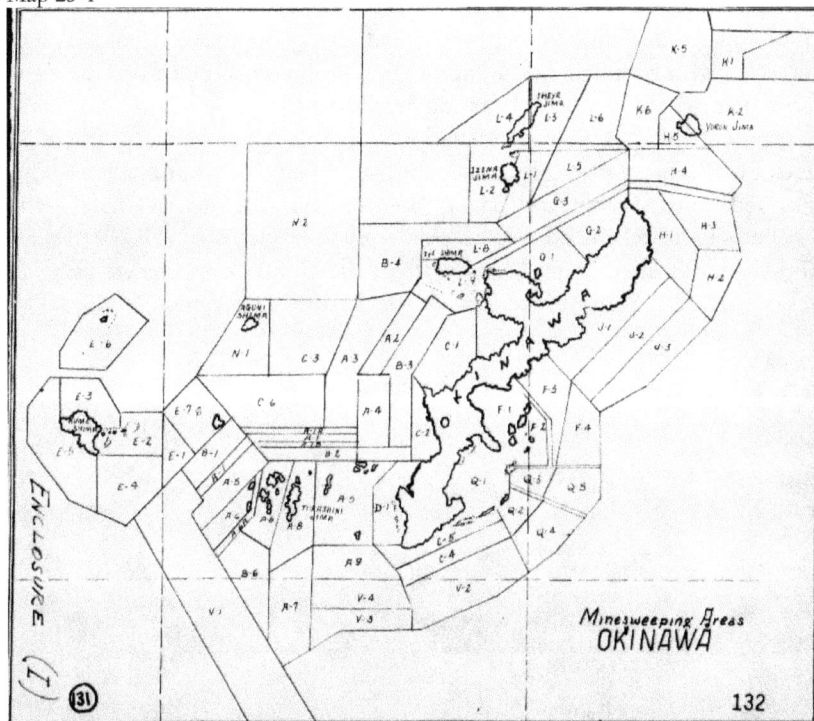

Minesweeping Areas at Okinawa
CTF 52.2 Report of Capture of Okinawa Gunto – Phases One and Two, 23 July 1945

A fifth minesweeper, *Dorsey* (DMS-1), joined Sweep Unit Two upon her arrival on 25 March. In late morning that day, she began exploratory sweeping operations with the other DMSs for moored and acoustic mines in areas A-3 and N-2. This work was completed the following day, results negative.[5]

Sweep Unit Two proceeded at dawn on 27 March for Kerama Retto for logistics. Only one day earlier, the U.S. 77th infantry division had captured the Kerama Islands (located twenty miles southwest of Okinawa) for use as a seaplane anchorage and advanced base in support of the planned invasion. At 0616, *Dorsey* observed an enemy plane shot down by friendly ships five miles distant. Two minutes later, as three Japanese planes approached her starboard bow, *Dorsey*'s guns opened fire. The aircraft split, two passing across her starboard bow. The third (a "Val," Aichi D3A Navy type 99 carrier bomber), some 1,500 yards away, continued down her starboard side, before crossing astern, turning sharply to the right and starting a suicide dive at 800 feet above the water from directly astern.[6]

Lt. John M. Hayes, USN, *Dorsey's* commanding officer, ordered right full rudder, and all engines ahead flank, 25 knots. The Val was repeatedly hit by automatic weapons fire, as evidenced by a bomb or gas tank being shot off and bursting into flames. The plane, however, continued downward, and crashed into the port side of the ship's main deck, just aft of the galley deckhouse. Hayes ordered all engines stopped, and life rafts put over for men thrown into the sea by the impact. The fire started was quickly put out by damage control parties, and wounded men were provided treatment. Of eight men missing in action, five were later picked up by the *Macomb* and *Forrest*.[7]

Photo 23-2

Minelayer USS *Dorsey* (DMS-1) on 6 February 1943. Commissioned a destroyer (DD-117) on 16 September 1918, she was reclassified as a highspeed minesweeper, DMS-1, on 19 November 1940.
Naval History and Heritage Command photograph #NH 83214

Dorsey opened fire on an enemy bomber approaching from her port quarter at 0626, as it passed overhead. Two small dropped bombs struck the water on the ship's starboard bow. The "Betty" was later downed by fire from other ships, and *Dorsey* suffered only minor damage as a result of the attack. Disaster struck that autumn when, she and other Navy ships were driven aground at Okinawa by typhoons—*Dorsey* on 9 October 1945. Judged a total loss, she was decommissioned and her battered hulk destroyed 1 January 1946.[8]

MINESWEEPING FLOTILLA ARRIVES

A day behind the arrival of Sweep Unit Two at Okinawa came a hundred or more minecraft, six days out of Ulithi. The massive flotilla was split into two groups, with Rear Admiral Sharp in tactical command of one, and Capt. William G. Beecher Jr., USN, commanding the other. The minesweepers sighted Kerama Retto at 0600 on 25 March and, with guns at the ready, set about their appointed duties.[9]

Sweep Unit Two completed its swept path parallel to the southeast shores. Unit Three—*Ellyson* (DMS-19), *Hambleton* (DMS-20), *Rodman* (DMS-21), *Emmons* (DMS-22), and *Lindsey* (DM-32)—and Unit Four—*Jeffers* (DMS-27), *Harding* (DMS-28), *Butler* (DMS-29), *Gherardi* (DMS-30), and *Aaron Ward* (DMS-34)—then extended it westward to a point fifteen miles south of Kerama Retto, then northward and eventually east again to the Haguchi beaches.[10]

The AMs of Sweep Unit Five—*Heed* (AM-100), *Champion* (AM-314), *Defense* (AM-317), *Devastator* (AM-318), *Ardent* (AM-340), and *Adams* (DM-27)—plus Sweep Unit Eight—*Gladiator* (AM-319), *Impeccable* (AM-320), *Spear* (AM-322), *Triumph* (AM-323), *Vigilance* (AM-324), *Shea* (DM-30), and *PC-1598*—swept a long, southeast-northwest strip, thirty miles south of Kerama Retto. Its purpose was to provide safe passage for large ships swinging round those islands en route to the Haguchi beachhead—landing site for the forthcoming Okinawa assault.[11]

On 28 March, Unit Seven—*Sheldrake* (AM-62), *Skylark* (AM-63), *Starling* (AM-64), *Swallow* (AM-65), *Henry A. Wiley* (DM-29), and *PC-1179*—plus Unit 8 proceeded to a point 100 miles northwest of Okinawa; and began sweeping a path for transports heading down to the Haguchi beaches.[12]

Rear Admiral Sharp's sweep plan also called for clearance of the fire support areas off the beachheads, and a sweep through the eastern islands so bombardment groups could enter Chimu Wan and Nakagusuku Wan in support of ground troops. Next would come a sweep along the entire western coast of Okinawa in order to stage additional surprise landings if necessary. Finally, the minesweepers were to clear a path round the entire 75-mile-long island, so that patrol craft would have free rein to intercept possible Japanese reinforcements.[13]

FIRST KAMAKAZI TO ATTACK A MINECRAFT

The day before the kamikaze crashed into *Dorsey*, the *Robert H. Smith* (DM-23) was patrolling astern of Sweep Units 12 and 13 (comprised of YMSs) in area B-6, just south of Kerama Retto. At 0659, a Val broke through the overcast approximately 2,500 yards distant, closing the starboard side of the ship at 2,500 feet altitude, before initiating a suicide dive. Aboard the *Smith*, engine speed was increased to flank, and fire opened with all batteries—5"/38, 40mm, and 20mm guns. (The 40mm appeared to be the most effective.) As the plane passed over the minelayer's superstructure just forward of the mast, it made a sudden sharp dive into the water on her port beam.[14]

Apparently either the pilot had been killed and the plane was out of control, or the kamikaze realized that he had overshot the target and was making a desperate effort for a hit by pushing the stick all the way forward. Two almost simultaneous explosions occurred as the plane impacted the sea, throwing two large columns of water up over the bridge and gun director. No casualties to ship's force resulted, although several pieces of sheet metal from the aircraft were found on board.[15]

Day and night, suicide planes rained out of the sky, but by 26 March, as ordered, the minesweepers opened the door to Kerama Retto and the 77th Infantry Division landed on schedule. It captured 350 Japanese suicide boats (Shinyo) there. These small special attack boats utilized an explosive charge in their bow by ramming the side of the intended victim. The motor boats were collected in special attack basins along the coast or carried on mother ships. After the war, it was learned that the suicide craft were manned by boys of 15 and 16 years of age. A supply of volunteer pilots was obtained because of special privileges, early responsibility, fast promotion, and the promise of a posthumous monetary award to the volunteers' parents.[16]

Photo 23-3

Painting "Kerama Retto" by C.W. Smith, 1945, depicts a destroyer firing on what may be suicide boats in the roadstead.
Naval History and Heritage Command photograph #NH 92591-KN

SKYLARK LOST TO MINE STRIKES

We were sweeping in formation as follows—Skylark–Sage–Revenge–Pursuit. *The* Requisite *had just lost her gear and proceeded to fall out of line to stream more gear. Then it happened—what we all had feared most—the lead ship hit a mine. I guess the picture of that ship will stick with me as long as I live. I guess it was about 11:20 AM when she got hit because chow was called down a little after, but somehow food didn't interest me. The* Sage *and us were ordered to sweep as close to the burning ship as possible and continue on. She was hit on the port side about amid ships. Fire engulfed the ship from one side to the other. She leaned up as if she was lifted up about 5 feet and then settled down listing fairly heavy to port. Smoke and fire was pouring out pretty heavy from no. 1 stack so I guess that's where she got hit. The guys in her forward engine room didn't have a chance.*

—Entry from the diary of Donald D. Panek, a crewmember aboard
the minesweeper USS *Revenge* (AM-110) at Okinawa[17]

Photo 23-4

Minesweeper USS *Skylark* (AM-63), location and date unknown.
Naval History and Heritage Command photograph #NH 89202

On 28 March, *Skylark* was sweeping in a starboard echelon formation (with *Sage*, *Revenge*, and *Pursuit*) about four miles off the west coast of Okinawa Jima.   Many Japanese mines had been cut that morning by AMs of Sweep Units Six and Seven (60 per the most recent report) and there were many adrift.   Accordingly, commander, Sweep Unit Six, who was in charge of both the Six and Seven Units, ordered Lt. Comdr. George M. Estep, USN—the commanding officer of *Skylark* and commander, Mine Division 15—to take charge of such ships as were available and to continue to sweep area C-2.   After one north-to-south pass through adjoining water that had been swept the previous day, the minelayer *Tolman* (DM-28), acting as collaborating and fire support vessel, advised finishing a small portion of unswept water remaining in area B-2 near the southern end of C-2. This was completed and B-2 reported swept.[18]

The echelon, with *Skylark* leading, then proceeded northward in waters believed safe. In late morning, *Skylark* hit a mine, portside, amidships. The force of the explosion vented upward through her engine room and living compartment amidships, and propulsion was lost immediately. The ship also immediately took a port list, partly because the boat on the starboard side had been blown overboard. *Skylark* had been at general quarters, and since the dressing station was located on the mess decks, in the middle of the ship, the doctor and pharmacist mate were both burned, as were most of the men whose battle stations were on the main deck, port side. The wounded were taken to the forecastle.[19]

The forward engine room was a mass of flames. Attempts made to flood magazines and connect hoses to fight the fire were futile, because there was no water pressure due to lack of power and broken piping. It was also not possible to reach CO_2 controls to flood the engine room due to the intensity of the fire. An immediate check was made to ensure that the depth charges were set on safe. *Skylark* settled slowly, but any hope of saving her vanished when she drifted into a second mine and began sinking rapidly. The order was given to abandon ship, and the wounded were taken off on rafts alongside the fo'c'sle, which by then was quite low in the water.[20]

Five sailors perished. Small boats from the *Tolman* and *PC-1179* picked up the survivors (ship's force and embarked Mine Division 15 staff) and took them to *Tolman*, where they were given medical assistance, clothing, and other care. They were then transferred in succession to the patrol craft *Amherst* (PCER-853), and attack transports *Pitt* (APA-223), *Hinsdale* (APA-120), and *Wayne* (APA-54). The Mine Division staff was returned to ComMinPac headquarters at Pearl Harbor, and the surviving officers and men of *Skylark* delivered to the Receiving Ship, San Francisco (an installation), for survivors leave.[21]

ADAMS KNOCKED OUT OF ACTION

During the sweeping period before Love Day [landings at Okinawa], the ADAMS drove off a great number of potential attacking planes before they could commit themselves to a definite attack. However, every plane which did attack the unit attacked the ADAMS. *Six of these were shot down, five of the six were very close aboard. It is believed that every attempt should be made in similar future operations to operate two sweep units together whenever possible. The DM's could then operate in pairs and would have a much greater chance of successfully accomplishing their missions without being damaged. This type ship can handle a single bogey quite successfully, and the* ADAMS *on two occasions handled two simultaneously. But when enemy planes gang up beyond that point and make simultaneous attacks, the chance of being hit skyrockets rapidly. On the morning when the* ADAMS *was finally put out of action by a near miss, there were three Jap planes plus a PBM [Mariner patrol bomber] not showing IFF [Identify Friend or Foe] to contend with. A pair of DM's instead of a single ship would, it is believed, have been able to successfully handle this situation.*

—Excerpt from a report by Capt. John H. Sides, USN, commander, Mine Division Eight, who was embarked aboard the minelayer *Adams* (DM-27) during the attacks described.[22]

The new 2,200-ton destroyer minelayers present at Okinawa operated with various kinds of minesweepers. Their primary function was to protect the sweep unit to which they were assigned against air, surface, and shore attack, and to provide a communications capability for the unit. Secondary functions were: (1) Laying of Mk 6 Mine Case buoys, and Dan buoys, (2) Salvage and Rescue, and (3) Destruction of mines cut adrift by AM and YMS sweepers when this did not interfere with the maneuverability of the DMs. The assigned sub-chaser, patrol craft, or designated YMS ordinarily did the destruction work. However, when operating with the higher speed DMSs, capable of cutting much greater numbers of mines, the destruction of unencumbered "ship killers" became of paramount importance.[23]

Providing protection for a low speed formation of AMs or YMSs against air attack required a fine sense of judgement, as to the use of high speed and maneuverability for their own protection vice the necessity of remaining within supporting distance of the unit. All DM commanding officers realized that they were not required to maintain the speed of the sweep unit during an air attack, but ordering flank speed, as would otherwise be common practice, might take the destroyer minelayer four thousand yards from the protected unit in three minutes.[24]

SEQUENCE OF ATTACKS

Adams (Comdr. Henry J. Armstrong Jr., USN) was first damaged on 23 March during a night attack by planes. An explosion off her starboard quarter injured thirteen men and killed two, but the planes were driven away and the minelayer remained in action. Sadly, these casualties were due to the premature detonation of a 5"/38 AA Common projectile within the first few feet of travel after leaving the muzzle of the gun. It was believed this was perhaps the first casualty of its kind in the many thousands of such rounds fired during the war.[25]

At around sunrise on 28 March, four enemy aircraft were picked up visually about ten miles to the west. *Adams* took evasive action and unmasked her main battery as one of the suiciders came in. The first salvo was fired when the plane was 8,000 yards distant, and the 40mm and 20mm guns opened when within range. Hit multiple times, and completely out of control, the Val passed over the ship and crashed about 200 feet off her starboard quarter.[26]

Five minutes later, a second aircraft made a low-level run on the *Adams*, headed directly for her bridge. All guns that could be brought to bear opened, but the plane did not break up until twenty-five feet off the ship's port bow. Pieces of it (also a Val) were found spread all over the fore part of the ship, but mostly under Mount No. 1. There was also considerable gasoline and oil about the decks, but no fire started and there were no personnel casualties aboard ship from either attack.[27]

On 1 April, *Adams* was hit for the third and last time, following two unsuccessful attacks earlier that day. Shortly after midnight, her combat information center reported a contact, sixty-five miles to the north-northeast. It proved to be a raid of two planes, which was driven off northward by the ship's main battery operating in full radar control. At 0220, *Adams* came under a second attack by two aircraft of unknown type. Her 5"/38 guns opened at 10,000 yards; by 6,000 yards, the planes had apparently had enough—they also retired to the north.[28]

Later that morning, approximately eight miles west of Kuba Shima, the minelayer was attacked at 0613 by a "Tony" (Kawasaki Ki-61 Hien Army type 3 fighter) making a low-level run from the *Adams'* port beam directly toward her bridge. The ship took evasive action, using 25 knots and increasing left rudder to full left. The plane was only 3,000 yards away and all guns that could bear commenced firing. Soon badly damaged and out of control, it crashed in close vicinity to the fantail, port side. Its two bombs exploded just under the stern, jamming the rudders from full left to full right, and leaving the *Adams* turning in a tight circle to starboard.[29]

Five minutes later, *Adams* came under attack by another suicide plane, a Zero (Mitsubishi A6M Navy type carrier fighter) making a long steep dive on the ship. The minelayer's main battery brought it down about 500 yards on the ship's starboard bow. A second suicider in the vicinity was brought down by the destroyer *Mullany* (DD-528), patrolling to the east about 5,000 yards away.[30]

Photo 23-5

Escort carrier USS *Sangamon* (CVE-26) being attacked by a "Tony" kamikaze aircraft off Okinawa on 4 May 1945.
National Archives photograph #80-G-334505

Photo 23-6

A "Zero" fighter about to crash into the sea after an unsuccessful kamikaze run on the carrier USS *Essex* (CV-9), on 15 May 1945.
National Archives photograph #80-G-324121

Due to the considerable damage to her rudders, the *Adams* was no longer in effective fighting condition. Even with her starboard engine ahead and port one backing, she could not proceed on a straight course. The DM was taken by the fleet tug *Tekesta* (AT-93) to the North Anchorage at Kerama Retto and moored alongside the repair ship *Endymion* (ARL-9). Temporary repairs were made by freeing her starboard rudder, and securing it in a midship position, and cutting off most of the mangled port rudder. The minelayer left Kerama Retto on the morning of 7 April for Guam as part of the screen of a task unit.[31]

Some questions were raised, as a result of damage done to *Adams* and other DMs, regarding the viability of the role of destroyer minelayers at Okinawa, given they were frequent targets of kamikaze. Commander, Amphibious Group One noted about these concerns:

> This command considers that DM's in support of smaller minecraft serve very important functions in addition to the morale effect they may have. Their fire power and superior navigational and communication facilities are valuable and necessary adjuncts to the meager facilities available in the minesweeping units composed of the smaller types of vessels. It is not believed that in the absence of DM's the enemy planes would necessarily have habitually ignored the lesser craft to seek more valuable targets.[32]

COMMANDING OFFICER AWARDED A SILVER STAR

Armstrong was promoted to Captain and awarded a Silver Star for heroism. The text of his medal citation follows:

> The President of the United States of America takes pleasure in presenting the Silver Star to Captain Henry Jacques Armstrong, Jr., United States Navy, for conspicuous gallantry and intrepidity in action as Commanding Officer of the U.S.S. *ADAMS* (DM-27), in support of a group of minesweepers at Okinawa Gunto during the period 23 March 1945 to 1 April 1945. His inspiring leadership consistently maintained the fighting efficiency of his ship at a high peak and made possible an outstanding record of destruction to enemy aircraft and the driving off of many air attacks upon his unit, with the result that their mission was successfully accomplished without loss of, or damage to, the ships being supported. His courage, initiative, and professional skill were at all times in keeping with the highest traditions of the United States Naval Service.

MINESWEEPING CONTINUES

By 1 April, minesweepers had cleared 141 mines and *Skylark* and the destroyer *Halligan* (DD-584) were both on the bottom as a result of enemy mines. The painstakingly slow, dangerous work continued, generally by smaller YMSs and AMs, capable of shallow water work, while the larger destroyer minesweepers began taking up radar picket stations. Duty there was equally dangerous as DMSs were not properly gunned for such tasking. *Emmons* (DMS-22) was sunk with massive casualties, and others suffered damage with men killed and wounded.[33]

For protection against kamikaze attack, most of the invasion fleet withdrew out to sea from Okinawa in groups each night; and returned at dawn. To provide these ships some advance warning of air attacks, fifteen radar picket stations ringed the island, some out as far as 150 miles. Ships on the picket line detected incoming aircraft on their radar screens and radioed warnings. Not surprisingly, the Japanese wanted to knock out the pickets, so that they could proceed in as far as possible undetected. For heavily gunned destroyers, picket duty was extremely dangerous—for destroyer minelayers with smaller batteries, it could be sheer murder.[34]

Except for assault sweeps, most of the sweeping after 1 April was performed by the 136-foot YMSs. The DMSs served as radar pickets, furnished shore bombardment and fire support, and served as escort vessels. The 220-foot AMs mostly served in the outer anti-submarine screen, carrying out 15-knot patrols involving seemingly endless time on station. The smaller 180-foot AMs were assigned to the inner anti-submarine screen and to anti-surface craft patrol. Many of the AMs operated at sea for all but two days of each of their three months at Okinawa, with no significant down time for repairs and upkeep.[35]

YMS-103 SUNK BY A MINE

The YMS-103 *was one of the oldest ships of her type in the Pacific Fleet and was manned by a well-trained, confident and experienced crew.*

—Lt. (jg) Leslie M. Thornton, commanding officer of *YMS-103*.[36]

The gallantry of YMS 103 *in rendering assistance to the stricken* PGM 18 *is typical of the steadfast courage and fortitude shown by all minecraft engaged in the hazardous task of clearing* NAKAGUSUKU WAN *of enemy mines.*

—Commander, Amphibious Group One, in his endorsement of USS *YMS-103*'s report on the loss of the ship to a mine.[37]

Photo 23-7

USS *YMS-103* under way, location and date unknown.
Courtesy of NavSource (from the collection of SM1c Howard Rider)

On 6 April, *YMS-103* entered Nakagusuku Wan with Sweep Unit Seventeen to function as its mine disposal vessel. (This large bay on the southern coast of Okinawa, would later be termed "Buckner Bay" by U.S. military personnel in honor of Lt. Gen. Simon Bolivar Buckner Jr. He was killed on 18 June, during the closing days of the Battle of Okinawa by enemy artillery fire.) The minesweeper destroyed seven enemy mines of the contact type that day by 20mm. Optimum range was 200 yards, as all explosions were of high order. Four of the mines appeared to be drifters, not cut by the sweep unit. The following day, *YMS-103* accounted for three mines, one lying outside swept waters.[38]

At about 1700 that evening, the entire sweep unit was taken under fire by enemy gun installations to the south-southwest, located on a peninsula on the southeastern tip of Okinawa. *YMS-103* returned counterfire from 6,000 yards distant with her main batteries, until ordered by the task group commander to disengage. During retirement, she received a direct hit, port side at frame 90, and while taking evasive action was straddled by another five rounds close aboard. The enemy battery appeared to be about 75mm.[39]

On 8 April, *YMS-103* again entered Nakagusuku Wan, this time in position four in a starboard echelon formation consisting of *YMS-1*, *YMS-299*, *YMS-92* and herself, in that order. Type "O" gear was streamed using 250 fathoms scope and a 40-foot pendant, with 700 yards distance between ships. *PGM-18* was 500 yards astern of the formation, positioned ten degrees off *YMS-103*'s port quarter to serve as mine disposal ship. Aboard *YMS-103*, per standard practice, all hatches were "dogged tight" (closed) to increase compartmentation—survivability of the ship in the event of damage—and all personnel above decks were wearing life jackets and helmets.[40]

At about 0743 as the formation was coming right to a new course, a mine exploded under *PGM-18* amidships, lifting her stern clear of the water. The gunboat came down with a 20-degree starboard list, and with her decks aflame, then made a slow turn to the right while rolling completely over on her starboard side. The task group commander ordered *YMS-103* to slip her sweep gear and to aid survivors.[41]

The minesweeper carried out this tasking with all possible speed, entering swept water and making an approach upwind of the gunboat. At 0800, *YMS-103* stopped both screws, so as not to harm survivors along both sides, and crewmembers were stationed along each life rail to aid in the rescue of those in the water. The wherry was put overboard to pick up wounded and lifelines passed to groups of survivors. *PGM-18* sank at 0805. Three minutes later, a mine blew off *YMS-103*'s bow at the forward edge of the bridge. Several men on her 3"/50-gun platform were blown overboard by the blast. One was killed immediately, and it was believed that seven of the gunboat's fourteen casualties were killed by the force of the underwater explosion.[42]

Steps were being taken aboard *YMS-103* to try to save the ship, her crew was jettisoning topside weight and had brought a "handy-billy" into use to pump the after-crew's quarters, when the detonation of a second mine sheared the minesweeper as far aft as the generator compartment. This explosion injured several men. Abandon ship was ordered and, aided by Dan buoys and life rafts already in the water, all men were able to make their way to safe positions. Lt. (jg) Leslie M. Thornton, the commanding officer, left the ship at 0820, having first inspected all compartments and machinery rooms for survivors. Boats from the *YMS-103*, *YMS-92*, and *Buoyant* (AM-153) picked up survivors in the water. *Buoyant* recovered ten men, including the badly wounded executive officer of the minesweeper.[43]

After seeing to the treatment of survivors taken to the destroyer minelayer *Lindsey* (DM-32), Thornton returned aboard *YMS-103*, which was under tow by the fleet tug *Cree* (ATF-84). He retrieved all registered, secret and confidential publications, as well as the pay accounts and service records of the crew. The hulk of his former command was beached in an area held by U.S. ground forces, and machinery spares and supplies were salvaged as time permitted. Personnel casualties were one officer and four men killed in action.[44]

Thornton was later promoted to lieutenant and awarded a Silver Star for heroism, as described in a synopsis of his award citation:

Lieutenant Leslie M. Thornton, Jr., United States Navy, was awarded the Silver Star for gallantry in action as Commanding Officer of the U.S.S. *YMS 103*, in action against the Japanese at Okinawa, on 8 April 1945. His gallant actions and dedicated devotion to duty, without regard for his own life, were in keeping with the highest traditions of military service and reflect great credit upon himself and the United States Naval Service.[45]

After the rescue work, sweeping began anew, but the unit's troubles were not quite over. Later that afternoon, *YMS-92* hit a mine and, minus her stern, was towed to Kerama Retto—the refuge for wrecked ships where Service Force units patched them up enough to keep going. A short time later, the clearing of Nakagusuku Wan was completed without further incident, and large gunfire support ships were able to enter the anchorage to better provide bombardment for troops ashore.[46]

EMMONS SENT TO BOTTOM BY KAMIKAZES

Photo 23-8

USS *Emmons* (DD-457) at anchor, circa 1942. Commissioned on 5 December 1941, the *Gleaves*-class destroyer was later converted to a high-speed minesweeper (DMS-22) on 22 November 1944.
Naval History and Heritage Command photograph #NH 107417

Many books have been written about the invasion and occupation of Okinawa. Descriptions of the activities of the mine warfare vessels, and heroic actions of their crews, could fill an additional one. It is not possible to provide herein details about the dozens of minecraft that suffered damage and/or loss of life. The accounts of damage vested on ships by kamikaze attacks, mines, and shore batteries serve as representative examples. Three more examples follow before turning to the activities of Task Force 57 (the British Pacific Fleet) and U.S. Mobile Explosive Investigation Unit No. 4 at Okinawa. The *Emmons* and *Swallow* were lost to kamikaze attack and, proving that no ship was safe at Okinawa, the flagship *Terror* also suffered significant damage and large numbers of casualties owing to a suicide plane. The four U.S. Navy minesweepers lost at Okinawa were among the twenty-five total mine warfare ships lost in the Pacific in World War II, to enemy action, enemy mines, or in one case to avoid capture. One of these ships, a minesweeper, was lost during ensuing post-war mine clearance. A list of all these ships may be found near the end of the following chapter.

U.S. Navy Mine Warfare Ships Lost at Okinawa
(Does not include ones so severely damaged as to warrant scraping)

28 Mar 45	*Skylark* (AM-63)	Sunk by a mine off Okinawa, Ryukyu Islands
6 Apr 45	*Emmons* (DMS-22)	Sunk after being hit by five kamikaze aircraft off Okinawa
8 Apr 45	*YMS-103*	Sunk by a mine off Okinawa
22 Apr 45	*Swallow* (AM-65)	Sunk after hit by a kamikaze off Okinawa

On 6 April, while *YMS-103* was serving as a mine disposal vessel in Nakagusuku Wan, the destroyer-minesweepers *Rodman* (DMS-11) and *Emmons* (DMS-22) were supporting operations of Sweep Unit Eleven in area G-3. These waters in the northern Okinawa area lay to the east of Ie Shima. In mid-afternoon, the minesweepers received a report from *Macomb* (DMS-23), operating in an adjacent area, of enemy planes inbound their location. Aboard *Emmons*, all guns were brought to bear on a group of aircraft closing her port quarter, but *Rodman* positioned between her and the threat, fouled the range.[47]

One of the planes crashed the fo'c'sle of *Rodman*, another carried out an unsuccessful bombing attack on her and, as the range cleared, *Emmons* opened fire on the remaining plane with negative results. The sailors aboard *Rodman* quickly brought the fire started by the kamikaze under control, as *Emmons* began circling her in an effort to provide protection. Numerous bogies were reported to be in the area (later developments showed them to number from 50 to 75) as well as many units of the Combat Air Patrol. Aero "dog fights" were plentiful, and many enemy planes were "splashed," although it appeared to witnesses that the enemy did not choose to engage U.S. aircraft, and was instead intent on making attack runs on surface vessels.[48]

Many attacks were directed at *Emmons*; Tonys, Vals and Zekes were visually identified, and she definitively splashed six planes before suffering the first of five hits. Four other attacks missed her by a matter of a few yards. The five hits occurred in rapid succession, almost instantaneously beginning at 1732, and appeared well coordinated. The first kamikaze hit was taken on *Emmons'* fantail; the second on the starboard side of her pilot house; the third on the port side of the Combat Information Center; the fourth on the starboard side of number three 5-inch gun; and the fifth forward near the waterline at frame 30, starboard side. It is believed that all of the suicide planes carried bombs and strafed during their approach. Target speed of the aircraft was calculated to be 235 knots.[49]

MASSIVE DAMAGE AND CASUALTIES RESULT

> *All of the hull aft of frame 115 was entirely missing and serious damage was inflicted on the port screw rendering it inoperative. The entire bridge structure was destroyed and fire raged in all spaces from frame 67 forward to gun one, from the main deck up. Little or nothing remained of the decks from the main deck to the bridge overhead in that area. Several small fires were started in the after part of the ship and would break out again after being extinguished. The proportions of the fire, together with the exploding of 20 mm ready [ammunition] boxes, made firefighting extremely difficult forward although the handling rooms of both forward guns were sprinkled [sprinklers energized].*

> —Lt. John J. Griffin Jr., USNR, acting commanding officer of
> USS *Emmons* (DMS-22), reporting her loss off Okinawa.[50]

Emmons' gunnery officer, Lt. John J. Griffin Jr., USNR, assumed command during the attacks as a result of the commanding officer, Lt. Comdr. Eugene Noble Foss, II, USNR, being blinded and the executive officer missing in action. At 1800, Griffin made a decision to abandon ship because of several reasons:

- The port main engine was inoperative, and there was no means of steering the ship, because the rudder had been blown off
- The fire forward could not be controlled due to its intensity and low water pressure on the firemain
- The ship was continuing to take additional seawater through hull breeches and settle perceptibly
- There was imminent danger of explosion in both magazines and fuel tanks

- All gun batteries with the exception of two of the 20mm guns were inoperative[51]

Six officers and fifty-seven men remained aboard the *Emmons* until the last, expending every effort to combat damage. About 1930, the gunboat *PGM-11* came alongside to render assistance. As wounded men were being transferred to her, a heavy explosion occurred in the handling room of number two gun. Lieutenant Griffin then ordered those still aboard to abandon. Although the minelayer was still afloat when abandoned, there was no danger of her falling into enemy hands, as friendly units would be able to sink her should she start to drift toward the enemy beach. The hulk was sunk by the *Ellyson* (DMS-19) the following morning.[52]

PGM-11 brought sixty-three survivors to the attack transport *Gosper* (APA-17), where the wounded were treated. Additional survivors arrived aboard other vessels. The following day, six officers and 144 men were transferred to the *Wayne* (APA-54). The wounded remained aboard the *Gosper* and *Crescent City* (APA-21). *Emmons* had borne massive casualties, fifty-seven personnel killed or missing in action, and presumed dead.[53]

Griffin was awarded the Silver Star. His report on the loss of the *Emmons* reflects the rank of lieutenant and not lieutenant (jg), as indicated in the synopsis of his award citation. Of course, he may have been immediately meritoriously promoted for his actions during the kamikaze attacks and in trying to save his temporary command.

> Lieutenant, Junior Grade John J. Griffin, Jr., United States Navy, was awarded the Silver Star for gallantry in action as Gunnery Officer of the U.S.S. *EMMONS* (DD-457), in action against the Japanese off Okinawa, on 6 April 1945. His gallant actions and dedicated devotion to duty, without regard for his own life, were in keeping with the highest traditions of military service and reflect great credit upon himself and the United States Naval Service.[54]

SWALLOW LOST TO A KAMIKAZE ATTACK

> *The ship was struck by a Japanese suicide plane at approximately 1858. Plane hit the starboard side amidships just above the water line. Both engine rooms flooded and ship immediately took a forty-five-degree list to starboard. All power was lost and there was no pressure on fire main. Ship was abandoned at approximately 1901. At approximately 1905 she was capsized and sank shortly thereafter in about 85 fathoms of water.*

—Lt. Comdr. Whitefield F. Kimball, USN, commanding officer
USS *Swallow* (AM-65) reporting on loss of the minesweeper.[55]

Photo 23-9

USS *Swallow* (AM-65) passing mail to a destroyer somewhere in the South Pacific.
Photo from September 1943 issue of *All Hands* magazine

The fleet tug *Molala* (ATF-106) had observed the kamikaze crash into
the starboard side of *Swallow*, followed by the minesweeper veering
sharply to starboard and begin capsizing. Released by the destroyer
Isherwood (DD-520), for whom she'd been sent out from Kerama Retto
to provide assistance, the tug went to the aid of *Swallow* survivors. *Molala*
picked up fifteen personnel directly, and another ten transferred to her
by the *Gayety* (AM-239). The tug's motor whaleboat retrieved another
twelve and delivered them to the destroyer *Paul D. Bates* (DE-643).
Molala entered Kerama Retto after midnight on 23 April and transferred
those aboard her to the *Netrona* (APA-214).[56]

Gayety, operating near *Swallow* (also assigned to patrol duties in the
outer screen) had sighted the enemy plane at 1853, closing rapidly, and
had opened with all automatic weapons that could be brought to bear.
Hit twice, the plane had turned and executed a suicide dive into *Swallow*,
about 8,000 yards on the *Gayety*'s port quarter.[57]

FLAGSHIP *TERROR* HIT, WITH MANY CASUALTIES

During this month [April 1945] we have gone to general quarters 93 times. The shortest period was seven minutes and the longest, six and one-half hours. The enemy is making full use of high speed, weak IFF and running lights in their air attacks. The use of our air search radar for close in detection and ranging has been seriously handicapped by the fact that we are almost totally surrounded at close range by land.

—Comdr. Horace W. Blakeslee, USN, commanding officer
USS *Terror* (CM-5), in a report on damage and casualties
caused by a kamikaze hit on the ship on 1 May 1945
while she lay anchored in Kerama Retto.[58]

Photo 23-10

Kamikaze attack on shipping at Kerama Retto anchorage, at dusk, on 6 April 1945.
Minelayer and flagship USS *Terror* (CM-5) is in the left center, putting up anti-aircraft
fire like other ships present.
Naval History and Heritage Command photograph #80-G-311870

The mission of the *Terror* (CM-5) at Okinawa was to serve as flagship for Rear Admiral Sharp, who as commander, Task Group 52.2 was in charge of all minesweeping, mine destruction, and anti-submarine net laying activities.   To support the Mine Force, *Terror* carried various minesweeping, radio, radar, sonar and diesel spares as well as fresh and dry provisions for issue to ships of the task group. When Kerama Retto had earlier been secured by Army infantry ashore, the flagship had entered the anchorage there, and acted as a tender for minecraft.[59]

In early morning on 1 May, the *Terror* lay at anchor, as she had for about six weeks, when at 0356, a kamikaze hit her superstructure amidships. The plane had come in on her port beam, passed astern of her ship, then reversed course and made the attack from her starboard quarter. From the resultant damage, it was believed that the aircraft carried two 500lb high explosive bombs and also at least three thermite bombs. Only one of the latter ignited, burning in the wardroom and adding greatly to the difficulty of fighting fire and rescuing casualties.[60]

When the plane crashed into the ship's communication platform, one of its 500lb bombs immediately exploded. The other penetrated the main deck below and also exploded. The aircraft's engine tore through ship's bulkheads to land in the wardroom. Fire that flared up in the superstructure was soon controlled and, within two hours, extinguished. However, the kamikaze exacted a terrible toll aboard *Terror*—171 casualties: 41 dead, 7 missing, and 123 wounded.[61]

The flagship was moored the following day alongside the attack transport *Natrona* (APA-214) for emergency repairs. Made sufficiently seaworthy, *Terror* got under way on 8 May to rendezvous with a convoy in passage to Saipan. Following arrival at the northern Mariana island, and determination that her damage was too great to be repaired in a forward area, *Terror* proceeded via Eniwetok and Pearl Harbor to the U.S. west coast. She reached San Francisco on 1 June 1945, offloaded ammunition, and began an overhaul.[62]

Valor and sacrifice was abundant aboard Mine Flotilla ships at Okinawa. Mine warfare ships awarded a Presidential Unit Citation, or Navy Unit Commendation for heroism are identified at chapter's end. The eight ships thus honored were all minelayers, frequent targets off Okinawa. Discussion of action aboard these ships, and damage and casualties suffered, have been limited to the *Adams* and *Robert H. Smith*; two of the many ships hit, some sunk. The following photograph of the destroyer minelayer USS *Lindsey* (DM-32), hit by two kamikazes on 12 April 1945, serves as a vivid reminder of the collective sacrifices of the Flotilla at Okinawa.

Photo 23-11

Minelayer USS *Lindsey* (DM-32) at Kerama Retto, after losing 60 feet of her bow when crashed by two kamikaze aircraft on 12 April 1945.
Australian War Memorial photograph 302655

THE BRITISH PACIFIC FLEET (TASK FORCE 57)

When a kamikaze hits a US carrier it means six months of repair at Pearl [Harbor]. When a kamikaze hits a Limey carrier it's just a case of 'Sweepers, man your brooms.'

—Comment made by the USN liaison officer aboard the British fleet aircraft carrier HMS *Indefatigable*.[63]

In March and April 1945, while supporting the invasion of Okinawa, the British Pacific Force (formed on 22 November 1944 under Adm. Bruce Fraser) was responsible for suppressing Japanese air activity in the Sakishima Islands. Gunfire and air attack were used against potential kamikaze staging airfields that might otherwise be used to support attacks against U.S. Navy vessels at Okinawa. The Sakishimas, located to the southwest of Okinawa, are a part of the same Ryukyu Islands.[64]

The BPF began operations at the end of March 1945, attached to the U.S. Fifth Fleet. The BPF (Task Force 57) initially consisted of some twenty-five surface combatants including four fleet carriers and a Fleet Train of some thirty ships designated Task Force 112. The fighting core of the task force was the 1st Aircraft Carrier Squadron, under Rear Adm. Philip Vian, which included all six *Illustrious*-class carriers, although only four were in action at any one time. The British fleet carriers were subjected to heavy and repeated kamikaze attacks off the Sakishimas. However, boasting armoured flight decks, they were quite resilient and returned to action relatively quickly.[65]

At the time of the Japanese surrender, the BPF had expanded to some 270 ships including 18 aircraft carriers, 4 battleships, 11 cruisers, 29 submarines, over 80 destroyers and escorts, 32 minesweepers, and a Fleet Train comprising more than 100 ships. Among the mostly British force were a few Royal Canadian and Royal New Zealand Navy ships. Australia provided six destroyers, and the eighteen *Bathurst*-class corvettes which became the 21st and 22nd Minesweeping Flotillas under the British Pacific Fleet.[66]

Photo 23-12

Adm. Sir Bruce Fraser accompanied by Sub. Lt. Paton Forster RANVR, inspecting 21st Minesweeping Flotilla at Watsons Bay, New South Wales, Australia, in February 1945. Australian War Memorial photograph 018080

The duties of the Australian minesweeping corvettes primarily involved escorting the Fleet Train as it moved from base to base and within sea replenishment areas. The destroyers and corvettes listed below all earned Battle Honour – OKINAWA 1945. The corvettes played an important support role with the BPF from its inception and in the immediate post war period in Hong Kong and China. Four were present at Tokyo Bay for the Japanese surrender on 2 September 1945.[67]

Royal Australian Navy Destroyers and Corvettes

Destroyers	Corvettes
Quiberon, *Quickmatch*, *Napier*, *Nepal*, *Nizam*, and *Norman*	*Ballarat*, *Bendigo*, *Burnie*, *Cairns*, *Kalgoorlie*, *Launceston*, *Lismore*, *Pirie*, and *Whyalla*[68]

Photo 23-13

Able Seaman Bob Skinner, RAN, greets the crew of HMAS *Ballarat* mustered on the forecastle as the corvette moors at Yokosuka Naval Dockyard, 3 September 1945. Australian War Memorial photograph 019168

MOBILE EXPLOSIVE INVESTIGATIVE UNIT NO. 4

As battle was waged on land and in the skies over Okinawa, members of MEIU #4 were ashore searching for Japanese explosives to ship back to Pearl Harbor for examination/intelligence collection. The unit had been formed on 4 June 1944 under commander, Service Forces Pacific, and subsequently transferred to commander, Minecraft Pacific Fleet, following the establishment of this new command. For purposes of administration, messing and berthing, MEIU #4 was assigned to Mine Assembly Base, West Loch Oahu, in Hawaii.[69]

In April 1945, Bomb and Mine Disposal Teams were formed with an officer in charge of each. The teams were sent where most needed, and many participated in the invasion and occupation of Okinawa. Assignments included duty with: (1) Beach Garrison Parties responsible for removing ordnance from island beachheads; (2) Beachmaster Groups; (3) Minesweeper flagships; (4) Advance Base Construction Units; and (5) Group Pacific, a part of the Service Force[70]

Team #27, under the command of Lt. Robert B. Whittemore and Lt. (jg) Donald F. Annen, returned over 130 tons of ordnance materiel to Pearl Harbor, via transport aboard the cargo ship USS *Virgo* (AKA-20), for examination and distribution. Recovery of Japanese weapons and ordnance on Okinawa included the first acquisition of piloted baka bombs, a suicide weapon involving a single flight for the volunteer steering it toward a target. After the bomber carrying the manned-bomb had released it, the pilot of the suicide weapon would glide toward the target then activate rockets or a jet engine to dive into the target and explode its one-ton warhead.[71]

Photo 23-14

Baka piloted-glide bomb carried to within 12 miles of the target by a medium bomber. Courtesy of Ron Swart

Scores of individuals aboard mine warfare ships at Okinawa received medals for personal valor, and their associated citations could fill an entire chapter. The final two pages of this chapter cite only the unit awards garnered by minelayers.

MINELAYER UNIT AWARDS AND BATTLE STARS

Presidential Unit Citation

Okinawa

Ship	Award Period	Commanding Officer
Harry F. Bauer (DM-26)	24 Mar-11 Jun 45	Comdr. Richard Claggett Williams Jr., USN
Harry A. Wiley (DM-29)	23 Mar-24 Jun 45	Comdr. Paul Henrik Bjarnason, USN

Vicinity of Okinawa

Aaron Ward (DM-34)	3 May 45	Comdr. William Henry Sanders Jr., USN

Navy Unit Commendation

Okinawa

Adams (DM-27)	24 Mar-1 Apr 45	Comdr. Henry Jacques Armstrong Jr., USN
Gwin (DM-33)	24 Mar-24 Jun 45	Comdr. Frederick Samuel Steinke, USN
Shea (DM-30)	24 Mar-4 May 45	Comdr. Charles Cochran Kirkpatrick, USN
Robert H. Smith (DM-23)	24 Mar-24 Jun 45	Comdr. Henry Farrow, USN Comdr. Wilber Haines Cheney Jr., USN
J. William Ditter (DM-31)	24 Mar-6 Jun 45	Comdr. Robert Roy Sampson, USN

BATTLE STARS AWARDED MINELAYERS

Okinawa Operation: Assault and Occupation of Okinawa

Monadnock (CM-9)	10 Apr- 27 May 45	Lt. Comdr. Frederick O. Goldsmith, USNR; Lt. Comdr. John E. Cole, USNR
Salem (CM-11)	26 Mar- 4 Apr 45	Lt. Comdr. George Carlton King, USN
Terror (CM-5)	25 Mar- 8 May 45	Comdr. Horace William Blakeslee, USN
Weehawken (CM-12)	10 Apr- 30 Jun 45	Comdr. Walter Patrick Wrenn Jr., USNR
Aaron Ward (DM-34)	25 Mar- 11 Jun 45	Comdr. William Henry Sanders Jr., USN
Adams (DM-27)	1-7 Apr 45	Comdr. Henry Jacques Armstrong Jr., USN
Breese (DM-18)	25 Mar- 30 Jun 45	Lt. George W. McKnight, USNR
Gwin (DM-33)	1 Apr- 30 Jun 45	Comdr. Frederick Samuel Steinke, USN
Henry A. Wiley (DM-29)	25 Mar- 30 Jun 45	Comdr. Paul Henrik Bjarnason, USN
Henry F. Bauer (DM-26)	25 Mar- 11 Jun 45	Comdr. Richard Claggett Williams Jr., USN
J. William Ditter (DM-31)	30 Jun 45	Comdr. Robert Roy Sampson, USN
Lindsey (DM-32)	16-19 Feb 45	Comdr. Thomas Edward Chambers, USN
Robert H. Smith (DM-23)	25 Mar- 30 Jun 45	Comdr. Henry Farrow, USN; Comdr. Wilber Haines Cheney Jr., USN
Shannon (DM-25)	25 Mar- 30 Jun 45	Comdr. Edward Lee Foster, USN; Comdr. William Thomas Ingram II, USN
Shea (DM-30)	25 Mar- 16 May 45	Comdr. Charles Cochran Kirkpatrick, USN
Thomas E. Fraser (DM-24)	25 Mar- 30 Jun 45	Comdr. Ronald Joseph Woodman, USN
Tolman (DM-28)	1 Apr- 28 Jun 45	Comdr. Clifford Arthur Johnson, USN
Tracy (DM-19)	25 Mar- 28 Jun 45	Lt. Comdr. Richard Edward Carpenter, USNR

24

Third Fleet Operations against Japan

After nearly three weeks of replenishment in Leyte Gulf, subsequent to their support of the Okinawa operation, the fast carrier forces of Admiral Halsey's Third Fleet, comprising the greatest mass of sea power ever assembled, proceeded northward on 1 July toward Japan. This huge armada was to complete the destruction of the Japanese fleet, conduct a pre-invasion campaign of destruction against every industry and resource contributing to Japan's ability to wage war, and maintain maximum pressure on the Japanese in order to lower their will to fight.

—Fleet Adm. Ernest J. King, from *United States Navy at War Final Official Report to the Secretary of the Navy covering the period March 1, 1945, to October 1, 1945.*

On 1 July 1945, after organized enemy resistance on Okinawa had ceased, Task Force 38 (the fast carrier task force of the Third Fleet) proceeded northward from Leyte Gulf to operate close to Japan; and there it stayed during final phases of the War in the Pacific until Japan surrendered on 15 August. Task force aircraft struck at the city of Tokyo on 10 July and then attacked the islands of Honshu and Hokkaido on 14-15 July. Moving southward, the Third Fleet, joined by units of the British Pacific Fleet under Vice Adm. Sir Henry Bernard Hughes Rawlings, on July 17, made the first combined American-British bombardment of the Japanese homeland. Later that month, the same force made extended air strikes in the Inland Sea area and on the Japanese naval base at Kure, returning to harass Tokyo, for the third time in three weeks, on 30 July.[1]

The Third Fleet had been formed on 15 March 1944 under the command of Adm. William F. Halsey and had previously operated in and around the Solomon Islands, the Philippines, Formosa, Okinawa and the Ryukyu Islands. As Allied air power struck Japanese targets, a flotilla of American mine warfare ships at Okinawa carried out tasking to clear a suspected mine barrier located to the northwest of Kume-shima in the East China Sea. This area was designated "Juneau."[2]

MINESWEEPING OPERATIONS

This group will clear area JUNEAU of moored mines to a depth of ninety feet giving first priority to the area where the suspected enemy minefield is believed to be located. Since visual and radar fixes cannot be obtained in this area, a system of buoys, reference vessels, and possible Loran lines [electronic navigation] must be used to supplement celestial navigation and dead reckoning in order that acceptable accuracy may be obtained.

On completion of sweeping JUNEAU area, a check sweep will be made along the axis of any located minefield. The check sweep will follow immediately after the first sweep, or, at a later date as circumstances dictate.

DD's, PGM's [gunboats], DM's and DMS's accompanying sweep units will act as Mine Destruction, Fire Support, Salvage and Rescue vessels.

—Commander, Minecraft U.S. Pacific Fleet
Operation Order No. 7-45

In executing these orders, the Mine Force swept a total of 343 mines, and destroyed 61 floaters in the East China Sea in July, with no losses of, or significant damage to vessels.[3]

Map 24-1

Kume-shima, one of the most beautiful of the Okinawa Islands, lies west of Okinawa.
U.S. National Atlas 1970

Most significant was the numbers of minecraft comprising Task Group 39.11 under Capt. Wayne R. Loud. Immediately following the war, they would comprise the bulk of mine forces sweeping Philippine, Japanese, Korean and Chinese waters. During the war, and afterward, the unheralded officers and men of the U.S. Mine Force were mostly reservists. A badge depicting a mine awash, being struck by lightning bolts, symbolized the dangerous work in which they engaged.

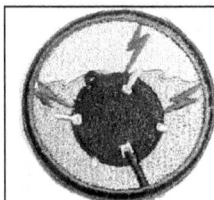

The U.S. Navy Mine Forces uniform shoulder patch, worn by sailors as a badge of a hazardous trade which paid no extra compensation, was small enough recompense for those who risked their lives and ships against an enemy they seldom saw until it was too late. Top Navy brass "deep-sixed" the one visible mark that distinguished Minecraft sailors from other people on 1 July 1946.[4]

Sweep Unit One: Capt. T. W. Davison (CominRon 15)

Ship	Commanding Officer
Hazard (AM-240)	Lt. C. B. Tibbals, USNR
Ransom (AM-283)	Lt. Comdr. W. N. McMillen, USNR
Execute (AM-232)	Lt. R. E. Brenkman, USNR
Recruit (AM-285)	Lt. Wade O. Hankinson, USNR
Reform (AM-286)	Lt. Francis Worcester Jr., USNR
Strategy (AM-308)	Lt. V. A. Brown, USNR
Success (AM-310)	Lt. R. N. Hall, USNR
Shelter (AM-301)	Lt. S. D. Letourneau, USNR
YMS-183	Lt. S. C. Klein, USNR
YMS-271	Lt. (jg) C. J. Verplank, USNR
YMS-360	Lt. R. F. Donovan, USNR
PGM-19	Lt. L. E. Burger, USNR
PGM-21	No name found

Sweep Unit Two: Lt. Comdr. R. R. Forrester (CominDiv 40)

Ship	Commanding Officer
Density (AM-218)	Lt. Comdr. R. R. Forrester Jr., USNR
Design (AM-219)	Lt. L. A. Young, USNR
Device (AM-220)	Lt. R. Remage Jr., USNR
Gayety (AM-239)	Lt. R. B. Harrell, USNR
Diploma (AM-221)	Lt. A. B. Baxter, USNR
Notable (AM-267)	Lt. Nathaniel William Roe, USNR
Rebel (AM-284)	Lt. F. S. Wooster, USNR
YMS-140	Lt. L. A. Agar, USNR
YMS-176	Lt. L. H. Countrymen, USNR
YMS-410	Lt. (jg) L. G. Knoles, USNR
PGM-10	Lt. (jg) B. W. Clemens, USNR
PGM-32	No name found

Sweep Unit Three: Comdr. Wetmore (CominRon 11)	
Ship	**Commanding Officer**
Phantom (AM-273)	Lt. James Edward Lambert, USNR
Nimble (AM-266)	Lt. Francis Robert Kitchell Jr., USNR
Pivot (AM-276)	Lt. Robin Max Hartmann, USNR
Pirate (AM-275)	Lt. Comdr. Gilbert Louis Leindecker, USNR
Inflict (AM-251)	Lt. Raymond Charles Davidson Jr., USNR
Pledge (AM-277)	Lt. Thomas Knotts, USNR
Dunlin (AM-361)	Lt. Tom Terrance Wuerth, USNR
YMS-324	Lt. H. B. Silveria, USNR
YMS-331	Lt. E. D. Dougherty, USNR
YMS-398	Lt. W. A. Latta, USNR
PGM-23	Lt. A. B. Carlisle, USNR
PGM-25	Lt. C. S. Munson Jr., USNR

Two members of Sweep Unit Three—the USS *Pirate* (AM-275) and *Pledge* (AM-277)—would survive the war, but later be sunk by mines at Wonsan Harbor during the Korean War.

Photo 24-1

Painting by Richard DeRosset, titled "Hidden Menace at Sin-Do Island," is the cover art for the author's book, *Wooden Ships and Iron Men: The U.S. Navy's Coastal and Motor Minesweepers, 1941-1953.*

Sweep Unit Four: Comdr. G. W. Allen (CominRon 5)

Ship	Commanding Officer
Strive (AM-117)	Lt. Comdr. Elliott Burris "Buzz" Knowlton, USN
Steady (AM-118)	Lt. Robert David Faber, USN
Sustain (AM-119)	Lt. Comdr. James Edward Lindeman Jr., USNR
Dextrous (AM-341)	Lt. Comdr. Stanley Sherman Trotman, USNR
Pioneer (AM-105)	Lt. Leroy Edward Rogers Jr., USNR
Prevail (AM-107)	Lt. Comdr. William McCarmick Mark, USNR
YMS-325	Lt. J. N. Ferguson, USNR
YMS-327	Lt. G. A. Ashland, USNR
YMS-401	Lt. A. P. Hess Jr., USNR
PGM-30	Lt. S. F. Whitman, USNR
PGM-31	No name found

Sweep Unit Five: Lt. Comdr. D. D. Long (CominDiv 24)

Ship	Commanding Officer
Pochard (AM-375)	Lt. Comdr. D. D. Long Jr., USNR
Requisite (AM-109)	Lt. Comdr. H. R. Pierce Jr., USNR
Token (AM-126)	Lt. W. T. Hunt, USNR
Tumult (AM-127)	Lt. L. W. Sharp, USNR
Velocity (AM-128)	Lt. G. J. Buyse, USNR
Zeal (AM-131)	Lt. Comdr. E. W. Woodhouse, USNR
YMS-362	Lt. T. N. I. Davis, USNR
YMS-388	Lt. R. E. Crowley, USNR
YMS-434	Lt. (jg) D. Birdwell, USNR
PGM-20	Lt. H. T. Scott Jr., USNR
PGM-29	Lt. N. H. Rumbaugh, USNR

Sweep Unit Six: Lt. Comdr. Robert W. Costello (CominDiv 11)

Ship	Commanding Officer
Gladiator (AM-319)	Lt. Comdr. Robert W. Costello, USNR
Competent (AM-316)	Lt. A. S. Furtwangler Jr., USNR
Impeccable (AM-320)	Lt. Comdr. B. H. Smith Jr., USNR
Triumph (AM-323)	Lt. Comdr. C. R. Cunningham, USNR
Surfbird (AM-383)	Lt. R. H. Nelson Jr., USNR
Toucan (AM-387)	Lt. Comdr. Stephen Henry Squibb, USNR
YMS-371	Lt. (jg) C. J. Verplank, USNR
YMS-341	Lt. H. O. Arend, USNR
YMS-430	Lt. R. C. Bates, USNR
PGM-11	Lt. E. H. George, USNR
PGM-16	Lt. D. E. Wilson, USNR

Fire Support Unit: Capt. Wayne R. Loud (CominRon 20)

Ellyson (DMS-19)	Lt. Comdr. R. W. Mountrey, USNR
Hambleton (DMS-20)	Comdr. George Albert O'Connell Jr., USN
Gherardi (DMS-30)	Lt. Comdr. William Wade Gentry, USNR
Jeffers (DMS-27)	Lt. Comdr. Robert D. Elder Jr., USNR
Dorsey (DMS-1)	Lt. John M. Hayes, USNR
Southard (DMS-10)	Lt. Comdr. John Edward Brennan, USNR
Robert H. Smith (DM-23)	Comdr. Wilber Haines Cheney Jr., USN

Buoy Unit: Capt. Henry J. Armstrong Jr., USN (CominDiv 8)

Breese (DM-18)	Lt. George W. McKnight, USNR
	Lt. David James Pikkaart, USNR
Gwin (DM-33)	Comdr. Frederick Samuel Steinke, USN
Henry A. Wiley (DM-29)	Comdr. Paul Henrik Bjarnason, USN
Thomas E. Fraser (DM-24)	Comdr. Ronald Joseph Woodman, USN
Shannon (DM-25)	Comdr. William Thomas Ingram II, USN

Reserve Unit: 180- and 220-foot AMs, and YMSs

Sheldrake (AM-62)	Lt. Burt E. Taylor, USN
Revenge (AM-110)	Lt. Comdr. James Leland Jackson, USN
Sage (AM-111)	Lt. Daisy Louis Brantley, USNR
Spear (AM-322)	Lt. Comdr. Arthur Mix Savage, USNR
Vigilance (AM-324)	Lt. Jackson Leonard Morton, USNR
Devastator (AM-318)	Lt. William Francis Remington, USNR
Dour (AM-223)	Lt. William Vandiver Byrd, USNR
Opponent (AM-269)	Lt. George Edward Schaaf, USNR
Signet (AM-302)	Lt. Comdr. Coburn Louis Grabenhorst, USNR
Skirmish (AM-303)	Lt. Bruce Leverington Hyatt, USNR
Scurry (AM 304)	Lt. Charles Edward Dunston, USNR
Specter (AM-306)	Lt. J. Chevalier, USNR
Staunch (AM-307)	Lt. Comdr. John Charles Kettenring, USNR
Serene (AM-300)	Lt. James Edward Calloway, USNR

YMS-89	Lt. (jg) W. L. Ford, USNR	*YMS-376*	No name found
YMS-93	Lt. (jg) A. H. Bofinger, USNR	*YMS-388*	Lt. R. E. Crowley, USNR
YMS-94	Lt. J. R. Stanfield, USNR	*YMS-390*	Lt. E. Wigglesworth, USNR
YMS-147	Lt. (jg) J. M. Prokop, USNR	*YMS-398*	Lt. W. A. Latta, USNR
YMS-176	Lt. L. H. Countrymen, USNR	*YMS-401*	Lt. A. P. Hess Jr., USNR
YMS-243	Lt. (jg) W. R. Spencer, USNR	*YMS-403*	No name found
YMS-271	Lt. (jg) C. J. Verplank, USNR	*YMS-415*	No name found
YMS-283	Lt. J. W. Holmes, USNR	*YMS-418*	No name found
YMS-299	Lt. (jg) O. Sipari, USNR	*YMS-426*	Lt. M. R. Ball Jr., USNR
YMS-311	Lt. G. M. Hardy, USNR	*YMS-443*	Lt. (jg) John P. Hanna, USN
YMS-323	Lt. J. R. Thompson, USNR	*YMS-449*	No name found
YMS-327	Lt. G. A. Ashland, USNR	*YMS-461*	No name found
YMS-331	Lt. E. D. Dougherty, USNR	*YMS-467*	No name found
YMS-341	Lt. H. O. Arend, USNR	*YMS-468*	Lt. (jg) Ed Leach, USNR
YMS-362	Lt. T. N. I. Davis, USNR	*YMS-471*	No name found
YMS-371	Lt. John L. Grace Jr., USNR	*YMS-475*	Lt. W. B. Beardon Jr., USNR
YMS-372	No name found	*YMS-478*	Lt. R. D. Tuohy, USNR[5]

The U.S. Navy's commitment of large numbers of minesweepers to expeditiously clear area Juneas was a precursor to what was to follow. The first priority upon the cessation of hostilities would be to sweep channels leading into ports necessary for the evacuation of Allied prisoners of war, and for landing occupational troops ashore. Next would be the clearance of additional channels and harbor facilities to allow movement of shipping, and resupply of occupational troops.

Finally, the sweeping of sea lanes around Japan would be undertaken. Generally, British and Australian minesweeping forces were to clear Hong Kong, South China and the islands and ports south of the Philippines, whilst the U.S. (and Japanese) cleared the Philippines, Korea, North China, and Japan.[6]

MINELAYERS' SUPPORT OF JUNEAU OPERATIONS

Of the six minelayers supporting sweep operations in area Juneau, only USS *Breese* was one of the old "warhorses" (ex-World War I vintage destroyers converted to minelayers) so desperately needed in the early years of the war. (*Breese* would earn 10 battle stars in WWII.) Boasting only 3-inch guns as their main armament, three of her sister ships— *Tracy*, *Montgomery*, and *Preble*—had mined the "Tokyo Express" at Guadalcanal on 1 February 1943, and they and other DMs had gone wherever dispatched to lay mines. These missions often involved braving Japanese-controlled waters to help shield Allied forces from attack by enemy ships and submarines.

The other five minelayers were newly built, powerful destroyers, converted to a new mission in their builders' yards and commissioned in 1944. Five of the DMs collectively comprised the Buoy Unit, whose mission involved planting buoys to mark areas to be swept, and then retrieving them once a particular area had been determined to be clear of mines. The also served as mine disposal vessels, destroying cut mines floating on the ocean's surface. The most interesting duty was that of *Robert H. Smith* (DM-23), drafted to be a member of the Fire Support Unit (under direct control of Capt. Wayne R. Loud, USN, commander, Mine Squadron Twenty and Task Group 39.11).

The mission of the Fire Support Unit was to provide protection to the minesweepers working the approximately 120-mile-long by 57-mile-wide area. Air coverage consisted of a twelve-plane combat air patrol (CAP) provided by escort carriers of Task Force 32, located eastward of the area. Additionally, a dusk patrol of twelve F4U Corsair fighters was provided by Task Force 99. The day CAP was on station from dawn to 1830 and the dusk CAP from 1830 to sunset. *Robert H. Smith* (with a special Fighter Director team aboard) was the Primary Fighter Director Ship, and as such, controlled flash conditions (issuance of warnings of enemy aircraft) for the task group. Other fighter director ships in the task group were *Henry A. Wiley* (DM-29) and *Thomas E. Fraser* (DM-24).[7]

Minesweeping operations began on 5 July. Over the next several days, each morning *Robert H. Smith* took station astern of the sweep units after their work had begun, and as the day's operations completed, joined cruising disposition MS-1 as the minesweepers retired to the

southeast.   At 0225 on 12 July, the minelayer detected a "bogie"
(unknown aircraft) thirty-nine miles distant. It remained in the area until
0425, with the closest range being three miles. This pattern repeated
itself the following night when a bogie was detected at 0323.   This
contact was reported to Okinawa, and was intercepted by a night fighter,
which identified it as a "Betty," and splashed it.[8]

By the end of that day's work, 229 mines had been swept and 29
floaters sighted and destroyed. It had been determined that these mines
were part of a barrier located in the approximate position of the
suspected barrier.   In control of the twelve-plane daytime CAP, *Smith*
maintained a two-division "high CAP" at altitudes of 10,000 and 20,000
feet; the remaining division was split into two-plane sections, with one
assigned to *Wiley* and one to *Fraser* for visual control. At 0655 on 23
July, a bogie was detected by the *Smith*, fifty-five miles distant.   Upon
inquiry, Task Force 32 reported splashing a "Dinah" (Mitsubishi Ki-46
Army reconnaissance aircraft).[9]

Sweep operations concluded on 30 July, and the task force entered
Buckner Bay, Okinawa, the following day.[10]

UNIT AWARDS

The six minelayers received a battle star for the period 5-31 July 1945.
The ship commissioning dates in the below table provide indication of
the relative age of the vessels.

3rd Fleet Operations against Japan (5-31 July 1945)		
Breese (DM-18)	23 Oct 1918	Lt. George W. McKnight, USNR
		Lt. David James Pikkaart, USNR
Gwin (DM-33)	30 Sep 1944	Comdr. Frederick Samuel Steinke, USN
Henry A. Wiley (DM-29)	31 Aug 1944	Comdr. Paul Henrik Bjarnason, USN
Robert H. Smith (DM-23)	4 Aug 1944	Comdr. Wilber Haines Cheney Jr., USN
Shannon (DM-25)	8 Sep 1944	Comdr. William Thomas Ingram II, USN
Thomas E. Fraser (DM-24)	4 Aug 1944	Comdr. Ronald Joseph Woodman, USN
		Comdr. Nevett Brooke Atkins, USN

25

War's End

'Well,' said the pilot, 'it's all over.'
Goldy tried to focus. 'What is?'
'The war,' said the pilot, and moved on to spread the news.
Goldy stared through the window to the dull, brown landscape below.
'Over,' he said. 'It's over.'

—Lt. Comdr. Leon Goldsworthy, RANVR, bored and dehydrated,
travelling in a RAF transport plane over the Syrian Desert on
15 August 1945, reacting to the news that the war was finally
over. From Hector Donohue's article,
"Unpicking the Goldsworthy Myths."

Following extensive Third Fleet Operations against the Japanese
Empire in July and early August 1945, Japan was finished as a war-
making nation, in spite of its four million men still under arms. But
Japan's leaders were determined to fight on, because, seemingly, to not
lose "face" was more important than the hundreds of thousands of lives
that would be lost. To continue was no longer a question of military
strategy, but an aspect of Japanese culture and psychology.

As American President Harry S. Truman was wrestling with the
decision to drop an atomic bomb on the city of Hiroshima on Honshu
Island, the Japanese government was mobilizing the entire population
of Japan to impose massive casualties on any American invasion.
Operation KETSU-GO called for the use of the Civilian Volunteer
Corps—comprised of all boys and men ages fifteen to sixty and all girls
and women seventeen to forty, except for those exempted as unfit.
These men and women received weapons training in the use of hand
grenades, swords, sickles, knives, fire hooks, and bamboo spears. Led
by regular forces, they were to actively defend a few selected beach
areas, and then to mass reserves for an all-out counterattack if the
invasion forces succeeded in winning a beachhead. A Japanese slogan
in 1945 was "The sooner the Americans come, the better.... One
hundred million [Japanese will] die proudly."[1]

Diagram 25-1

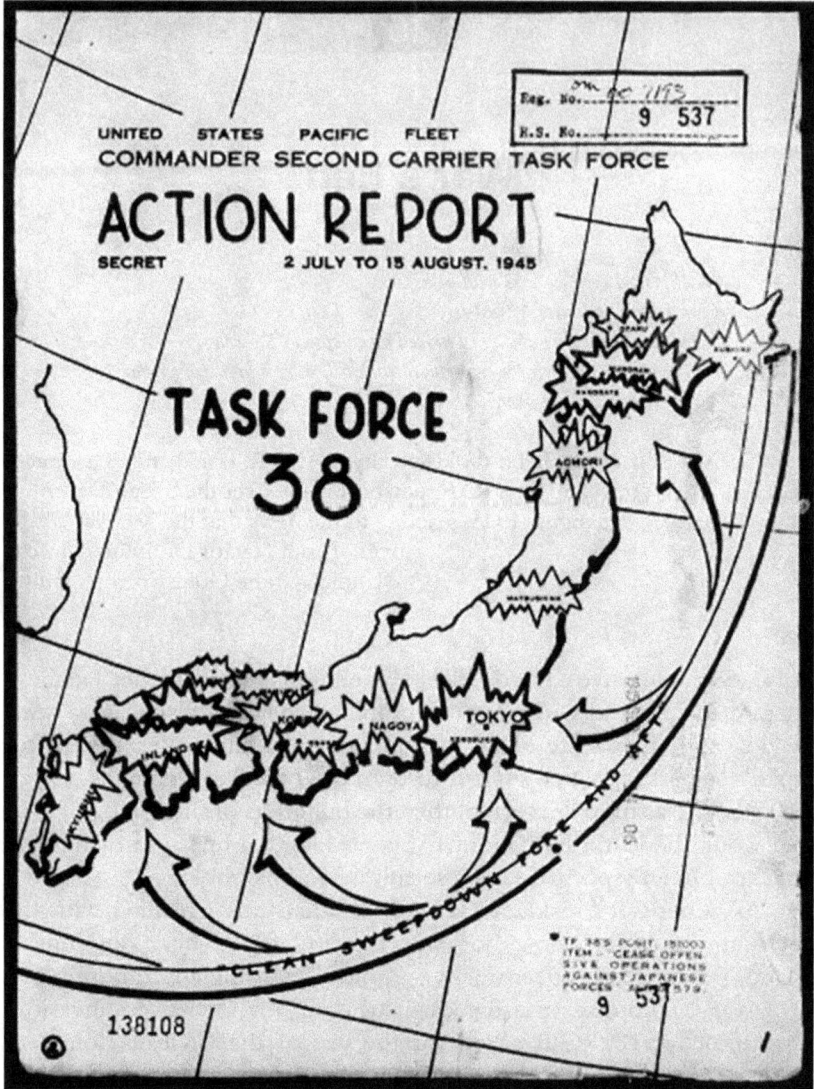

UNITED   STATES   PACIFIC   FLEET
COMMANDER SECOND CARRIER TASK FORCE

ACTION REPORT

SECRET   2 JULY TO 15 AUGUST, 1945

TASK FORCE 38

The cover of a report by Vice Adm. John S. McCain, USN, commander, Task Force 38, on the operations from 2 July-15 August 1945 of his fast aircraft carriers against Japan, graphically depicts the last naval actions in the Pacific.

USE OF ATOMIC BOMBS ON JAPAN ENDS WAR

On 6 August 1945, the American B29 bomber "Enola Gay" dropped a five-ton bomb over Hiroshima, reducing four square miles of the city to ruins and immediately killing 80,000 people. Tens of thousands more died in the following weeks from wounds suffered and from radiation poisoning. Despite these horrific civilian casualties, the Japanese government refused to capitulate. As a result, a bomb was dropped three days later on the city of Nagasaki, on the northwest coast of the island of Kyushu, killing nearly 40,000 more people. Thereafter, Emperor Hirohito accepted the inevitable defeat of Japan. At noon on 15 August, a recording of his formal announcement of surrender was transmitted from the Imperial Palace in Tokyo all over the Japanese Empire. In the broadcast, Hirohito stated in part:

> Moreover, the enemy has begun to employ a new and most cruel bomb, the power of which to do damage is, indeed, incalculable, taking the toll of many innocent lives. Should we continue to fight, not only would it result in an ultimate collapse and obliteration of the Japanese nation, but also it would lead to the total extinction of human civilization.[2]

That same day, 15 August, Adm. William F. Halsey, commander, Third Fleet, received orders from Fleet Adm. Chester W. Nimitz, commander in chief Pacific Fleet, to cease all operations against Japan. The cease fire order was received too late to stop the first of the day's air strikes against Tokyo, but the second strike which had been launched, was recalled in time. Two weeks later, Nimitz arrived in Tokyo Bay aboard a PB2Y Coronado seaplane, in preparation for Japan's formal surrender aboard the battleship *Missouri* (BB-63).[3]

SURRENDER CEREMONY IN TOKYO BAY

> *It is my earnest hope, and indeed the hope of all mankind, that from this solemn occasion a better world shall emerge out of the blood and carnage of the past—a world dedicated to the dignity of man and the fulfillment of his most cherished wish for freedom, tolerance and justice.*

—Closing remark made by Gen. Douglas MacArthur during a speech at the Surrender Ceremony of Japan aboard the battleship USS *Missouri* (BB-63) on 2 September 1945.

Photo 25-1

Fleet Adm. Chester W. Nimitz, USN, arrives at Tokyo Bay in a seaplane on 29 August 1945. The battleship *Missouri* (BB-63) is in the background. U.S. Naval History and Heritage Command Photograph #NH 96809

Photo 25-2

Nimitz signs the formal instrument of Japanese surrender as representative of the United States.
Source: Fleet Adm. Ernest J. King, *United States at War Final Official Report to the Secretary of the Navy covering the period March 1, 1945, to October 1, 1945*

Present in Tokyo Bay on that day were 255 Allied naval ships and merchant vessels. Among them were 31 mine warfare ships; of which four were Royal Australian Navy minesweepers.

Light Minelayer (DM)

USS *Gwin* (DM-33)	USS *Thomas E. Fraser* (DM-24)

Minesweeper, High Speed (DMS)

USS *Ellyson* (DMS-19)	USS *Hopkins* (DMS-13)
USS *Fitch* (DMS-25)	USS *Jeffers* (DMS-27)
USS *Gherardi* (DMS-30)	USS *Macomb* (DMS-23)
USS *Hambleton* (DMS-20)	

Minesweeper (AM)

HMAS *Ballarat* (J184)	USS *Pheasant* (AM-61)
HMAS *Cessnock* (J175)	USS *Pochard* (AM-375)
HMAS *Ipswich* (J186)	USS *Revenge* (AM-110)
HMAS *Pirie* (J189)	USS *Token* (AM-126)
	USS *Tumult* (AM-127)

Motor Mine Sweeper (YMS)

YMS-177	YMS-343	YMS-390	YMS-441
YMS-268	YMS-362	YMS-415	YMS-461
YMS-276	YMS-371	YMS-426	YMS-467

Auxiliary Mine Layer (ACM)

USS *Picket* (ACM-8)[4]

Photo 25-3

HMAS *Ballarat*, part of the 21st Minesweeping Flotilla, was the last Australian ship to leave Singapore before the surrender of the British Colony on 15 February 1942. Australian War Memorial photograph 101157

LEST WE FORGET

The following twenty-nine mine warfare ships (26 USN, 3 RAN) were scuttled by their crews to prevent capture; or were lost to mines or as a result of combat in the Pacific Theater, or during post-war mine clearance operations. Vessels lost to collisions at sea, groundings, or which foundered in typhoons are not included in the tables.

U.S. Navy Minelayers

Date	Ship	Location/Disposition
17 Oct 44	*Montgomery* (DM-17)	Scrapped after being damaged by a mine off Palau, Caroline Islands
18 Feb 45	*Gamble* (DM-15)	Damaged by aircraft bombs off Iwo Jima on 18 February, and scuttled off Saipan, 16 July 1945

U.S. Navy Minesweepers

8 Dec 41	*Penguin* (AM-33)	Sunk by Japanese aircraft off Guam, M.I.
10 Dec 41	*Bittern* (AM-36)	Sunk by aircraft bombs at Cavite, Luzon, P.I.
10 Apr 42	*Finch* (AM-9)	Sunk by Japanese aircraft off Corregidor, Luzon
4 May 42	*Tanager* (AM-5)	Sunk by shore batteries off Corregidor
5 May 42	*Quail* (AM-15)	Scuttled off Corregidor
13 Sep 44	*Perry* (DMS-17)	Sunk by a mine off Palau, Caroline Islands
24 Sep 44	*YMS-19*	Sunk by a mine off Palau
6 Jan 45	*Hovey* (DMS-11)	Sunk after being torpedoed by Japanese aircraft in Lingayen Gulf, Luzon, Philippine Islands
6 Jan 45	*Long* (DMS-12)	Sunk by kamikaze attack in Lingayen Gulf
7 Jan 45	*Palmer* (DMS-5)	Sunk by Japanese aircraft in Lingayen Gulf
14 Feb 45	*YMS-48*	Sunk by shore batteries in Manila Bay, Luzon
28 Mar 45	*Skylark* (AM-63)	Sunk by a mine off Okinawa, Ryukyu Islands
3 Apr 45	*YMS-71*	Sunk by a mine off Brunei, Borneo
6 Apr 45	*Emmons* (DMS-22)	Sunk after being hit by five kamikaze aircraft off Okinawa
8 Apr 45	*YMS-103*	Sunk by a mine off Okinawa
22 Apr 45	*Swallow* (AM-65)	Sunk after hit by a kamikaze off Okinawa
2 May 45	*YMS-481*	Sunk by shore batteries off Tarakan, Borneo
8 Jun 45	*Salute* (AM-294)	Sunk by a mine off Brunei, Borneo
18 Jun 45	*YMS-50*	Sunk by a mine off Balikpapan, Borneo
26 Jun 45	*YMS-39*	Sunk by a mine off Balikpapan
26 Jun 45	*YMS-365*	Sunk by a mine off Balikpapan
8 Jul 45	*YMS-84*	Sunk by a mine off Balikpapan
1 Oct 45	*YMS-385*	Sunk by a mine off Ulithi, Caroline Islands
29 Dec 45	*Minivet* (AM-371)	Sunk by a mine in Tsushima Strait, Japan[5]

Royal Australian Navy Minesweepers

1 Dec 42	*Armidale* (J240)	Sunk by Japanese aircraft off Timor
22 Jan 43	*Patricia Cam*	Sunk by Japanese aircraft off Wessel Island, N.T.
13 Sep 47	*Warrnambool* (J202)	Sank by mine near Cockburn Reef, Australia[6]

Postscript

Lt. Comdr. Leon Verdi Goldsworthy, GC, DSC, GM, MID, RANVR.
Australia War Memorial photograph 081383

The authors have endeavored to provide readers a broad swath of information about Allied mine warfare efforts against the Japanese in the Pacific Theater. To do so in detail would require a whole series of books, vice only *Nightraiders*. In order to keep the book to a reasonable length, little material was devoted to U.S. Army Air Force and Royal Air Force minelaying efforts—the former was incredibly important in crippling Japan in the latter stages of the war—and Royal Australian Navy minelaying and minesweeping likewise received scant attention.

With forewords by a distinguished Naval historian, and three flag officers representing Australia, Great Britain, and the United States providing perspective on Allied mine warfare generally, it is unnecessary to use the postscript to offer added information about material already covered. Instead it will be used to introduce readers to, and honor Australian lieutenant commanders Leon V. Goldsworthy and Palgrave E. Carr. Goldsworthy ("Goldy") was a Render Mines Safe officer, and the most highly decorated Australian Naval officer in the war. Carr's diverse service was particularly interesting; readers may recall his association with PBY Catalinas of the RAAF covered in Chapter 17. "Pally," as he was known, was the only serving officer in the Royal Australian Navy to be awarded the Distinguished Flying Cross.

LEON GOLDSWORTHY

Awarded (from left to right) the George Cross, Distinguished Service Cross, George Medal, and Mention in Despatches

Leon Goldsworthy was a mining engineer from Broken Hill, New South Wales prior to the war. After war broke out, his initial attempt to enlist in the Navy was thwarted by his small stature, but his second attempt resulted in his acceptance in March 1941 as a probationary sub-lieutenant in the Royal Australian Naval Volunteer Reserve (RANVR). After being sent to the UK to complete officer training, he volunteered for the Rendering Mines Safe Section of HMS Vernon.[1]

Vernon was a shore station ("stone frigate") at Portsmouth on the site of the old Gunwharf, responsible for mine disposal and mine countermeasures. Her officers and scientific staff achieved several coups involving the capture of mines and the development of countermeasures. One of the earliest of these was the rendering safe and recovery of the first German magnetic mine (Type GA) at

Shoeburyness on 23 November 1939. For this deed, Comdr. John G. D. Ouvry, RN, was decorated with the DSO by King George VI at a ceremony on HMS Vernon's parade ground on 19 December 1939. Others decorated at the same time for this, and other tasks where mines were rendered safe for recovery and examination, were Lt. Comdr. Roger C. Lewis (DSO), Lt. John E. M. Glenny (DSC), CPO Charles E. Baldwin (DSM) and AB Archibald L. Vearncombe (DSM). Of particular note, these were the first Royal Naval decorations of the Second World War.[2]

Photo Postscript-2

19 Dec 1939 – Lt. Comdr. John Ouvry at HMS Vernon showing King George VI the German magnetic ground mine he rendered safe on the mudflats at Shoeburyness on 23 November 1939. The projections were intended to prevent the cylindrical mine rolling across the seabed. Courtesy of Rob Hoole

In August 1941, Goldsworthy joined the Admiralty Mine Disposal
Section based in London; which became known as the Land Incident
Section, which dealt with German mines dropped as bombs from the
air. During these duties, he rendered safe nineteen mines and also
qualified as a diver. As German air attacks on English cities began to
decline, Goldsworthy transferred to the Enemy Mining Section, based
at HMS Vernon, in January 1943.[3]

Photo Postscript-3

Lowestoft, June 1943. Vernon Enemy Mining Group posing behind a recently rendered
safe German aerial mine lying on a beach. Lt. Comdr. John Mould (2nd from left) and
Lt. Leon Goldsworthy (far right).
Australian War Memorial PO3434.023

In early 1943, he helped Lt. Comdr. John S. Mould, RANVR, to
develop a diving suit with an independent gas supply, suitable for mine
disposal work. Mould then formed and trained Port Clearance Parties
("P Parties") to clear liberated ports in Europe of munitions dumped in
harbors by retreating German forces in an effort to prevent their use.
Goldsworthy remained at Vernon, working closely with its mine
recovery flotilla, auxiliary vessels fitted out and employed for mine
location and recovery work.[4]

In September and October 1943, Goldsworthy defused two mines,
one of which had been lying under a Southampton wharf for two years
and the other in the River Thames. For this he received the George
Medal. In August 1944 he received a Mention in Despatches for 'Great
courage and undaunted devotion to duty.'   In September 1944,
Goldsworthy earned a George Cross, for his skill and courage of a high

order in recovering and defusing three magnetic and one acoustic ground mine between June 1943 and April 1944. This medal was awarded by the Crown "for acts of the greatest heroism or for most conspicuous courage in circumstance of extreme danger."[5]

Photo Postscript-4

British mine recovery vessel *Esmeralda* at Cherbourg, France, August 1944. Australian War Memorial PO3434.030

Shortly after the Allied invasion of France, Goldsworthy (based aboard *Esmeralda* of the Mine Recovery Flotilla) joined the P Parties to undertake mine disposal, underwater demolition, and other diving tasks off the Normandy coast. There, he rendered safe the first Katey mine in Cherbourg Harbor, and three mines on the British assault beaches. The former, particularly sinister weapon, appeared innocuous: a metal rod tripod supporting a single "hertz horn" mounted on top of a concrete block containing ten kilos of explosive. A snag line (so called because it was intended to ensnare a passing vessel) pulled a lever which broke the horn's acid vial. To approach the mine, it was necessary for Goldsworthy to swim through a hundred yards of giant weed. The water was intensely cold and the rumbles of distant underwater explosions did not add to comfort. Despite these obstacles and the very real possibility of booby traps, the mine was rendered safe.[6]

For his actions, he was awarded a Distinguished Service Cross in January 1945, "for gallantry and distinguished services in the work of mine-clearance in the face of the enemy." As the P Parties followed the advance of Allied forces in Europe, Goldsworthy remained with the flotilla, dealing with mines in the waters around the English coast.[7]

TRANSFER TO THE PACIFIC

Goldsworthy and Geoffrey John "Jack" Cliff, also RANVR, were both promoted acting lieutenant commanders in September 1944, and sent in October to the Pacific as British Naval Liaison and Intelligence officers. Attached to the U.S. Navy's Mobile Explosive Investigation Unit (MEIU) No. 1, their task was to learn about U.S. search, recovery and disposal techniques, and to forward samples of Japanese ordnance, particularly torpedoes and mines, to the UK.[8]

This action came about following a request from the U.S. Embassy to the Director of Torpedoes and Mining at the Admiralty that two RANVR officers serving in England visit the United States to exchange information on mine disposal. At completion, they were to proceed for duty as liaison officers to assist MEIU No. 1 in the Pacific with mine recovery operations. In approving this petition, Capt. Gervase B. Middleton, RN, noted in his letter reply:

> LCDR Cliff and LCDR Goldsworthy have been given special courses in underwater weapons and their recovery and are well qualified for the work required of them.
>
> Cliff now serving in Land Incident Section and can be spared without relief. Goldsworthy is serving in Enemy Mining Section Vernon and will have to be relieved.
>
> DTM welcomes this request for the attachment of these 2 officers to the US MEIU for they will make liaison a great deal easier and ensure that details of enemy weapons recovered by the US Forces are available in this country sooner than they have been in the past. The experience in this country will also be of great value to US Authorities.[9]

Goldsworthy and Cliff reported for duty on 4 February 1945 to MEIU No. 1, based in Australia on the southern side of the Brisbane River. Soon after the Render Mines Safe officers engaged in intelligence gathering, and "hands-on" work defusing Japanese mines and booby-traps in the Philippines and New Guinea area. On 9 May 1945 in connection with landings in the Borneo area, Goldsworthy boarded USS *YP-421* at the small Philippine island of Nalunga, across the Sulu Sea from Borneo. The former Massachusetts beam trawler commanded by Walter E. Baker, USN, was assigned to support MEIU No. 1.[10]

BORNEO CAMPAIGN

The jungle fighters of the AIF had long since learned to shoot first and ask questions afterwards. We made a quick detour to the quartermaster to be knitted out with a couple of slouch hats.

—Lt. Comdr. Leon Goldsworthy remarking on the necessity to be identifiable to Australian army patrols, after he and Lt. Nicholas H. Dosker Jr., USNR (MEIU No. 1), found themselves under fire, after getting in front of a patrol, during an investigation of the coast for Japanese weapons or ordnance.[11]

The objectives of the Allied Borneo Campaign of 1945 (1 May-15 August) were to deny Japan the continued fruits of its conquests in the Netherlands East Indies, and use of the approaches to those areas. To achieve these aims, an Australian-led force captured Tarakan Island to provide an airfield for support of an assault on Balikpapan, and seized Brunei Bay for an advanced fleet base that could protect resources in the area. The final phase was to occupy Balikpapan for its naval air and logistic facilities as well as its petroleum installations.[12]

Many sea mines were encountered in the campaign, and during the final amphibious assault of the last major Allied campaign in the Southwest Pacific area in World War II, twenty-four American minesweepers earned Presidential Unit Citations for extraordinary heroism. The bravery of the crewmen of the "sweeps" at Balikpapan was particularly notable because it came on the heels of heavy losses and damage to minesweepers by mines and Japanese artillery fire at Tarakan Island and Brunei Bay. Due to the perseverance of the American minesweepers and Underwater Demolition Team 11, the amphibious landing was made at Balikpapan on 1 July as scheduled.[13]

After an intense two-hour bombardment, assault waves moved ashore, landing without a single casualty in spite of enemy artillery, mortar and small arms fire. The troops met with increased resistance as they progressed inland and fire support was provided by cruisers day and night, and continued through 7 July. A subsequent landing was made at Cape Penajam on 2 July without casualties. However, the minesweepers paid a heavy price off Balikpapan.[14]

During the next fifteen days, some twenty "sweeps" accounted for another twenty-six mines, one of which sent *YMS-84* to the bottom— the last surface ship of the U.S. Navy to be sunk in that distant corner of the Pacific. During the Borneo Campaign, the U.S. Pacific Mine

Force lost *Salute* (AM-294), *YMS-39, 50, 84, 365,* and *481,* and suffered damage to twelve others—*YMS-10, 47, 49, 51, 314, 329, 334, 335, 339, 363, 364,* and *368.*[15]

Map Postscript-1

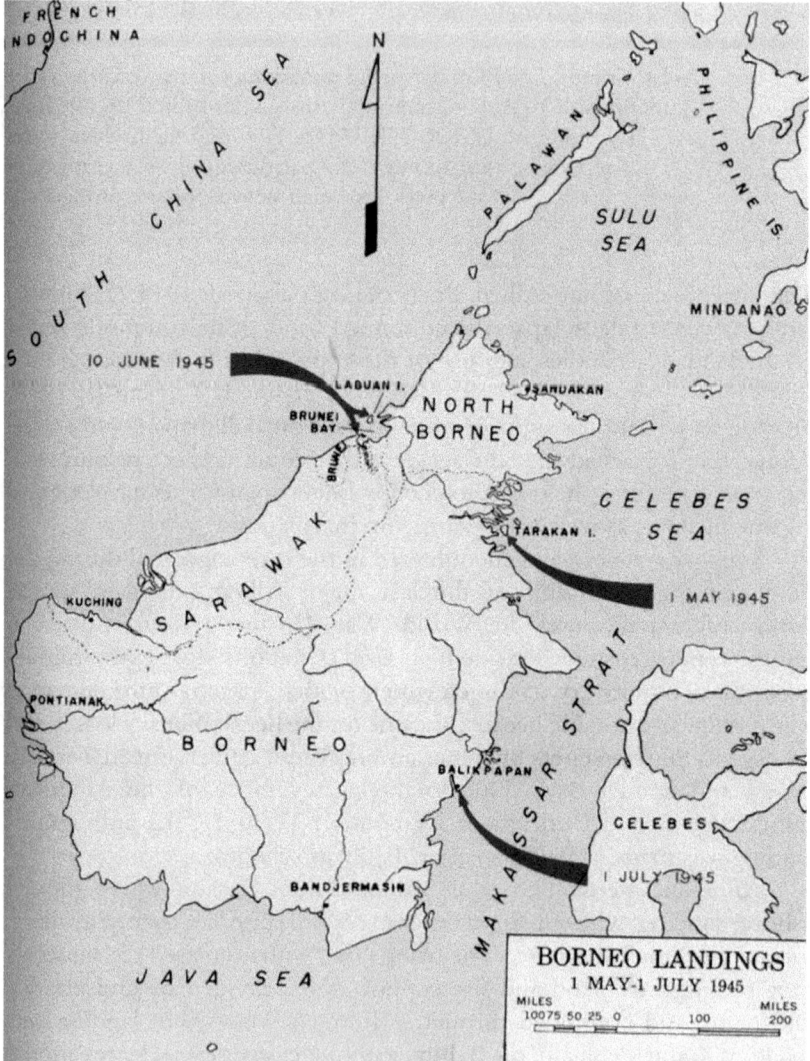

Movement of Australian assault forces en route to landings at Tarakan Island, at Labuan Island in Brunei Bay, and at Balikpapan during the 1945 Borneo Campaign.
www.lib.utexas.edu/maps/historical/engineers_v1_1947/borneo_landings_1945.jpg

It is unclear how many mines were disarmed by EOD experts at Balikpapan. Once Goldy got ashore, he wasted no time in examining enemy ordnance and armament. Having been cut off at Balikpapan by an Allied blockade of the port, the Japanese had constructed make-shift defenses. One of the most intriguing contraptions was a torpedo launching rig in lieu of normal torpedo tubes. A double-railed track had been constructed to move torpedoes from a storage pit through a tunnel in the sand. This was done through the use of a homemade wooden-wheeled trolley carrying a 21-inch torpedo to which two empty 40-gallon drums were lashed with woven grass rope. A section of each lashing was laid over a wooden block, above which was mounted a sharp, but primitive axe on a pivot.[16]

Apparently, the trolley was started rolling down the tracks by a vigorous push. At water's edge, the torpedo, with its affixed drums, would float free of the trolley. Swimmers would then point the torpedo in the direction of its intended course, and start the weapon's compressed air propulsion motors. Away would go the torpedo, at least in theory, under its own power, until the line tightened, tripping the axe handle and severing the drum lashings.[17]

Goldsworthy returned to the United Kingdom in August 1945 to close down the P Parties. In December, he was appointed to the British Technical Mission to Japan to assist in compiling a report on enemy underwater weapons. He returned home to Australia aboard HMS *Formidable* in April 1946 and separated from the Navy the following month, as part of the demobilization of Australian military personnel. By his account, Goldsworthy rendered safe thirty-three ground mines, the Katey mine off Cherbourg, and a large number of moored mines and armed conical floats during the war.[18]

ROYAL AUSTRALIAN NAVY BOMB AND MINE (BMD) DISPOSAL OPERATIONS IN THE PACIFIC THEATER

While Goldsworthy was out of country, RAN BMD personnel in the Pacific theater supported Allied operations in the islands to Australia's north. Whilst much has been written about the few RANVR officers who conducted BMD activities in the UK and Europe (and their bravery and well-deserved awards, given to 12 officers – including four GCs - and two sailors, is acknowledged), little has been written about the men who conducted BMD in the Australian theater.[19]

Following Japan's initiation of war in the Pacific, the RAN introduced Render Mines Safe (RMS) training at Flinders Naval Depot in early 1942, and RAN personnel were trained by the Army in bomb disposal. BMD personnel were deployed to vital ports and eventually in support of operations in the islands to the north of Australia. While RMS operations by RANVR officers in Europe dealt mostly with sophisticated German ground mines, BMD in support of Allied offensives focused on Japanese unexploded ordnance and booby traps. Accordingly, those involved lived in the field, often close to fierce fighting, as evidenced by Billman landing with the Sixth Army at Leyte.[20]

Photo Postscript-5

Lt. (Sp) Sydney Allan Arnold RANVR, and Lt. Maurice Samuel Batterham, RANVR, inspecting a Japanese JA mine, June 1943. Australian War Memorial photograph 015093

Photo Postscript-6

A Japanese mine in New Britain around November 1946. Batterham is on the left and Lt. Charles George Croft, RANVR, on the right. Australian War Memorial photograph 015094

No lists of personnel qualified in RMS and bomb disposal during the War were kept, but in excess of 30 RMS officers have been identified from Navy Lists and there would have been at least a similar number of sailors.   Awards for RAN members who carried out BMD operations in the Pacific during WWII and in the immediate post-war period were sparse; four officers and two sailors as shown below.

Name	Award	Received
Comdr. (Sp) Maurice Samuel Batterham, RANVR	OBE	5 Jun 1952
Lt. (Sp) Harold Leon Billman, RANVR	MID	26 Jun 1945
	DSC	6 Nov 1945
Lt. Charles George Croft, RANVR	MBE	28 Sep 1943
Lt. (Sp) Francis Nankivell, RANVR	MBE	26 Jun 1945
Able Seaman Reginald Frank Peel, RANR	BEM	26 Jun 1945
Able Seaman Victor William Turner, RAN	BEM	1 Jan 1952[21]

BEM: British Empire Medal
DSC: Distinguished Service Cross
MBE: Member of the Order of the British Empire
MID: Mention in Despatches
OBE: Officer of the Order of the British Empire

Photo Postscript-7

Lt. Charles George Croft, RANVR, 21 January 1946.
Australia War Memorial photograph 099763

PALGRAVE CARR

A great and unbelievably courageous Australian.

—Tribute paid Palgrave Ebden Carr at his funeral in August 1990 by
Air Commodore William Henry Garing, CBE, DFS, RAAF (Rtd.)[22]

**Earned RAAF pilot wings; awarded
the Distinguished Flying Cross**

Palgrave Carr was selected in 1922 to attend the Royal Australian Navy
College at Jervis Bay, on the south coast of New South Wales. There,
he excelled as Chief Cadet Captain, was the King's Medalist winner as
the best all round cadet, and won the Governor General's Cup for the
Best Individual All Round Sportsman. After graduating with distinction
in 1925, "Pally" was sent to the UK, where he spent four years at sea
and in various courses of instruction, returning to Australia in 1930 as a
sub-lieutenant.[23]

Volunteers were sought for secondment (temporary assignment) to
the RAAF, training as pilots. Carr offered his services, was accepted
and spent two years flying seaplanes at Point Cook, Melbourne, the
birthplace of the Royal Australian Air Force. Unfortunately, Service
rules precluded RAN officers from taking up the duties of RAAF pilots,
who were then flying the RAN's seaborne aircraft. A lieutenant in the
RAN, Carr was given the honorary rank in the RAAF of flight lieutenant
for which he dressed during the day. He qualified for his pilot's Brevet
wings in 1932, and in 1933 flew with Air Commodore Garing.[24]

"Pally" returned to sea service in 1935 as a fully qualified Naval pilot.
Unable to fly with the RAN, the Naval Board offered him an Observer's
course in the UK, which he accepted. Following graduation, he spent
four years serving in the aircraft carriers HMS *Furious*, *Courageous*, and
Ark Royal. In July 1939, Carr joined HMAS *Perth* en route to Australia.
During the voyage, war became imminent, and the light cruiser had no
planes. Accordingly, while at Jamaica, Carr transferred to HMS *Orion*,
also a light cruiser, but which had two Seafox floatplanes embarked and
he flew them for six months.[25]

Photo Postscript-8

British cruiser HMS Orion at a U.S. southern port, circa 1937.
Seaplane is a Fairey "Seafox" serial #K-8571.
Naval History and Heritage Command photograph #NH 57582

In 1940, "Pally" returned to Australia to serve briefly aboard the armed merchant cruiser HMAS *Manoora* and the auxiliary cruiser HMAS *Westralia*, and for a short period at Pearce RAAF base, in Bullsbrook to the north of Perth in Western Australia. At year's end, Carr was sent back to Britain on loan to the RN. Following duty at Naval Air Stations at Arbroath and Crail, he was appointed to the aircraft carrier HMS *Hermes*, operating in the Indian Ocean. One day after he departed her for Australia, *Hermes* was sunk by aircraft from a Japanese carrier.[26]

DUTY IN AUSTRALIA WITH PBY CATALINAS

Following his arrival in Australia, Carr joined the heavy cruiser HMAS *Australia*, taking part in the Guadalcanal Campaign and Battle of Savo Island. At the end of 1942, he was again attached to the RAAF, and served briefly with No. 100 Torpedo Bomber Squadron based at Milne Bay, New Guinea. "Pally" was posted next to the RAAF command at Allied headquarters in Brisbane. There, he initiated, with the USN, the highly-secret offensive minelaying campaign, using PBYs of No.11 and 12 Squadrons RAAF. These operations are described in Chapter 17.[27]

In September 1943, a Catalina in which Carr was flying was shot down off Pomala, in the Celebes. His mission had been to test under operational conditions, a new American torpedo. "Pally" was able to save one other crewman from the downed PBY, but he expired within thirty-six hours of being brought ashore. Carr was captured by the Japanese and shipped to the secret Ofuna interrogation center, inland from Yokosuka. There, prisoners thought to hold critical intelligence— mostly officers and any men who were submariners and airmen—were placed under a strict regimen designed to make them break. Along with routine beatings, Japanese interrogators used solitary confinement, sleep deprivation, meager rations, poor medical care, uncomfortable positions and other techniques to make prisoners talk.[28]

Photo Postscript-9

Entrance to the main building at the Allied POW camp at Ofuna, 30 August 1945. Australian War Memorial photograph 121297

In this hell hole, "Pally" endured eight months of his two years as a POW without breaking. Following Japan's defeat, he was repatriated to Australia and thereafter retired from the service. His bravery and devotion to duty were recognized by his award of the DFC.[29]

Appendix A: Japanese Ships Sunk by Allied Mines

Japanese Merchant and Naval Vessels Sunk by Australian Mines

Date	Ship	Displ. (tons)	Location
24 Jul 43	Cargo ship *Mie Maru*	2,913	2°45'S, 133°40'E
11 Sep 43	Minesweeper *No. 16*	492	6°08'S, 119°20'E
16 Sep 43	Converted gunboat *Seiksi Maru*	2,633	2°30'S, 150°48E
4 Nov 43	Cargo ship *Ryuosan Maru*	2,455	2°40'S, 150°40'E
4 Nov 43	Surveying ship *Tsukushi*	2,000	2°40'S, 150°40'E
23 Jul 44	Cargo ship *Takasan Maru*	1,428	4°35'S, 122°17'E
26 Oct 44	Cargo ship *Seito Maru*	2,219	3°14'S, 116°13'E
29 Oct 44	Cargo ship *Kokko Maru*	2,863	1°17'S, 116°48'E
3 Mar 45	Miscellaneous auxiliary *Hario*	4,000	18°10'N, 109°40'E[1]

Japanese Merchant and Naval Vessels Sunk by Dutch Mines

Date	Ship	Displ.	Location
18 Dec 41	Destroyer *Shinonome*	1,950	4°24'N, 114°00'E
2 Feb 42	Minesweeper *No. 9*	630	3°42'S, 128°10'E[2]

Japanese Merchant and Naval Vessels Sunk by British Mines

Date	Ship	Displ.	Location
Oct 42	Submarine *I-30*	2,212	1°00'N, 105°00'E
9 Sep 44	Special submarine chaser *No. 8*	100	3°54'N, 98°44'E
10 Sep 44	Tanker *Takekun Maru*	3,029	3°54'N, 98°43'E
1 Jan 45	Cargo ship *Kyokko Maru*	593	12°26'N, 98°38'E
23 Jan 45	Tanker *Hozan Maru No. 1*	868	4°08'N, 98°15'E
23 Jan 45	Cargo ship *Nikkaku Maru*	1,946	4°08'N, 98°15'E
24 Feb 45	Cargo ship *Kyuryu Maru*	1,339	13°45'N, 100°35'E
27 Mar 45	Special minelayer *No. 1*	500	3°52'N, 98°45'E
30 Apr 45	Tanker *Yuno Maru*	2,345	0°58'S, 104°31'E
29 May 45	Cargo ship *Etsunan Maru* (1)	880	10°30'N, 99°24'E[3]

Note: (1) Sunk by Army mine.

Japanese Vessels Sunk by American Mines

Date	Ship	Displ.	Location
20 Dec 42	Unknown Maru (cargo ship)	4,000	35°45'N, 140°55'E
29 Dec 42	Passenger-cargo ship *Fukken Maru*	2,558	20°04'N, 109°18'E
17 Apr 43	Cargo ship *Tatsunan Maru*	6,417	6°50'S, 155°45'E
8 May 43	Destroyer *Kuroshio*	1,900	8°05'S, 156°55'E
8 May 43	Destroyer *Oyashio* (1)	1,900	8°05'S, 156°55'E
8 May 43	Destroyer *Kagero* (1)	1,900	8°05'S, 156°55'E
27 Jul 43	Cargo ship *Teikin Maru*	1,972	19°57'N, 109°05'E
26 Oct 43	Cargo ship *Shozan Maru* (2)	2,937	20°52'S, 106°40'E
12 Dec 43	Cargo ship *Kanjo Maru* (2)	2,197	22°37'S, 120°15'E
15 Feb 44	Passenger-cargo ship *Ryoka Maru* (2)	5,307	31°16'S, 121°45'E
22 Feb 44	River gunboat *Francis Garnier* (2)	639	10°30'N, 108°00'E
8 Mar 44	Cargo ship *Juyo Maru* (2)	5,457	13°10'N, 100°50'E
23 Apr 44	Destroyer *Amagiri*	1,950	2°10'S, 116°45'E
26 May 44	Cargo ship *Akishima Maru* (2)	1,993	33°48'N, 131°33'E
1 Jul 44	Cargo ship *Nikko Maru*	3,098	5°39'S, 119°28'E
10 Jul 44	Cargo ship *Iyang Maru* (2)	943	31°00'N, 122°30'E
11 Aug 44	Tanker *Ikuta Maru* (2)	1,018	2°56'S, 104°56'E
29 Aug 44	Passenger-cargo *Tientsin Maru* (2)	2,349	31°23'N, 121°30'E
30 Aug 44	Cargo ship *Iwaishima Maru* (2)	695	33°07'N, 119°59'E
1 Sep 44	Cargo ship *Namba Maru* (2)	2,300	31°10'N, 121°50'E
6 Sep 44	Cargo ship *Eiji Maru* (2)	6,968	22°19'N, 120°30'E
7 Sep 44	Cargo ship *Kokka Maru* (2)	5,300	25°12'N, 121°45'E
9 Sep 44	Passenger-cargo *Sainan Maru* (2)	3,232	31°23'N, 121°30'E
10 Sep 44	Tanker *Hoei Maru No. 2* (2)	834	31°23'N, 121°37'E
6 Oct 44	Gunboat *Saga* (2)	990	22°17'N, 114°10'E
9 Oct 44	Cargo ship *Hato Maru* (2)	880	31°10'N, 122°25'E
25 Nov 44	Submarine *RO-100*	525	7°00'S, 156°00'E
8 Jan 45	Cargo ship *Seikai Maru No. 2* (2)	550	26°35'N, 142°08'E
18 Jan 45	Passenger-cargo *Reizan Maru* (2)	3,426	30°11'N, 115°07'E
26 Jan 45	Cargo ship *Tamon Maru No. 15*	6,925	13°34'N, 109°17'E
27 Jan 45	Cargo ship *Hsin Yang Maru* (2)	1,674	29°40'N, 115°52'E
27 Jan 45	Passenger-cargo *Ryuzan Maru* (2)	2,482	29°46'N, 116°52'E
6 Feb 45	Tanker *Oei Maru* (2)	2,858	1°00'N, 103°36'E
22 Feb 45	Tanker *Tatekawa Maru No. 2* (4)	10,000	11°08'N, 108°44'E
11 Mar 45	Cargo ship *Koko Maru* (2)	1,520	31°22'N, 121°34'E
17 Mar 45	Cargo ship *Bansei Maru* (2)	3,120	31°19'N, 121°45'E
19 Mar 45	Tanker *Sarawak Maru* (2)	5,135	1°20'N, 104°34'E
19 Mar 45	Cargo ship *Kozan Maru* (2)	3,923	32°05'N, 119°56'E
19 Mar 45	River gunboat *Suma* (2)	625	32°00'N, 120°00'E
3 Apr 45	Cargo ship *Yamabishi Maru* (2)	880	34°00'N, 131°00'E
4 Apr 45	Cargo ship *Hozan Maru* (2)	1,175	33°45'N, 131°46'E
4 Apr 45	Submarine *RO-64* (2)	988	34°14'N, 132°16'E
4 Apr 45	Submarine *RO-67* (2)	988	34°00'N, 133°00'E
4 Apr 45	Frigate *Mokuto RO-64* (2)	900	33°53'N, 131°03'E
5 Apr 45	Cargo ship *Nichinan Maru* (2)	880	33°58'N, 131°02'E
6 Apr 45	Passenger-cargo *Fushimi Maru* (2)	1,229	33°59'N, 130°52'E

6 Apr 45	Passenger-cargo *Hsing Yun Maru* (2)	3,414	32°05'N, 119°56'E
7 Apr 45	Cargo ship *Hatsukari Maru* (2)	875	33°55'N, 130°50'E
8 Apr 45	Cargo ship *Tamon Maru No. 12* (2)	2,220	31°16'N, 121°29'E
10 Apr 45	Cargo ship *Maruko Maru* (2)	3,000	31°15'N, 121°29'E
13 Apr 45	Cargo ship *Shinro Maru* (2)	877	33°58'N, 131°02'E
15 Apr 45	Cargo ship *Sannan Maru* (2)	873	33°58'N, 131°02'E
17 Apr 45	Cargo ship *Sonjo Maru* (2)	800	31°13'N, 121°52'E
18 Apr 45	Cargo ship *Teizui Maru* (2)	8,426	34°03'N, 130°50'E
20 Apr 45	Cargo ship *Sanko Maru* (2)	847	33°48'N, 131°00'E
20 Apr 45	Cargo ship *Yamamitsu Maru* (2)	876	34°32'N, 128°44'E
27 Apr 45	Cargo ship *Kaiho Maru* (2)	2,563	34°00'N, 130°50'E
28 Apr 45	Cargo ship *Gakujo Maru* (2)	2,220	34°00'N, 130°50'E
4 May 45	Tanker *Yaei Maru No. 2* (2)	834	2°00'N, 105°00'E
5 May 45	Cargo ship *Manshu Maru* (2)	5,266	33°47'N, 131°35'E
6 May 45	Cargo ship *Sagamigawa Maru* (2)	6,880	34°00'N, 130°50'E
7 May 45	Minesweeper *No. 29* (2)	630	34°02'N, 130°54'E
7 May 45	Cargo ship *Harushio Maru* (2)	1,989	34°06'N, 130°47'E
7 May 45	Cargo ship *Kashima Maru* (2)	870	34°02'N, 130°54'E
7 May 45	Passenger-cargo ship *Teika Maru* (2)	8,009	34°06'N, 130°47'E
7 May 45	Passenger-cargo *Shofuku Maru* (2)	1,768	33°59'N, 130°52'E
8 May 45	Passenger-cargo *Kotobuki Maru* (2)	18,765	34°30'N, 126°30'E
10 May 45	Passenger-cargo *Tatsuwa Maru* (2)	6,332	34°05'N, 132°27'E
11 May 45	Passenger-cargo *Kitsurin Maru* (2)	6,783	34°39'N, 135°11'E
11 May 45	Cargo ship *Tamae Maru* (2)	1,282	34°06'N, 130°47'E
12 May 45	Cargo ship *Brazil Maru* (2)	5,859	34°40'N, 135°12'E
12 May 45	Cargo ship *Nissho Maru No. 1* (2)	1,362	34°06'N, 130°47'E
12 May 45	Cargo ship *Manbo Maru* (2)	1,253	34°30'N, 135°10'E
13 May 45	Cargo ship *Gyoryu Maru* (2)	2,180	34°40'N, 135°10'E
13 May 45	Cargo ship *Kinoto Maru* (2)	1,561	34°39'N, 135°11'E
13 May 45	Tanker *Takasago Maru* (2)	834	34°39'N, 135°11'E
14 May 45	Cargo ship *Miyajima Maru* (2)	2,000	34°20'N, 134°50'E
15 May 45	Cargo ship *Anko Maru* (2)	2,821	34°30'N, 134°47'E
15 May 45	Cargo ship *Mishima Maru* (2)	1,934	34°30'N, 135°10'E
16 May 45	Cargo ship *Yamanami Maru* (2)	873	34°36'N, 135°08'E
17 May 45	Cargo ship *Koan Maru* (2)	4,305	34°38'N, 135°11'E
17 May 45	Passenger-cargo *Mikazuki Maru* (2)	1,412	32°05'N, 119°56'E
17 May 45	Cargo ship *Tairyu Maru* (2)	1.912	34°27'N, 135°11'E
18 May 45	Cargo ship *Nanko Maru No. 5* (2)	834	33°58'N, 130°58'E
22 May 45	Passenger-cargo *Uwajima Maru* (2)	744	34°20'N, 134°20'E
23 May 45	Cargo ship *Kimigayo Maru* (2)	670	33°06'N, 129°43'E
23 May 45	Cargo ship *Shinei Maru No. 2* (2)	882	33°58'N, 131°02'E
24 May 45	Cargo ship *Fukuei Maru* (2)	1,868	33°06'N, 129°43'E
24 May 45	Cargo ship *Kaishin Maru* (2)	870	34°01'N, 130°54'E
24 May 45	Tanker *Nanko Maru No. 7* (2)	834	33°55'N, 131°20'E
24 May 45	Cargo ship *Tatsuine Maru* (2)	873	33°58'N, 131°02'E
24 May 45	Cargo ship *Kiyokawa Maru* (2)	6,862	32°44'N, 129°52'E
25 May 45	Cargo ship *Ginzan Maru* (2)	2,220	33°58'N, 131°02'E
25 May 45	Cargo ship *Ginsei Maru* (2)	2,218	33°59'N, 130°52'E
25 May 45	Cargo ship *Hiwaka Maru* (2)	891	33°58'N, 131°02'E

25 May 45	Cargo ship *Kinryuzan Maru* (2)	1,370	34°00'N, 130°50'E
25 May 45	Cargo ship *Koei Maru* (2)	882	33°58'N, 131°01'E
25 May 45	Cargo ship *Shinragi Maru* (2)	3,032	33°58'N, 131°02'E
25 May 45	Cargo ship *Toyu Maru No. 3* (2)	834	33°55'N, 131°20'E
25 May 45	Cargo ship *Tobi Maru* (2)	878	33°58'N, 130°52'E
25 May 45	Cargo ship *Nitcho Maru* (5)	549	34°01'N, 130°54'E
26 May 45	Special sub-chaser No. *172* (2)	100	36°48'N, 137°05'E
26 May 45	Cargo ship *Kaishin Maru No. 9* (2)	875	34°00'N, 130°50'E
26 May 45	Cargo ship *Nissan Maru No. 1* (2)	6,889	33°58'N, 130°52'E
26 May 45	Cargo ship *Miyakawa Maru* (2)	872	33°58'N, 131°02'E
26 May 45	Cargo ship *Mogi Maru* (2)	884	33°43'N, 131°38'E
26 May 45	Cargo ship *Kunuki Maru* (2)	884	34°39'N, 131°11'E
26 May 45	Cargo ship *Washin Maru* (2)	1,870	33°56'N, 131°10'E
26 May 45	Cargo ship *Yoshin Maru* (2)	2,820	34°21'N, 132°27'E
26 May 45	Cargo ship *Shiokubi Maru* (2)	3,747	33°56'N, 131°11'E
26 May 45	Cargo ship *Shozan Maru* (2)	6,890	33°57'N, 131°46'E
27 May 45	Cargo ship *Chizan Maru* (2)	884	34°39'N, 135°11'E
27 May 45	Transport *Kongo Maru* (2)	7,081	33°36'N, 130°25'E
27 May 45	Cargo ship *Kozan Maru* (2)	882	33°30'N, 130°30'E
28 May 45	Coast defense vessel *No. 29* (2)	800	33°07'N, 129°44'E
29 May 45	Cargo ship *Uwatsu Maru* (2)	882	34°00'N, 130°50'E
30 May 45	Cargo ship *Fujitama Maru* (2)	2,218	34°38'N, 135°11'E
30 May 45	Cargo ship *Hakuun Maru* (2)	594	33°36'N, 130°25'E
30 May 45	Passenger-cargo *Jindai Maru* (2)	1,095	33°59'N, 130°52'E
30 May 45	Tanker *Takasago Maru No. 14* (2)	834	35°15'N, 133°44'E
31 May 45	Cargo ship *Peking Maru* (2)	3,011	34°01'N, 130°45'E
31 May 45	Cargo ship *Yoro Maru No. 2* (2)	2,231	33°58'N, 130°56'E
31 May 45	Cargo ship *Yumihari Maru No. 3* (2)	1,154	34°00'N, 130°50'E
31 May 45	Passenger-cargo *Tensho Maru* (2)	601	34°35'N, 135°15'E
1 Jun 45	Cargo ship *Abukumagawa Maru* (2)	6,887	34°35'N, 134°14'F
1 Jun 45	Cargo ship *Goko Maru* (2)	873	33°58'N, 130°41'E
1 Jun 45	Cargo ship *Kasumi Maru* (2)	1,274	33°58'N, 130°59'E
1 Jun 45	Cargo ship *Kenkon Maru No. 7* (2)	880	33°58'N, 130°56'E
1 Jun 45	Cargo ship *Kishun Maru* (2)	1,864	33°56'N, 130°56'E
1 Jun 45	Cargo ship *Seishu Maru* (2)	1,223	34°45'N, 137°05'E
1 Jun 45	Cargo ship *Myosei Maru* (2)	6,886	33°58'N, 131°03'E
1 Jun 45	Tanker *Yoko Maru* (2)	956	34°30'N, 135°10'E
2 Jun 45	Cargo ship *Kashima Maru* (2)	873	33°58'N, 130°42'E
3 Jun 45	Cargo ship *Konei Maru* (2)	884	33°33'N, 129°58'E
3 Jun 45	Cargo ship *Momo Maru* (2)	884	34°20'N, 134°20'E
3 Jun 45	Cargo ship *Taiei Maru* (2)	6,923	33°56'N, 131°11'E
4 Jun 45	Cargo ship *Kifune Maru* (2)	1,606	37°57'N, 137°05'E
4 Jun 45	Cargo ship *Sawa Maru* (2)	1,264	33°59'N, 130°52'E
5 Jun 45	Cargo ship *Taisho Maru* (2)	1,122	33°58'N, 130°52'E
6 Jun 45	Cargo ship *Gassan Maru* (2)	887	34°00'N, 130°50'E
6 Jun 45	Special sub-chaser No. *195* (2)	100	37°10'N, 137°05'E
7 Jun 45	Cargo ship *Yumihari Maru No. 2* (2)	1,186	34°00'N, 130°00'E
9 Jun 45	Cargo ship *Inaura Maru* (2)	873	34°00'N, 131°00'E
12 Jun 45	Cargo ship *Aizan Maru* (2)	2,221	33°38'N, 130°22'E

12 Jun 45	Transport *Shimonoseki Maru* (2)	527	33°36'N, 130°19'E
12 Jun 45	Cargo ship *Fumitsuki Maru* (2)	875	33°57'N, 130°44'E
13 Jun 45	Converted gunboat *Hiyoshi Maru* (2)	1,287	33°57'N, 131°02'E
13 Jun 45	Cargo ship *Matsuo Maru* (2)	1,911	33°56'N, 131°11'E
13 Jun 45	Cargo ship *Koun Maru* (2)	1,665	23°30'N, 113°30'E
13 Jun 45	Cargo ship *Shirahi Maru* (2)	6,872	33°57'N, 130°55'E
14 Jun 45	Tanker *Koryu Maru* (2)	974	33°55'N, 131°05'E
14 Jun 45	Cargo ship *Nissho Maru No. 8* (2)	1,174	33°54'N, 131°07'E
14 Jun 45	Cargo ship *Uwajima Maru No. 18* (2)	885	33°39'N, 130°15'E
16 Jun 45	Cargo ship *Taikyu Maru* (2)	2,128	33°58'N, 130°34'E
18 Jun 45	Passenger-cargo ship *Bisan Maru* (2)	3,420	33°58'N, 130°44'E
18 Jun 45	Cargo ship *Nissho Maru* (2)	1,276	34°00'N, 130°50'E
19 Jun 45	Cargo ship *Kaisei Maru* (2)	873	34°22'N, 130°51'E
20 Jun 45	Cargo ship *Huashan Maru* (2)	2,103	33°38'N, 130°22'E
20 Jun 45	Cargo ship *Kenan Maru* (2)	2,220	33°59'N, 130°48'E
20 Jun 45	Cargo ship *Nanki Maru No. 1* (2)	834	33°58'N, 131°01'E
22 Jun 45	Cargo ship *Yuba Maru* (2)	884	34°00'N, 131°00'E
22 Jun 45	Cargo ship *Shimotsuki Maru* (2)	875	33°58'N, 131°01'E
22 Jun 45	Cargo ship *Taigen Maru* (2)	884	34°00'N, 130°50'E
22 Jun 45	Cargo ship *Takasegawa Maru* (2)	879	33°55'N, 131°20'E
23 Jun 45	Cargo ship *Goshu Maru* (2)	2,211	34°00'N, 130°50'E
23 Jun 45	Cargo ship *Kocho Maru* (2)	6,888	33°50'N, 130°30'E
23 Jun 45	Cargo ship *Sagami Maru* (2)	877	34°00'N, 131°00'E
24 Jun 45	Cargo ship *Katsura Maru* (2)	884	34°19'N, 133°35'E
24 Jun 45	Cargo ship *Kenkon Maru No. 8* (2)	882	33°58'N, 130°52'E
24 Jun 45	Cargo ship *Wakamatsu Maru* (2)	2,722	33°48'N, 131°32'E
24 Jun 45	Cargo ship *Shinka Maru* (2)	873	33°28'N, 130°01'E
25 Jun 45	Cargo ship *Anri Maru* (2)	1,668	33°58'N, 130°50'E
25 Jun 45	Cargo ship *Ungatsu Maru* (2)	887	33°59'N, 130°52'E
27 Jun 45	Cargo ship *Kaishin Maru No. 7* (2)	873	33°50'N, 130°50'E
27 Jun 45	Cargo ship *Mibunesan Maru* (2)	872	34°35'N, 134°15'E
27 Jun 45	Cargo ship *Mifuku Maru* (2)	1,211	37°06'N, 137°02'E
27 Jun 45	Cargo ship *Tsuki Maru No. 1* (2)	875	33°58'N, 131°01'E
28 Jun 45	Cargo ship *Daikokuzan Maru* (2)	692	34°30'N, 130°30'E
28 Jun 45	Cargo ship *Reian Maru* (2)	1,936	33°59'N, 130°52'E
28 Jun 45	Tanker *Takasago Maru No. 12* (2)	834	34°23'N, 130°56'E
28 Jun 45	Destroyer *Enoki* (2)	1,260	35°26'N, 135°44'E
29 Jun 45	Cargo ship *Nichiyu Maru No. 1* (2)	6,873	34°36'N, 130°46'E
29 Jun 45	Cargo ship *Soshu Maru* (2)	1,219	35°41'N, 136°05'E
30 Jun 45	Cargo ship *Taruyasu Maru* (2)	19,254	37°07'N, 137°04'E
1 Jul 45	Cargo ship *Eijun Maru* (2)	2,156	34°02'N, 130°57'E
1 Jul 45	Cargo ship *Naoshima Maru* (2)	873	34°30'N, 134°17'E
1 Jul 45	Cargo ship Yamamichi Maru (2)	6,850	34°28'N, 135°08'E
2 Jul 45	Special sub-chaser *No. 188* (2)	100	35°59'N, 130°52'E
3 Jul 45	Tanker *Hoei Maru* (2)	784	34°22'N, 126°25'E
5 Jul 45	Tanker *Tosei Maru No. 5* (2)	523	37°57'N, 139°04'E
5 Jul 45	Cargo ship *Toyokawa Maru* (2)	5,123	33°56'N, 130°53'E
6 Jul 45	Cargo ship *Tokai Maru No. 5* (2)	879	33°59'N, 130°52'E
6 Jul 45	Cargo ship *Toyo Maru* (2)	3,718	33°58'N, 131°00'E

6 Jul 45	Cargo ship *Ujina Maru* (2)	2,218	37°56'N, 139°04'E
6 Jul 45	Cargo ship *Take Maru* (2)	1,265	33°52'N, 131°15'E
6 Jul 45	Cargo ship *Shori Maru* (2)	2,080	34°06'N, 130°47'E
7 Jul 45	Cargo ship *Kinyu Maru No. 1* (2)	834	34°00'N, 135°00'E
7 Jul 45	Cargo ship *Nachizan Maru* (2)	880	33°55'N, 130°56'E
7 Jul 45	Transport *Tairi Go* (2)	1,832	33°43'N, 131°38'E
7 Jul 45	Destroyer *Nara* (2)	1,000	35°54'N, 130°49'E
8 Jul 45	Cargo ship *Shinto Maru* (2)	2,211	33°06'N, 129°43'E
9 Jul 45	Cargo ship *Gakuzyo Maru* (2)	1,500	33°59'N, 130°48'E
9 Jul 45	Cargo ship *Michi Maru* (2)	1,383	33°41'N, 130°14'E
10 Jul 45	Cargo ship *Chikuma Maru* (2)	9,951	33°56'N, 131°00'E
10 Jul 45	Cargo ship *Nippu Maru* (2)	891	33°06'N, 129°43'E
11 Jul 45	Cargo ship *Takechi Maru No. 3* (2)	870	34°34'N, 134°21'E
11 Jul 45	Destroyer *Sakura* (2)	1,000	35°50'N, 135°20'E
12 Jul 45	Cargo ship *Kojun Maru* (2)	2,177	37°57'N, 139°04'E
12 Jul 45	Cargo ship *Mitsu Maru* (2)	5,682	33°38'N, 135°03'E
12 Jul 45	Cargo ship *Tatsutsuyu Maru* (2)	2,220	34°24'N, 131°12'E
12 Jul 45	Conv. salvage vessel *Nasu Maru* (2)	694	37°55'N, 139°04'E
13 Jul 45	Cargo ship *Agata Maru No. 7* (2)	873	33°56'N, 130°56'E
13 Jul 45	Cargo ship *Hayahi Maru* (2)	6,919	34°00'N, 130°57'E
13 Jul 45	Cargo ship *Korasan Maru* (2)	2,300	33°56'N, 130°56'E
13 Jul 45	Cargo *Yamabishi Maru No. 20* (2)	880	33°58'N, 131°01'E
14 Jul 45	Cargo ship *Chizu Maru* (2)	2,304	33°58'N, 130°59'E
14 Jul 45	Passenger-cargo ship *Kyuko Maru* (2)	3,222	35°00'N, 129°43'E
15 Jul 45	Cargo ship *Saikai Maru No. 5* (2)	880	34°00'N, 131°00'E
15 Jul 45	Cargo ship *Nichiyu Maru No. 5* (2)	873	33°38'N, 135°24'E
16 Jul 45	Cargo ship *Koi Maru* (2)	882	33°54'N, 131°08'E
16 Jul 45	Passenger-cargo ship *Nachi Maru* (2)	1,606	33°59'N, 130°52'E
16 Jul 45	Cargo ship *Nannin Maru* (2)	2,486	33°56'N, 130°52'E
17 Jul 45	Cargo ship *Kitakata Maru* (2)	2,218	33°56'N, 130°56'E
17 Jul 45	Cargo ship *Nissho Maru* (2)	2,284	33°59'N, 130°52'E
17 Jul 45	Cargo *Shokai Maru No. 2* (2)	883	33°59'N, 130°52'E
19 Jul 45	Cargo ship *Daikoku Maru* (2)	598	37°57'N, 139°04'E
20 Jul 45	Cargo ship *Fuei Maru* (2)	2,822	34°00'N, 134°00'E
20 Jul 45	Cargo ship *Tatsutakawa Maru* (2)	873	33°59'N, 130°52'E
22 Jul 45	Cargo ship *Banei Maru No. 3* (2)	1,778	39°53'N, 139°52'E
22 Jul 45	Passenger-cargo ship *Choyo Maru* (2)	2,225	42°30'N, 129°45'E
22 Jul 45	Cargo ship *Hakutetsu Maru No. 5* (2)	799	34°40'N, 138°47'E
22 Jul 45	Cargo ship *Katori Maru* (2)	2,218	37°57'N, 139°04'E
22 Jul 45	Tanker *Yuyo Maru* (2)	10,045	34°16'N, 133°38'E
23 Jul 45	Cargo ship *Daishin Maru No. 2* (2)	1,153	33°54'N, 130°54'E
23 Jul 45	Cargo ship *Nissho Maru No. 6* (2)	642	34°00'N, 130°50'E
23 Jul 45	Cargo ship *Tsurukawa Maru No. 3* (2)	875	33°56'N, 130°54'E
23 Jul 45	Cargo ship *Shoko Maru* (2)	2,208	37°57'N, 139°04'E
23 Jul 45	Cargo ship *Taiha Maru No. 1* (2)	6,889	33°56'N, 130°30'E
23 Jul 45	Cargo ship *Taishin Maru No. 2* (2)	1,152	34°00'N, 130°00'E
24 Jul 45	Cargo ship *Himekawa Maru* (2)	873	33°34'N, 130°01'E
24 Jul 45	Cargo ship *Inari Maru* (2)	2,759	33°55'N, 130°56'E
24 Jul 45	Cargo ship *Koichi Maru* (2)	873	39°10'N, 127°26'E

24 Jul 45	Escort aircraft carrier *Kaiyo* (3)	17,000	33°21'N, 131°32'E
25 Jul 45	Cargo ship *Shokai Maru No. 5* (2)	882	34°35'N, 135°27'E
25 Jul 45	Cargo ship *Eian Maru* (2)	3,825	35°33'N, 133°14'E
25 Jul 45	Cargo ship *Hoshi Maru* (2)	2,853	34°35'N, 135°21'E
26 Jul 45	Cargo ship *Anette Fritzen* (2)	2,774	35°00'N, 128°00'E
27 Jul 45	Cargo ship *Meiko Maru* (2)	887	34°10'N, 130°55'E
27 Jul 45	Cargo ship *Unten Maru* (2)	1,025	33°56'N, 131°11'E
28 Jul 45	Cargo ship *Kinzan Maru No. 11* (2)	873	34°48'N, 134°38'E
28 Jul 45	Cargo ship *Kiyotada Maru* (2)	3,079	33°56'N, 131°15'E
28 Jul 45	Transport *Teiritsu Maru* (2)	9,877	35°32'N, 135°20'E
30 Jul 45	Passenger-cargo ship *Shokei Maru* (2)	3,620	35°32'N, 135°12'E
30 Jul 45	Destroyer *Hatsushimo* (2)	1,600	35°33'N, 135°12'E
2 Aug 45	Cargo ship *Santo Maru* (2)	1,890	37°57'N, 139°04'E
3 Aug 45	Dredge *Aga Maru* (2)	529	38°55'N, 139°45'E
3 Aug 45	Cargo ship *Dairetsu Maru* (2)	873	34°39'N, 135°12'E
3 Aug 45	Cargo ship *Hohsing* (2)	2,111	35°40'N, 136°05'E
3 Aug 45	Cargo ship *Shinei Maru No. 3* (2)	2,220	42°10'N, 130°15'E
4 Aug 45	Cargo ship *Koshin Maru* (2)	975	34°37'N, 135°18'E
6 Aug 45	Cargo ship *Chokai Maru No. 2* (2)	1,287	38°49'N, 137°04'E
6 Aug 45	Cargo ship *Isojima Maru* (2)	884	33°56'N, 130°56'E
6 Aug 45	Cargo ship *Kozan Maru No. 2* (2, 6)	864	33°53'N, 132°00'E
7 Aug 45	Cargo ship *Kibitsu Maru* (2)	9,575	34°37'N, 135°04'E
10 Aug 45	Cargo ship *Taishun Maru* (2)	2,857	41°46'N, 129°49'E
11 Aug 45	Cargo ship *Nisshin Maru No. 2* (2)	882	33°45'N, 131°30'E
12 Aug 45	Cargo ship *Tamon Maru No. 16* (2)	6,885	37°07'N, 137°02'E
14 Aug 45	Cargo ship *Mikasasan Maru* (2)	865	30°10'N, 126°28'E
14 Aug 45	Cargo ship *Yoko Maru* (2)	2,220	34°38'N, 135°24'E
14 Aug 45	Cargo ship *Kashima Maru* (2, 6)[4]	2,211	35°10'N, 129°00'E

Notes regarding Japanese ship casualties:
(1) Land-based aircraft after being damaged by mine(s).
(2) U.S. Army mine.
(3) Carrier-based aircraft after being damaged by mine(s).
(4) U.S. Army aircraft, Army mine.
(5) Army mine, probably sunk.
(6) Probably agent (cause).

Appendix B: RAAF Personal Awards

Number 11 Squadron RAAF
Distinguished Service Order

Redmond Forrest Michael Green	David Vernon

Distinguished Flying Cross

Colin Stuart Brown	Denis Russell Lawrence
Reginald Bruce Burrage	Robert Michael Seymour
Francis Blomfield Chapman	James Farrell Spears
Arthur John Cleland	Vernon Eric Townsend
Terence Lawless Duigan	Athol Galway Hope Wearne
Brian Hartley Higgins	William George Searle White
Stuart Austin Ikin	

Air Force Cross

John Leonard Grey	Michael Vaughan Mather
Charles Raymond Gurney	

Distinguished Flying Medal

Albert Bates	Reginald John Longhurst
Robert Mason Caldwell	James Alexander McWade
Donald Alvin Ferme	Clive William Miller
Ronald William Thompson Francis	George James Ping
Ronald Stanley Jordon	Donald Arthur Roberts
Leslie Allen Langdon	Max Cecil Schulz
Frederick Ross Liebelt	

Mention in Despatches

Alfred Frederick Burne	Albert Thomas Edmunds
Gordon Lord Dawson	Frank William Jeffrey
Terence Lawless Duigan	Alfred Henry Lanagan
Lewis Melvin Dunham	John Henry Willmott

Number 20 Squadron RAAF
Distinguished Service Order

Athol Galway Hope Wearne	

Distinguished Flying Cross

Harold Gordon MacMurray Brown	Gordon Neville Read
Alexander Ronald Emslie	Norman Valentine Robertson
Robert Maxwell Hirst	Henry Corbett Scott
Victor Allan Hodgkinson	Thomas Vincent Stokes
Leslie Harold Hokanson	Gilbert Robert Thurstun
James Lionel Mutch	

Air Force Cross

Godfrey Ellard Hemsworth	

Distinguished Flying Medal

Lionel Henry Campbell	Noel Morgan Hall
Robert Frederick Cox	Henry Angell Kirkhouse
John Patrick Lawless D'Arcy	Victor Norman Knowles
Robert Maxwell Draper	George William McMaster
Russell Thomas Anthony Fleming	Jack Stuart Riddell

Mentioned in Despatches

William Keith Bolitho	Bryden Melrose Pain
Vivian Palmer Chapman	John Perkins
Arthur John Moline	Arthur Kenneth Sandell
Allan Leslie Norman	John Percy Rushton Shields

Number 42 Squadron RAAF

Distinguished Flying Medal	Mention in Despatches
Harold Richard Longworth	Charles Thomas Spencer Bates

Number 43 Squadron RAAF
Distinguished Flying Cross

Anthony Noel Lee Atkinson	Reginald Joseph Marr
Noel Charles Edward Barr	Patrick John McMahon
Robert Trevor Clark	Ronald Nevill Damian Miller
Armand Andre Etienne	Brian Ortlepp
Robin Henry Gray	Charles Frederick Thompson
Lindley Maxwell Hurt	John William Thompson
John Kenneth Longmuir	Benjamin Alfred Titshall

Distinguished Flying Medal

Verdun William Cameron	Ronald John Hamilton Maclean
Thomas Allen Carman	Claude Arnold Ripper

Mention in Despatches

Arthur Berthold Girling	Frank Elvey Paramour
William Charles Andrew Hammond	Athol Lawrence Sewell
William Murray McLean	Donald Sowerby[1]

Bibliography/Notes

PUBLISHED WORKS

Barbey, Daniel E. *MacArthur's Amphibious Navy*. Annapolis, Md: Institute, 1969.

Bruhn, David D. *Battle Stars for the "Cactus Navy": America's Fishing Vessels and Yachts in World War II*. Berwyn Heights, Md: Heritage Books, 2014.

—*Eyes of the Fleet: The U.S. Navy's Seaplane Tenders and Patrol Aircraft in World War II*. Berwyn Heights, Md: Heritage Books, 2016.

—*MacArthur and Halsey's "Pacific Island Hoppers": The Forgotten Fleet of World War II*. Berwyn Heights, Md: Heritage Books, 2016.

—*Wooden Ships and Iron Men: The U.S. Navy's Coastal and Motor Minesweepers, 1941-1953*. Westminster, Md: Heritage Books, 2009.

Bruhn, David D., Rob Hoole. *Home Waters: Royal Navy, Royal Canadian Navy, and U.S. Navy Mine Forces Battling U-Boats in World War I*. Berwyn Heights, MD: Heritage Books, 2018.

Bulkley Jr., Robert J., *At Close Quarters: PT Boats in the United States Navy*. Washington, DC: Naval History Division, 1962.

Chandonnet, Fern. *Alaska at War, 1941-1945: The Forgotten War Remembered*. Fairbanks, Alaska: University of Alaska Press, 2007.

Cloe, John Haile. *The Aleutian Warriors: A History of the 11th Air Force and Fleet Air Wing 4*. Anchorage, AK and Missoula, MT: Anchorage Chapter, Air Force Association and Pictorial Histories Publishing Co., Inc., 1990.

Cowie, J. S. *Mines, Minelayers and Minelaying*. London: Oxford University Press, 1951.

Frank, Richard B. *Guadalcanal: The Definitive Account of the Landmark Battle*. New York: Penguin Group, 1990.

Jane, Fred T. *Jane's Fighting Ships 1939*. London: Macmillan, 1939.

Jersey, Stanley Coleman. *The Untold Story of Guadalcanal*. College Station, Texas: Texas A&M University Press, 2008.

Johnson, Ellis A., David A. Katcher, *Mines Against Japan*. Washington, DC: Government Printing Office, 1973.

Kemp, Paul. *The T-class submarine: the classic British design*. Annapolis, Md: Naval Institute Press, 1990.

Lott, Arnold S. *Most Dangerous Sea: A History of Mine Warfare, and an Account of U.S. Navy Mine Warfare Operations in World War II and Korea*. Annapolis, Md: Naval Institute, 1959.

Low, A. M. *Mine and Countermine*. London: Hutchinson, 1940.

Macklin, Robert. *One False Move: Bravest of the Brave The Australian Mine Defusers in World War II*. Sydney, Australia, Hachette Australia, 2012.

Melia, Tamara Moser. *Damn the Torpedoes" A Short History of U.S. Naval Mine Countermeasures, 1777-1991*. Washington, DC: Naval Historical Center, 1991.

Ministry of Defence. *British Mining Operations, 1939–1945, Vol. 1.* London: Director of Naval Warfare, 1973.

Morison, Samuel Eliot Morison. *The Struggle for Guadalcanal, August 1942 – February 1943, vol. 5 of History of United States Naval Operations in World War II.* Boston: Little, Brown and Company, 1958.

—*The Two-Ocean War.* Boston: Little, Brown, 1963.

Stille, Mark. *Guadalcanal 1942–43: America's first victory on the road to Tokyo.* Oxford, England: Osprey, 2015.

"Taffrail" (Henry Taprell Dorling). *Blue Star Line at War, 1939-45.* London: W. Foulsham & Co., 1973.

Toland, John. *The Rising Sun: The Decline and Fall of the Japanese Empire, 1936-1945.* New York: The Modern Library, 2003.

Turner, Mike, Hector Donohue. *Australian Minesweepers at War.* Canberra: Sea Power Centre - Australia, 2018.

U.S. Government Printing Office. *Building the Navy' Bases in World War II: History of the Bureau of Yards and Docks and the Civil Engineer Corps 1940-1946 vol. II.* Washington, DC: U.S. Bureau of Yards and Docks, 1947.

Womack, Tom. *The Allied Defense of the Malay Barrier, 1941-1942.* Jefferson, North Carolina, McFarland & Co., 2016.

PREFACE NOTES:

[1] Ellis A. Johnson and David A. Katcher, *Mines Against Japan* (Washington, DC: Government Printing Office, 1973), Preface.

[2] Tamara Moser Melia, *"Damn the Torpedoes" A Short History of U.S. Naval Mine Countermeasures, 1777-1991* (Washington, DC: Naval Historical Center, 1991), 9; A. M. Low, *Mine and Countermine* (London: Hutchinson, 1940), 83.

[3] Johnson and Katcher, *Mines Against Japan*, 16.

[4] Ibid, 23.

[5] Ibid.

[6] Ibid.

[7] Ibid, 30.

[8] Ibid, 23, 30.

[9] Ibid, 21.

[10] Ibid, 35.

[11] Ibid.

[12] Ibid.

[13] Ron Swart, 23 January 2018.

[14] Scott C. Truver, "An Act of War? The Law of Naval Mining" (https://warontherocks.com/2014/10/an-act-of-war-the-law-of-naval-mining/: accessed 20 January 2018); Ron Swart, 23 January 2018.

[15] J. S. Cowie, *Mines, Minelayers and Minelaying,* (London: Oxford University Press, 1951, 26-27.

[16] *British Mining Operations, 1939–1945, Vol. 1* (Ministry of Defence Director of Naval Warfare, 1973), 749, 756

[17] *British Mining Operations, 1939–1945, Vol. 1* (Ministry of Defence Director of Naval Warfare, 1973), 722, 769.

[18] *British Mining Operations, 1939–1945, Vol. 1*, 728, 752.

[19] Mike Turner and Hector Donohue, *Australian Minesweepers at War* (Canberra: Sea Power Centre - Australia, 2018), 70, 81, 289; "HMAS *Bungaree*" (http://www.navy.gov.au/hmas-bungaree: accessed 14 January 2018).

[20] Mike Turner and Hector Donohue, *Australian Minesweepers at War* (Canberra: Sea Power Centre - Australia, 2018), 70, 81, 289.

[21] Mike Turner and Hector Donohue, *Australian Minesweepers at War* (Canberra: Sea Power Centre - Australia, 2018), 70, 81, 289; Lott, *Most Dangerous Sea*, 215.

[22] "Rathmines – The Top Secret Catalina Operations" (https://www.catalinaflying.org.au/documents/Top%20secret%20Catalina%20ops.pdf: accessed 18 January 2018).

[23] Ibid.

[24] Lott, *Most Dangerous Sea*, 217.

[25] Lott, *Most Dangerous Sea*, 81; "*Prins van Oranje*-class," "*Pro Patria*-class" (http://www.netherlandsnavy.nl/: accessed 14 January 2018).

[26] Samuel Eliot Morison, *The Two-Ocean War* (Boston: Little, Brown, 1963), 86-87.

[27] Morison, *The Two-Ocean War*, 87-89.

[28] "Royal Dutch Navy (Netherlands) Mine Warfare Ships" (http://www.navypedia.org/ships/netherlands/nl_ms_pro_patria.htm); "Royal Netherlands Navy Warships of World War II" (http://www.netherlandsnavy.nl/: both accessed 13 January 2018).

[29] "*Smeroe*-class auxiliary minesweepers" (http://netherlandsnavy.nl/: accessed 13 January 2018).

[30] "HMAS *Abraham Crijnssen*" (http://www.navy.gov.au/hmas-abraham-crijnssen: accessed 14 January 2018).

[31] *British Mining Operations, 1939–1945, Vol. 1*, 715.

[32] Ibid, 715-716.

[33] Ibid, 716.

CHAPTER 1 NOTES:

[1] Interrogations of Japanese Officials - Vols. I & II United States Strategic Bombing Survey [Pacific], 81 (https://www.history.navy.mil/research/library/online-reading-room/title-list-alphabetically/i/interrogations-japanese-officials-voli.html#no16: accessed 5 November 2017).

[2] Cdr. Warren L. Craig, USNR (Ret.), "We Sank the Tokyo Express," an unpublished account dated 2 August 1991; *Montgomery* War Diary, January 1943.

[3] Cdr. Warren L. Craig, USNR (Ret.), "We Sank the Tokyo Express," an unpublished account dated 2 August 1991; *Montgomery* War Diary, January 1943.

[4] David D. Bruhn, *Battle Stars for the "Cactus Navy": America's Fishing Vessels and Yachts in World War II* (Berwyn Heights, Md: Heritage Books, 2014), 4.

[5] Bruhn, *Battle Stars for the "Cactus Navy,"* 4-5.

[6] Bruhn, *Battle Stars for the "Cactus Navy,"* 5.

[7] Bruhn, *Battle Stars for the "Cactus Navy,"* 22-23.

[8] Japanese Withdrawal, February 1943 (https://www.history.navy.mil/browse-by-topic/wars-conflicts-and-operations/world-war-ii/1942/guadalcanal/japanese-withdrawal.html: accessed 5 November 2017).

[9] Commander Mine Division 1, Guadalcanal Mine Fields, Report of Laying, 30 April 1943.

[10] Ibid.

[11] Commander Mine Division 1, Guadalcanal Mine Fields, Report of Laying, 30 April 1943; *Montgomery* and *Preble* Deck Logs, 1 February 1943.

[12] Logs of the Guadalcanal PT's August 1942-February 1943 (http://pt-king.gdinc.com/TokyoExpress1.html: accessed 6 November 2017).

[13] Richard B. Frank, *Guadalcanal: The Definitive Account of the Landmark Battle* (New York: Penguin Group, 1990), 585–586, 758; Samuel Eliot Morison, *The Struggle for Guadalcanal, August 1942 – February 1943, vol. 5 of History of United States Naval Operations in World War II* (Boston: Little, Brown and Company, 1958), 366; Stanley Coleman Jersey, *Hell's Islands: The Untold Story of Guadalcanal* (College Station, Texas: Texas A&M University Press, 2008), 392–393.

[14] Frank, *Guadalcanal: The Definitive Account of the Landmark Battle*, 587–588; Morison, *The Struggle for Guadalcanal, August 1942 – February 1943*, 367–368; Jersey, *Hell's Islands: The Untold Story of Guadalcanal*, 393–395; John Toland, *The Rising Sun: The Decline and Fall of the Japanese Empire, 1936-1945* (New York: The Modern Library, 2003), 429–430; Gene Kirkland, Operational Losses https://www.google.com/search?q=Pt+boats+1+February+1943+guadalcanal&ie=utf-8&oe=utf-8.

[15] Commander Mine Division 1, Guadalcanal Mine Fields, Report of Laying, 30 April 1943; *Montgomery* and *Preble* Deck Logs, 1 February 1943

[16] Three PT boats were destroyed off Guadalcanal on 1 Febuary: *PT-37* by Japanese destroyer *Kawakaze*; *PT-111* by destroyer gunfire, possibly that of *Kawakaze*; and *PT-123* by a Mitsubishi F1M reconnaissance floatplane. Commander Mine Division 1, Guadalcanal Mine Fields, Report of Laying, 30 April 1943; Kirkland, Operational Losses; IJN Kawakaze: Tabular Record of Movement (http://www.combinedfleet.com/kawaka_t.htm); *PT-123* (http://www.navsource.org/archives/12/05123.htm: both accessed 6 November 2017).

[17] Commander Mine Division 1, Guadalcanal Mine Fields, Report of Laying, 30 April 1943.

[18] Commander Mine Division 1, Guadalcanal Mine Fields, Report of Laying, 30 April 1943.

[19] Commander Mine Division 1, Guadalcanal Mine Fields, Report of Laying, 30 April 1943.

[20] Frank, *Guadalcanal: The Definitive Account of the Landmark Battle*, 561.

[21] Mark Stille, *Guadalcanal 1942–43: America's first victory on the road to Tokyo* (Oxford, England: Osprey, 2015), 87.

[22] Stille, *Guadalcanal 1942–43: America's first victory on the road to Tokyo*, 87; Japanese Withdrawal, February 1943 (https://www.history.navy.mil/browse-by-topic/wars-conflicts-and-operations/world-war-ii/1942/guadalcanal/japanese-withdrawal.html: accessed 5 November 2017).

[23] Commander South Pacific, Guadalcanal Mine Fields, Report of Laying, 31 May 1943.

[24] Commander in Chief, U.S. Pacific Fleet, Guadalcanal Mine Fields, Report of Laying, 14 July 1943.

[25] General Orders: Bureau of Naval Personnel Information Bulletin No. 321 (December 1943).

[26] Obituaries Thursday, January 14, 1999, Services scheduled at Pearl Harbor for Adm. Collis (http://archives.starbulletin.com/1999/01/14/news/obits.html: accessed 6 November 2017).

CHAPTER 2 NOTES:

[1] David Bruhn, *Home Waters: Royal Navy, Royal Canadian Navy, and U.S. Navy Mine Forces Battling U-Boats in World War I* (Berwyn Heights, MD: Heritage Books, 2018), 238-239.

[2] "NavSource, Minelayer (CM) Fleet Minelayer (MMF) Minelaying Submarine (SM) Index" (http://www.navsource.org/archives/11/06idx.htm: accessed 22 December 2017).

[3] "USS *Oglala* (CM-4, later ARG-1), 1917-1965" (http://www.shipscribe.com/usnaux/ww1/ships/cm4.htm; "Class: *Terror* (CM-5)" (http://www.shipscribe.com/usnaux/CM/CM05.html: both accessed 23 December 2017).

[4] "Class: *Terror* (CM-5)".

[5] "Class: *Terror* (CM-5)"; Robert S. Egan, "USS *Terror* and Her Family," *Warship International*, Vol 48, Issue 4, December 2011.

[6] "Class: *Terror* (CM-5)"; *Terror, DANFS*.

[7] *Keokuk, DANFS*.

[8] *Monadnock* and *Miantonomah, DANFS*.

[9] *Salem, DANFS*.

[10] *Weehawken, DANFS*.

[11] *Salem* War Diary, April, May, 1944; *Monadnock* War Diary, December 1941 to July 1943.

[12] "Light Minelayer (DM) Fast Minelayer (MMD) Index" (http://www.navsource.org/archives/11/08idx.htm: accessed 23 December 2017).

[13] Ibid.

[14] Ibid.

[15] "Class: *Chimo* (ACM-1)"
(http://www.shipscribe.com/usnaux/ACM/ACM01.html: accessed 23
December 2017).
[16] Ibid.
[17] Ibid.
[18] *Buttress, Monadnock, DANFS.*
[19] Johnson and Katcher, *Mines Against Japan*, 35.

CHAPTER 3 NOTES:
[1] Fraser G. Machaffie, "The Short Life and Sudden Death of HMS *Princess Victoria*," *Sea Breezes*, 18 January 2017
(USAhttp://www.seabreezes.co.im/index.php/features/ships/2370-the-short-life-and-sudden-death-of-hms-princess-victoria: accessed 4 January 2018).
[2] Machaffie, "The Short Life and Sudden Death of HMS *Princess Victoria*."
[3] Ibid.
[4] "Royal Navy Minelaying Operations" (http://www.naval-history.net/xGM-Ops-Minelaying.htm); "Commonwealth & Dominion Line Port Line" (https://web.archive.org/web/20120708035811/http://www.red-duster.co.uk/PORT7.htm: both accessed 4 January 2018).
[5] J. S. Cowie, *Mines, Minelayers and Minelaying* (London: Oxford University Press, 1951), 205-206; "*Adventure* class"
(https://uboat.net/allies/warships/class/76.html: accessed 3 January 2018).
[6] Cowie, *Mines, Minelayers and Minelaying*, 205-206; *British Mining Operations, 1939–1945, Vol. 1* (Ministry of Defence Director of Naval Warfare, 1973), 665-678; "Auxiliary Minelayers Class"
(https://uboat.net/allies/warships/class.html?ID=621&navy=HMS); "M 1 Class" (https://uboat.net/allies/warships/class/624.html: accessed 3 January 2018).
[7] "*Abdiel* class" (https://www.uboat.net/allies/warships/class/77.html: accessed 3 January 2018).
[8] Cowie, *Mines, Minelayers and Minelaying*, 205-206; "*Abdiel* class."
[9] "*E* class" (https://uboat.net/allies/warships/class.html?ID=13: accessed 3 January 2018).
[10] "*I* class" (https://uboat.net/allies/warships/class.html?ID=17: accessed 3 January 2018).
[11] Cowie, *Mines, Minelayers and Minelaying*, 205-206; "*E* class"
(https://uboat.net/allies/warships/class.html?ID=13: accessed 3 January 2018).
[12] Cowie, *Mines, Minelayers and Minelaying*, 205-206; "*I* class"
(https://uboat.net/allies/warships/class.html?ID=17: accessed 3 January 2018).
[13] "Royal Navy Minelaying Operations" (http://www.naval-history.net/xGM-Ops-Minelaying.htm); "*O* class"
(https://www.uboat.net/allies/warships/class/26.html: both accessed 4 January 2018); *British Mining Operations, 1939–1945, Vol. 1*, 116-117, 221-222,

and 453-454 cites mines laid by O-class minelayers in Operations BS86, 14 October 1942; CH, 8 April 1945; and Trammel, 22 April 1945.

[14] "Admiralty *S* class" (https://www.uboat.net/allies/warships/class/20.html: accessed 3 January 2018).

[15] "Royal Navy Minelaying Operations" (http://www.naval-history.net/xGM-Ops-Minelaying.htm: accessed 4 January 2018).

[16] *British Mining Operations, 1939–1945, Vol. 1*, 665-678.

[17] Cowie, *Mines, Minelayers and Minelaying*, 205-206; "*Plover* class" https://uboat.net/allies/warships/class.html?ID=622&navy=HMS; "*Plover* Class Coastal Minelayer Leader" (http://gb-navy-ww2.narod.ru/HTM-ML-plover.html: both accessed 1 January 2018).

[18] "*Corncrake* Class Coastal Minelayers" (http://gb-navy-ww2.narod.ru/HTM-ML-corncrake.html: accessed 1 January 2018).

[19] *Jane's Fighting Ships 1939* (London: Macmillan, 1939), 98.

[20] "*Linnet* class" (https://uboat.net/allies/warships/class.html?ID=623&navy=HMS: accessed 4 January 2018).

[21] "Miner Class Coastal Minelayer" (http://gb-navy-ww2.narod.ru/HTM-ML-miner.html: accessed 1 January 2018).

[22] Mike Turner and Hector Donohue, *Australian Minesweepers at War* (Canberra, Australia: Seapower Centre – Australia, 2018), 79, 89; "*Atreus* 1911 HMS - Minelayer Base Ship" (http://forums.clydemaritime.co.uk/viewtopic.php?t=25463); "Naval Trawlers" (https://www.battleships-cruisers.co.uk/naval_trawlers.htm: both accessed 4 March 2018).

[23] Rob Hoole, 9 January 2018.

[24] Cowie, *Mines, Minelayers and Minelaying*, 206; *British Mining Operations, 1939–1945, Vol. 1* (Ministry of Defence Director of Naval Warfare, 1973), 789; "Royal Navy Coastal Forces" (http://www.unithistories.com/units_british/RN_CoastalForces.html: accessed 6 January 2018).

[25] *British Mining Operations, 1939–1945, Vol. 1*, 725, 757; "Chronological Sequence of Submarine Minelaying Operations 1939-45" (http://www.naval-history.net/xGM-Ops-Minelaying.htm: accessed 6 January 2018); "HNMS *O 19* (N 54)" (https://www.uboat.net/allies/warships/ship/2890.html: accessed 8 January 2018).

[26] *British Mining Operations, 1939–1945, Vol. 1*, Table L.

[27] Cowie, *Mines, Minelayers and Minelaying*, 205-206; "*Porpoise* class" (https://uboat.net/allies/warships/class/49.html); "Chronological Sequence of Submarine Minelaying Operations 1939-45" (http://www.naval-history.net/xGM-Ops-Minelaying.htm: both accessed 6 January 2018).

[28] "Chronological Sequence of Submarine Minelaying Operations 1939-45" (http://www.naval-history.net/xGM-Ops-Minelaying.htm: accessed 6 January 2018).

[29] *British Mining Operations, 1939–1945, Vol. 1,* 788; "*T* class"
(https://uboat.net/allies/warships/class.html?ID=53&navy=HMS: accessed
6 January 2018).
[30] "Chronological Sequence of Submarine Minelaying Operations 1939-45"
(http://www.naval-history.net/xGM-Ops-Minelaying.htm: accessed 6
January 2018).
[31] *British Mining Operations, 1939–1945, Vol. 1,* 788; "S class"
(https://uboat.net/allies/warships/class.html?ID=52&navy=HMS: accessed
6 January 2018).
[32] Cowie, *Mines, Minelayers and Minelaying,* 205-206; *British Mining Operations,
1939–1945, Vol. 1,* 790.
[33] Ellis A. Johnson and David A. Katcher, *Mines Against Japan* (Washington,
DC: Government Printing Office, 1973), 35.

CHAPTER 4 NOTES:

[1] "RAN in the Second World War"
(http://www.navy.gov.au/history/feature-histories/ran-second-world-war);
"HMAS *Bungaree*" (http://www.navy.gov.au/hmas-bungaree: both accessed 7
January 2018).
[2] *British Mining Operations, 1939–1945, Vol. 1,* 674.
[3] Ibid,
[4] Ibid, 675.
[5] Ibid.
[6] "HMAS *Bungaree.*"
[7] "Conrad Emil Lambert Helfrich"
(https://www.britannica.com/biography/Conrad-Emil-Lambert-Helfrich);
"The conquest of Java Island, March 1942"
(http://www.dutcheastindies.webs.com/java.html: both accessed 22 January
2018).
[8] "Netherlands East Indies Naval Forces"
(http://pwencycl.kgbudge.com/N/e/Netherlands_East_Indies_Naval_Force
s.htm: accessed 22 January 2018).
[9] "*O-19* Class, Dutch Submarines"
(http://www.pwencycl.kgbudge.com/O/-/O-19_class.htm); "*O 19* class"
(https://uboat.net/allies/warships/class.html?ID=190&navy=HNMS: both
accessed 21 January 2018).
[10] "*Willem van der Zaan History*"
(http://www.netherlandsnavy.nl/Zaan_his.htm); "*Willem van der Zaan* class"
(https://www.uboat.net/allies/warships/class.html?ID=202&navy=HNMS:
both accessed 22 January 2018).
[11] "*Prins van Oranje*-class" (http://www.netherlandsnavy.nl/); "HNMS
Gouden Leeuw" (https://www.uboat.net/allies/warships/ship/2835.html);
"HNMS *Prins van Oranje*"
(https://www.uboat.net/allies/warships/ship/2834.html: all accessed 22
January 2018).

[12] "HNMS *Pro Patria*"
(https://www.uboat.net/allies/warships/ship/2838.html); "*Pro Patria*-class"
(http://www.netherlandsnavy.nl/: both accessed 22 January 2018).
[13] "HNMS *Krakatau*"
(https://www.uboat.net/allies/warships/ship/2829.html: accessed 22 January 2018).
[14] "*Rigel* class"
(https://www.uboat.net/allies/warships/class.html?ID=398&navy=HNMS);
"*Rigel*-class" (http://www.netherlandsnavy.nl/: both accessed 4 January 2018).
[15] "*Soemenep* class"
(https://www.uboat.net/allies/warships/class.html?ID=397&navy=HNMS);
"*Soemenep*-class" (http://www.netherlandsnavy.nl/: both accessed 22 January 2018).
[16] "*Bangkalan* class"
(https://www.uboat.net/allies/warships/class.html?ID=396&navy=HNMS);
"*Bangkalan*-class" (http://www.netherlandsnavy.nl/: both accessed 22 January 2018).
[17] Turner and Donohue, Australian Minesweepers at War, 289.
[18] "Royal Australian Navy Ship/Unit Approved Battle Honours"
(https://web.archive.org/web/20110614064156/http://www.navy.gov.au/w/images/Units_entitlement_list.pdf); "Britain's Navy
Fighting Ships - Operations – History"
(http://www.britainsnavy.co.uk/start.htm); "Escape from Soerabaja"
(http://www.netherlandsnavy.nl/Crijnssen.htm: all accessed 22 January 2018).

CHAPTER 5 NOTES:

[1] Terry Copp, "The Defence of Hong Kong"
(https://web.archive.org/web/20161229045704/https://web.archive.org/web/20060824135011/http:/arts.uwaterloo.ca/~ghayes/Copp.pdf: accessed 1 January 2018).
[2] *British Mining Operations, 1939–1945, Vol. 1* (Ministry of Defence Director of Naval Warfare, 1973), 665, 717-718.
[3] Ibid, 718.
[4] Ibid, 665.
[5] *British Mining Operations, 1939–1945, Vol. 1*, 665; "British Submarine HMS *Seal*" (http://www.historynet.com/british-submarine-hms-seal.htm: accessed 30 December 2017).
[6] *British Mining Operations, 1939–1945, Vol. 1*, 665-666; "HMS *Thracian* (D 86) - Old S-class Destroyer" (http://www.naval-history.net/xGM-Chrono-10DD-07T-HMS_Thracian.htm: accessed 31 December 2017).
[7] *British Mining Operations, 1939–1945, Vol. 1*, 665-666; J. S. Cowie, *Mines, Minelayers and Minelaying* (London: Oxford University Press, 1951), 117.
[8] Cowie, *Mines, Minelayers and Minelaying*, 92-93; "United Kingdom/Britain

Mines" (http://www.navweaps.com/Weapons/WAMBR_Mines.php: accessed 1 January 2018).
[9] *British Mining Operations, 1939–1945, Vol. 1*, 666.
[10] Ibid.
[11] *British Mining Operations, 1939–1945, Vol. 1*, 666; "Norwegian Merchant Fleet 1939 – 1945 Ships starting with Hø" (https://www.warsailors.com/freefleet/norfleeth4.html: accessed 30 December 2017).
[12] *British Mining Operations, 1939–1945, Vol. 1*, 666-667; "HMS *Redstart* (M 62)" (https://uboat.net/allies/warships/ship/13426.html: accessed 30 December 2017).
[13] Richard Walding, "Indicator Loops Royal Navy Harbour Defences – Hong Kong" (http://indicatorloops.com/hongkong.htm: accessed 31 December 2017).
[14] Michael Steemson, "H.M.S. Tamar R.N. Base Hong Kong" (http://www.royalnavyresearcharchive.org.uk/HMS_Tamar.htm#.Wkk3WDeIaM8: accessed 31 December 2017).
[15] *British Mining Operations, 1939–1945, Vol. 1*, 667.
[16] Ibid.
[17] Ibid, 667-668.
[18] Ibid, 668.
[19] Ibid.
[20] Ibid.
[21] Ibid, 669.
[22] Ibid, 672-673.
[23] Ibid, 670, 672-673.
[24] "Invasion of Malaya and Singapore 8 Dec 1941 - 15 Feb 1942" (https://ww2db.com/battle_spec.php?battle_id=47: accessed 25 January 2018).
[25] "General Tomoyuki Yamashita" (http://www.historynet.com/general-tomoyuki-yamashita.htm: accessed 24 January 2018).
[26] *British Mining Operations, 1939–1945, Vol. 1*, 719.
[27] Walding, "Indicator Loops Royal Navy Harbour Defences – Hong Kong."
[28] The War Office, January, 1948, Operations in the Far East, from 17th October 1940 to 27th December 1941, *Supplement to The London Gazette*, Of Thursday, the 20th of January, 1948, 22 January, 1948.
[29] Walding, "Indicator Loops Royal Navy Harbour Defences – Hong Kong".
[30] Walding, "Indicator Loops Royal Navy Harbour Defences – Hong Kong"; "Black Christmas - The Fall of Hong Kong – 1941" (http://www.combinedfleet.com/HongKong_t.htm: accessed 31 December 2017).
[31] "Black Christmas - The Fall of Hong Kong – 1941."
[32] Ibid.
[33] Ibid.

[34] "HMS *Redstart* - Yard No 263 - Indicator Loop Mine Layer - Royal Navy - Built 1938" (http://leithshipyards.com/ships-built-in-leith/1918-to-1939/313-hms-redstart-yard-no-263-indicator-loop-mine-layer-royal-navy-built-1938.html: accessed 31 December 2017).

[35] "HMS *Redstart* - Yard No 263 - Indicator Loop Mine Layer - Royal Navy - Built 1938"; Henry C. S. Collingwood-Selby, *In Time of War* (Hong Kong: Proverse, 2013), about the author.

[36] "Black Christmas - The Fall of Hong Kong – 1941."

[37] "Black Christmas - The Fall of Hong Kong – 1941"; "George Egan Able Seaman Royal Navy HMS *Thracian*" (http://wartimeheritage.com/storyarchive2/story_george_egan_hong_kong. htm: accessed 25 January 2018).

[38] Guy Walters, "Was this Japan's most heinous war crime? How 800 British PoW's were locked up and left to drown when their prison ship was torpedoed" (http://www.dailymail.co.uk/news/article-5010553/Was-sinking-Lisbon-Maru-Japan-s-worst-war-crime.html: accessed 25 January 2018).

[39] Ibid.

[40] "George Egan Able Seaman Royal Navy HMS *Thracian*" (http://wartimeheritage.com/storyarchive2/story_george_egan_hong_kong. htm); "Casualty Lists of the Royal Navy and Dominion Navies, World War 2" (http://www.naval-history.net/xDKCas1941-12DEC2.htm; HMS Cavalier Association (http://hmscavalier.org.uk/casualtySearch): all accessed 31 December 2017).

[41] *British Mining Operations, 1939–1945, Vol. 1,* 676.

[42] Diary of Admiral Layton, Commander-in-Chief, China (http://www.naval-history.net/xDKWD-EF1941ChinaStation.htm); "Naval Events, February 1942 (Part 1 of 2) Sunday 1st – Saturday 14th" (http://www.naval-history.net/xDKWW2-4202-42FEB01.htm: both accessed 26 January 2018); "Taffrail" (Henry Taprell Dorling), *Blue Star Line at War, 1939–45* (London: W. Foulsham & Co., 1973), 61-64.

[43] Diary of Admiral Layton, Commander-in-Chief, China; "HMS *Stronghold* (H 50) - Old S-class Destroyer" (http://www.naval-history.net/xGM-Chrono-10DD-05S-Stronghold.htm: accessed 26 January 2018).

[44] Diary of Admiral Layton, Commander-in-Chief, China; "Ernest John Spooner" (http://military.wikia.com/wiki/Ernest_John_Spooner: accessed 26 January 2018).

[45] Turner and Donohue, *Australian Minesweepers at War,* 45.

[46] Tom Womack, *The Allied Defense of the Malay Barrier, 1941–1942* (Jefferson, North Carolina, McFarland & Co., 2016), 287.

[47] "HMS *Li Wo* Crew members" (http://forcez-survivors.org.uk/biographies/listliwocrew.html); "Royal Navy Vessels Lost at Sea, 1939-45 – By Type" (http://www.naval-history.net/WW2BritishLosses1Major.htm); "Naval Events, February 1942 (Part 1 of 2) Sunday 1st – Saturday 14th" (http://www.naval-history.net/xDKWW2-4202-42FEB01.htm); "Royal Navy Ships, January 1942 (Part 4 of 4)" (http://www.naval-history.net/xDKWW2-4201-

40RNShips4Overseas.htm); "Casualty Lists of the Royal Navy and Dominion Navies, World War 2" (http://www.naval-history.net/xDKCasAlpha1939-45Ma.htm: all accessed 26 January 2018).

[48] Third Supplement to *The London Gazette* of Friday, 13 December 1946.

[49] Ibid.

[50] Ibid.

[51] Ibid.

[52] Diary of Admiral Layton, Commander-in-Chief, China (http://www.naval-history.net/xDKWD-EF1941ChinaStation.htm: accessed 26 January 2018).

CHAPTER 6 NOTES:

[1] "Tarakan," The Pacific War Online Encyclopedia (http://pwencycl.kgbudge.com/T/a/Tarakan.htm: accessed 1 July 2015); Morison, *The Rising Sun in the Pacific 1931-April 1942*, p. 272-273.

[2] David D. Bruhn, *Eyes of the Fleet: The U.S. Navy's Seaplane Tenders and Patrol Aircraft in World War II* (Berwyn Heights, Md: Heritage Books, 2016), 73.

[3] Bruhn, *Eyes of the Fleet*, 111, 114.

[4] "Escape from Soerabaja" (http://www.netherlandsnavy.nl/Crijnssen.htm); "Escape of Hr. Ms. Abraham Crijnssen" (http://www.go2war2.nl/artikel/2728/Escape-of-Hr-Ms-Abraham-Crijnssen.htm: both accessed 30 January 2018).

[5] "Escape from Soerabaja"; "Escape of Hr. Ms. Abraham Crijnssen".

[6] "Escape from Soerabaja"; "Escape of Hr. Ms. Abraham Crijnssen".

[7] "Escape from Soerabaja"; "Escape of Hr. Ms. Abraham Crijnssen"; Tom Womack, *The Allied Defense of the Malay Barrier, 1941-1942* (Jefferson, North Carolina, McFarland & Co., 2016), 282.

[8] "Escape from Soerabaja"; Womack, *The Allied Defense of the Malay Barrier, 1941-1942*, 282.

[9] "Escape from Soerabaja"; "Escape of Hr. Ms. Abraham Crijnssen".

[10] "Escape from Soerabaja"; "Escape of Hr. Ms. Abraham Crijnssen".

[11] "Escape from Soerabaja"; "Escape of Hr. Ms. Abraham Crijnssen".

[12] "Escape from Soerabaja".

[13] Ibid.

[14] Womack, *The Allied Defense of the Malay Barrier, 1941-1942*, 282; "Escape from Soerabaja".

[15] Womack, *The Allied Defense of the Malay Barrier, 1941-1942*, 282; "HNMS A" (https://www.uboat.net/allies/warships/ship/2843.html: accessed 1 February 2018).

[16] "World War II unit histories and officers" (http://www.unithistories.com/units_index/default.asp?file=../units_dutch/navy_abc.asp: accessed 31 January 2018).

[17] "Smeroe-class auxiliary minesweepers" (http://www.netherlandsnavy.nl/); "Smeroe auxiliary minesweepers (1941)" (http://www.navypedia.org/ships/netherlands/nl_ms_smeroe.htm: both

accessed 1 February 2018); Womack, *The Allied Defense of the Malay Barrier, 1941-1942*, 282.
[18] Womack, *The Allied Defense of the Malay Barrier, 1941-1942*, 282.
[19] Ibid.
[20] Letter from the Royal Netherlands Navy Office of Naval Attache "Stanhill", 34 Queen's Road, Melbourne (S.C.2) to the Director of Naval Intelligence, Department of the Navy, Navy Office, Melbourne (S.C.1) of 15 August 1956, serial: 721/285, subject: The Java Campaign – World War II.

CHAPTER 7 NOTES:
[1] David D. Bruhn, *Eyes of the Fleet: The U.S. Navy's Seaplane Tenders and Patrol Aircraft in World War II* (Berwyn Heights, Md: Heritage Books, 2016), 29.
[2] Bruhn, *Eyes of the Fleet: The U.S. Navy's Seaplane Tenders and Patrol Aircraft in World War II*, 29-30. [2] Bruhn, *Eyes of the Fleet: The U.S. Navy's Seaplane Tenders and Patrol Aircraft in World War II*, 29-30.
[3] Bruhn, *Eyes of the Fleet: The U.S. Navy's Seaplane Tenders and Patrol Aircraft in World War II*, 30.
[4] Bruhn, *Eyes of the Fleet: The U.S. Navy's Seaplane Tenders and Patrol Aircraft in World War II*, 30.
[5] Bruhn, *Eyes of the Fleet: The U.S. Navy's Seaplane Tenders and Patrol Aircraft in World War II*, 31.
[6] Bruhn, *Eyes of the Fleet: The U.S. Navy's Seaplane Tenders and Patrol Aircraft in World War II*, 31.
[7] "Ships and District Craft Present at Pearl Harbor, 0800 7 December 1941" (https://www.history.navy.mil/research/library/online-reading-room/title-list-alphabetically/s/ships-present-at-pearl-harbor.html: accessed 29 November 2017).
[8] "Coastal Minesweeper (AMc) Index" (http://www.navsource.org/archives/11/03idx.htm: accessed 30 November 2017); "The United States Navy, The Navy Afloat October 1, 1941" (http://www.fleetorganization.com/1941intro.html: accessed 27 November 2017).
[9] David D. Bruhn, *Battle Stars for the "Cactus Navy": America's Fishing Vessels and Yachts in World War II* (Berwyn Heights, Md: Heritage Books, 2014), 29.
[10] Bruhn, *Battle Stars for the "Cactus Navy": America's Fishing Vessels and Yachts in World War II*, 30-31.
[11] Commander-in-Chief, United States Pacific Fleet, Report of Japanese Raid on Pearl Harbor, 7 December, 1941, 15 February 1942.
[12] Lott, *Most Dangerous Sea*, 24-26; *Kingfisher*, *DANFS*.
[13] "The United States Navy, The Navy Afloat October 1, 1941" (http://www.fleetorganization.com/1941intro.html: accessed 27 November 2017).
[14] Commander Minecraft, Battle Force, Japanese Plane Attack on Pearl Harbor, Morning of Sunday, December 7, 1941, 7 December 1941.
[15] Lott, *Most Dangerous Sea*, 22-23.

[16] Commander Minecraft, Battle Force, Japanese Plane Attack on Pearl Harbor, Morning of Sunday, December 7, 1941, 7 December 1941; Bureau of Ships Navy Department, USS *Oglala* Torpedo & Bomb Damage, December 7, 1941, Pearl Harbor, 14 February 1942.

[17] Bureau of Ships Navy Department, USS *Oglala* Torpedo & Bomb Damage, December 7, 1941, Pearl Harbor, 14 February 1942.

[18] Commander Minecraft, Battle Force, Japanese Plane Attack on Pearl Harbor, Morning of Sunday, December 7, 1941, 7 December 1941; Bureau of Ships Navy Department, USS *Oglala* Torpedo & Bomb Damage, December 7, 1941, Pearl Harbor, 14 February 1942.

[19] Commander Minecraft, Battle Force, Japanese Plane Attack on Pearl Harbor, Morning of Sunday, December 7, 1941, 7 December 1941; Bureau of Ships Navy Department, USS *Oglala* Torpedo & Bomb Damage, December 7, 1941, Pearl Harbor, 14 February 1942.

[20] Commander Minecraft, Battle Force, Japanese Plane Attack on Pearl Harbor, Morning of Sunday, December 7, 1941, 7 December 1941.

[21] Commander Minecraft, Battle Force, Japanese Plane Attack on Pearl Harbor, Morning of Sunday, December 7, 1941, 7 December 1941.

[22] Executive Officer, USS *Breese*, Japanese Air Raid on Pearl Harbor, T.H., 7 December 1941, 9 December 1941.

[23] Commanding Officer USS *Breese*, Action, Pearl Harbor Air Raid, Report of -; 9 December 1941; Executive Officer, USS *Breese*, Japanese Air Raid on Pearl Harbor, T.H., 7 December 1941, 9 December 1941; Commanding Officer USS *Gamble*, Japanese Air Raid on Pearl Harbor, T.H., December 7, 1941, Report of, 13 December 1941.

[24] Commanding Officer USS *Breese*, Action, Pearl Harbor Air Raid, Report of -; 9 December 1941; Executive Officer, USS *Breese*, Japanese Air Raid on Pearl Harbor, T.H., 7 December 1941, 9 December 1941.

[25] Commanding Officer USS *Gamble*, Japanese Air Raid on Pearl Harbor, T.H., December 7, 1941, Report of, 13 December 1941.

[26] Executive Officer, USS *Breese*, Japanese Air Raid on Pearl Harbor, T.H., 7 December 1941, 9 December 1941.

[27] Commanding Officer USS *Gamble*, Japanese Air Raid on Pearl Harbor, T.H., December 7, 1941, Report of, 13 December 1941.

[28] Commanding Officer USS *Gamble*, Japanese Air Raid on Pearl Harbor, T.H., December 7, 1941, Report of, 13 December 1941.

[29] Commanding Officer USS *Breese*, Action, Pearl Harbor Air Raid, Report of -; 9 December 1941.

[30] Commanding Officer, USS *Ramsay*, Report of Offensive Measures During Air Raid Seven December 1941, 13 December 1941.

[31] Commanding Officer USS *Breese*, Action, Pearl Harbor Air Raid, Report of -; 9 December 1941.

[32] Commanding Officer USS *Gamble*, Japanese Air Raid on Pearl Harbor, T.H., December 7, 1941, Report of, 13 December 1941.

[33] Commanding Officer USS *Gamble*, Japanese Air Raid on Pearl Harbor, T.H., December 7, 1941, Report of, 13 December 1941.

[34] Commanding Officer USS *Gamble*, Japanese Air Raid on Pearl Harbor, T.H., December 7, 1941, Report of, 13 December 1941.

[35] Commanding Officer, USS *Montgomery*; Japanese Air Raid on Pearl Harbor, T.H., December 7, 1941 – report of; 12 December 1941.

[36] Commanding Officer, USS *Pruitt*, Engagement with Japanese Planes During Air Raid on Pearl Harbor, T.H., December 7, 1941, 10 December 1941.

[37] Commanding Officer, USS *Pruitt*, Engagement with Japanese Planes During Air Raid on Pearl Harbor, T.H., December 7, 1941, 10 December 1941.

[38] Commanding Officer, USS *Pruitt*, Engagement with Japanese Planes During Air Raid on Pearl Harbor, T.H., December 7, 1941, 10 December 1941.

[39] Commanding Officer, USS *Pruitt*, Engagement with Japanese Planes During Air Raid on Pearl Harbor, T.H., December 7, 1941, 10 December 1941.

[40] "Pearl Harbor Casualties" (http://www.pearlharbor.org/history/casualties/pearl-harbor-casualties/); "U.S. Navy Casualties" (https://www.nps.gov/valr/learn/historyculture/navy-casualties.htm: both accessed 8 December 2017).

[41] Commanding Officer, USS *Sicard*, Report of Action, December 7, 1941, 9 December 1941.

[42] Commanding Officer, USS *Sicard*, Report of Action, December 7, 1941, 9 December 1941.

[43] Commanding Officer, USS *Sicard*, Report of Action, December 7, 1941, 9 December 1941.

[44] Commanding Officer, USS *Preble*, Japanese Air Attack, Sunday, December 7, 1941, 10 December 1941.

[45] Commanding Officer, USS *Tracy*, Report of Action, December 7, 1941, 10 December 1941.

[46] Commanding Officer, USS *Tracy*, Report of Action, December 7, 1941, 10 December 1941; Commanding Officer, USS *Cummings*, Pearl Harbor Action of Sunday December 7, 1941 -- Report of, 18 December 1941.

[47] Commanding Officer, USS *Tracy*, Report of Action, December 7, 1941, 10 December 1941; Lott, *Most Dangerous Sea*, 23.

[48] Commanding Officer, USS *Tracy*, Report of Action, December 7, 1941, 10 December 1941.

[49] Commanding Officer, USS *Tracy*, Report of Action, December 7, 1941, 10 December 1941.

[50] Lott, *Most Dangerous Sea*, 20; Commanding Officer, USS *Zane*, Air Raid on Pearl Harbor, report of, 10 December 1941; Commanding Officer, USS *Trever*, Japanese Air Raid on Pearl Harbor, T.H., Sunday, December 7 1941; Report of, 12 December 1941.

[51] Commanding Officer, USS *Bobolink*, Offensive Measure Taken by the USS *Bobolink* Against Japanese Air Attack of December 7, 1941, 10 December

1941; Commanding Officer, USS *Tern*, December 7, 1941 Raid - report on, 5 December 1941; Lott, *Most Dangerous Sea*, 25-26.

[52] Lott, *Most Dangerous Sea*, 24, 84-85.

[53] Lott, *Most Dangerous Sea*, 85.

CHAPTER 8 NOTES:

[1] "Operation K" (http://fly.historicwings.com/2013/03/operation-k/: accessed 24 November 2017).

[2] "The Women's Air Raid Defense: Protecting the Hawaiian Islands" (http://www.historynet.com/the-womens-air-raid-defense-protecting-the-hawaiian-islands.htm: accessed 24 November 1943).

[3] "The Women's Air Raid Defense: Protecting the Hawaiian Islands" (www.historynet.com).

[4] "The Women's Air Raid Defense: Protecting the Hawaiian Islands" (www.historynet.com).

[5] "Operation K" (http://fly.historicwings.com); Operation K (https://pacificeagles.net/operation-k/: accessed 24 November 2017).

[6] "Operation K" (http://fly.historicwings.com).

[7] "Operation K" (http://fly.historicwings.com); Lott, *Most Dangerous Sea*, 31.

[8] "Operation K" (http://fly.historicwings.com).

[9] "Operation K" (http://fly.historicwings.com).

[10] Lott, *Most Dangerous Sea*, 30.

[11] U.S. Government Printing Office, *Building the Navy' Bases in World War II: History of the Bureau of Yards and Docks and the Civil Engineer Corps 1940-1946 vol. II* (Washington, DC: U.S. Bureau of Yards and Docks, 1947), 162.

[12] Operation K (https://pacificeagles.net); IJN Submarine *I-23*: Tabular Record of Movement (http://www.combinedfleet.com/I-23.htm: accessed 26 November 2017).

[13] CoMinDiv 1, *Tracy* War Diary, April 1942; Lott, *Most Dangerous Sea*, 32.

[14] CoMinDiv 1 War Diary, April 1942; Lott, *Most Dangerous Sea*, 32.

[15] CoMinDiv 1 War Diary, April 1942.

[16] *Sicard*, Marine Force 14th Naval District War Diary, April 1942.

[17] *Sicard* War Diary, April 1942.

[18] *Sicard* War Diary, April 1942.

[19] Commander in Chief, U.S. Pacific Fleet, [USS *Preble*] Ship's History – Forwarding of, 13 November 1945; VMF-222 War Diary, April 1942.

[20] Commander in Chief, U.S. Pacific Fleet, [USS *Preble*] Ship's History – Forwarding of, 13 November 1945.

[21] Commander in Chief, U.S. Pacific Fleet, [USS *Preble*] Ship's History – Forwarding of, 13 November 1945; Com U.S. NAS Midway Island War Diary, April 1942.

CHAPTER 9 NOTES:

[1] Commanding Officer, USS *Preble*, Destruction of *YP-277*, May 23, 1942, 25 May 1942; CoMinDiv 1 War Diary, May 1942.

[2] "Battle of Midway: 4-7 June 1942"
(https://www.history.navy.mil/research/library/online-reading-room/title-list-alphabetically/b/battle-of-midway-4-7-june-1942.html: accessed 26 November 2017).
[3] David D. Bruhn, *Battle Stars for the "Cactus Navy": America's Fishing Vessels and Yachts in World War II* (Berwyn Heights, Md: Heritage Books, 2014), 99.
[4] Bruhn, *Battle Stars for the "Cactus Navy": America's Fishing Vessels and Yachts in World War II*, 101.
[5] Ibid.
[6] Commander in Chief, U.S. Pacific Fleet, [USS *Preble*] Ship's History – Forwarding of, 13 November 1945; Bruhn, *Battle Stars for the "Cactus Navy": America's Fishing Vessels and Yachts in World War II*, 101.
[7] Bruhn, *Battle Stars for the "Cactus Navy": America's Fishing Vessels and Yachts in World War II*, 101-102.
[8] Bruhn, *Battle Stars for the "Cactus Navy": America's Fishing Vessels and Yachts in World War II*, 102.
[9] Commanding Officer, USS *Preble*, Destruction of *YP-277*, May 23, 1942, 25 May 1942; Robert J. Bulkley Jr., *At Close Quarters PT Boats in the United States Navy* (Washington, DC: Naval History Division, 1962), 9.
[10] Commanding Officer, USS *Preble*, Destruction of *YP-277*, May 23, 1942, 25 May 1942; *Preble* War Diary, May 1942.
[11] *Preble* War Diary, May 1942.
[12] "Battle of Midway: 4-7 June 1942"
(https://www.history.navy.mil/research/library/online-reading-room/title-list-alphabetically/b/battle-of-midway-4-7-june-1942.html: accessed 26 November 2017).
[13] "Battle of Midway: 4-7 June 1942" (www.history.navy.mil); [8] Naval Analysis Division Staff, *The Battle of Midway: The Campaigns of the Pacific War* (United States Strategic Bombing Survey (Pacific), 1946), 60.

CHAPTER 10 NOTES:

[1] *Sicard* War Diary, June 1942.
[2] Commander North Pacific Force, U.S. Pacific Fleet War Diary, July 1942.
[3] CoMinDiv 1 War Diary, June 1942.
[4] Naval Analysis Division Staff, *The Battle of Midway: The Campaigns of the Pacific War* (United States Strategic Bombing Survey (Pacific), 1946), 58.
[5] Naval Analysis Division Staff, *The Battle of Midway: The Campaigns of the Pacific War* (United States Strategic Bombing Survey (Pacific), 1946), 58.
[6] Bruhn, *Eyes of the Fleet*, 124.
[7] Bruhn, *Eyes of the Fleet*, 124-125.
[8] Bruhn, *Eyes of the Fleet*, 138.
[9] Bruhn, *Eyes of the Fleet*, 141-142.
[10] *Sicard* War Diary, June 1942.
[11] *Sicard* War Diary, June 1942; Lott, *Most Dangerous Sea*, 99; "Getting past Cape Mendocino: A coastwise passage south from Seattle" (http://www.oceannavigator.com/October-2009/Getting-past-Cape-

Mendocino-A-coastwise-passage-south-from-Seattle/: accessed 25 November 2017).

[12] CoMinDiv 1, *Sicard* War Diary, June 1942; BUPERS Circular Letter 205-43 of 12 October 1943.

[13] CoMinDiv 1 War Diary, June 1942.

[14] *Sicard* War Diary, June 1942.

[15] *Sicard* War Diary, July 1942.

[16] CoMinDiv 1 War Diary, July 1942; "USAT Atkins (FS 237) ex-USC&GS Explorer" (http://www.navsource.org/archives/12/179918.htm: accessed 25 November 2017)

[17] CoMinDiv 1 War Diary, July 1942; Lott, *Most Dangerous Sea*, 99.

[18] CoMinDiv 1 War Diary, July 1942.

[19] CoMinDiv 1 War Diary, July 1942; Lott, *Most Dangerous Sea*, 99-100.

[20] CoMinDiv 1 War Diary, September 1942.

[21] CoMinDiv 1 War Diary, September and October 1942.

[22] Robert J. Bulkley Jr., *At Close Quarters: PT Boats in the United States Navy* (Washington, DC: Naval History Division, 1962), 263; Lott, *Most Dangerous Sea*, 100; CoMinDiv 1 War Diary, October 1942; Ron Swart email, 5 December 2017.

[23] CoMinDiv 1 War Diary, November 1942; Bruhn, *Eyes of the Fleet*, 130.

[24] CoMinDiv 1 War Diary, November 1942.

[25] "The War in the Pacific: Japanese Suicide" (https://warinthepacific.wordpress.com/2013/02/08/colonel-yamazaki-yasuyo/: accessed 24 November 2017).

[26] The Aleutians Campaign (Office of Naval Intelligence, 1945), 11.

[27] USS *Pruitt*, Ship's History – Forwarding of, 13 November 1945; *Ramsay* War Diary, May 1943; Combat Narratives: The Aleutians Campaign (Office of Naval Intelligence, 1945), 79-81.

[28] Michael A. Mira, "Battle of Attu" (https://www.army.mil/article/20923/battle_of_attu; "Battle of Attu" https://www.revolvy.com/main/index.php?s=Battle%20of%20Attu&item_type=topic: both accessed 24 November 2017).

[29] John Haile Cloe, *The Aleutian Warriors: A History of the 11th Air Force and Fleet Air Wing 4* (Anchorage, AK and Missoula, MT: Anchorage Chapter, Air Force Association and Pictorial Histories Publishing Co., Inc., 1990), 293; ; "Battle of Attu" https://www.revolvy.com/main/index.php?s=Battle%20of%20Attu&item_type=topic: accessed 24 November 2017).

[30] Fern Chandonnet, *Alaska at War, 1941-1945: The Forgotten War Remembered* (Fairbanks: University of Alaska Press, 2007), 395.

[31] Lott, *Most Dangerous Sea*, 100.

CHAPTER 11 NOTES:

[1] Bruhn, *Eyes of the Fleet*, 203.

[2] Ibid, 203-204.

[3] Ibid, 204.

[4] *Ramsay* War Diary, February 1942; *Gamble* War Diary, March 1942.

[5] *Gamble, Ramsay* War Diary, March-April 1942.

[6] *Gamble* War Diary, April 1942.

[7] *Gamble, Ramsay* War Diary, April-May 1942.

[8] Ibid.

[9] *Ramsay* War Diary, May 1942.

[10] Ibid.

[11] *Montgomery* War Diary, May 1942.

[12] *Montgomery* War Diary, May-June 1942.

[13] *Montgomery* and *Ramsay* War Diary, June-July 1942.

[14] *Ramsay* War Diary, May 1942.

[15] *Ramsay* War Diary, May-June 1942.

[16] *Ramsay* War Diary, June 1942.

CHAPTER 12 NOTES:

[1] *Tracy* War Diary, August 1942.

[2] *Breese, Gamble, Tracy* War Diary, July 1942.

[3] *Breese, Gamble, Tracy* War Diary, July 1942.

[4] *Breese, Gamble, Tracy* War Diary, July 1942.

[5] *Breese* War Diary, July 1942.

[6] *Breese, Gamble* War Diary, July 1942.

[7] *Breese, Gamble* War Diary, July 1942; Lott, *Most Dangerous Sea*, 77.

[8] Lott, *Most Dangerous Sea*, 77.

[9] David D. Bruhn, *Battle Stars for the "Cactus Navy": America's Fishing Vessels and Yachts in World War II* (Berwyn Heights, Md: Heritage Books, 2014), 10-11.

[10] Ibid.

[11] Ibid, 11.

[12] Bruhn, *Battle Stars for the "Cactus Navy": America's Fishing Vessels and Yachts in World War II*, 18.

[13] Lott, *Most Dangerous Sea*, 78.

[14] *Tracy* War Diary, August 1942.

[15] Bruhn, *Battle Stars for the "Cactus Navy": America's Fishing Vessels and Yachts in World War II*, 4.

[16] Ibid, 4-5.

[17] Bruhn, *Eyes of the Fleet*, 174.

[18] Lott, *Most Dangerous Sea*, 92; Samuel Eliot Morison, *The Two-Ocean War*, 168-177.

[19] Bruhn, *Eyes of the Fleet*, 174; *Tracy* War Diary, August 1942.

[20] *Gamble* War Diary, August 1942.

[21] *Gamble* War Diary, August 1942; Commanding Officer, USS *Gamble* (DM15), Anti-submarine Action – USS *Gamble* on 29 August 1942, 29 August 1942.

[22] Commanding Officer, USS *Gamble* (DM15), Anti-submarine Action – USS *Gamble* on 29 August 1942, 29 August 1942.

[23] Ibid.

[24] Ibid.

[25] Ibid.

[26] Commanding Officer, USS *Gamble* (DM15), Anti-submarine Action – USS *Gamble* on 29 August 1942, 29 August 1942; *Gamble*, *Saratoga* War Diary, August 1942.

[27] "USS *Colhoun* (Destroyer # 85, later DD-85 and APD-2), 1918-1942" (http://www.ibiblio.org/hyperwar/OnlineLibrary/photos/sh-usn/usnsh-c/dd85.htm): accessed 9 February 2018; *Little, Gregory, DANFS*.

[28] Commander Amphibious Forces South Pacific: Serial 0122 (October 6, 1942).

CHAPTER 13 NOTES:

[1] Bruhn, *Eyes of the Fleet*, 185.

[2] Ibid.

[3] Ibid.

[4] Ibid, 185-187.

[5] Ibid, 189.

[6] Bruhn, *Battle Stars for the "Cactus Navy": America's Fishing Vessels and Yachts in World War II*, 207.

[7] Ibid.

[8] Lott, *Most Dangerous Sea*, 95.

[9] *Breese* War Diary, May 1943.

[10] *Gamble, DANFS*.

[11] *Gamble, DANFS*; *Breese* War Diary, May 1943.

[12] *Gamble, DANFS*; *Australia in the War of 1939–1945, Series 2 – Navy, Royal Australian Navy, 1942–1945 (Volume II)* 1st edition, 1968, 279.

[13] *Breese* War Diary, May 1943.

[14] Ibid.

[15] Ibid.

[16] *Gamble, DANFS*; "A Brief History of U.S. Navy Cruisers Part II - World War II (1941-1943)" (http://www.navy.mil/navydata/nav_legacy.asp?id=136: accessed 11 February 2018).

[17] "A Brief History of U.S. Navy Cruisers Part II - World War II (1941-1943)."

[18] "A Brief History of U.S. Navy Cruisers Part II - World War II (1941-1943)"; "USS *Gwin* DD433" (http://www.destroyerhistory.org/benson-gleavesclass/ussgwin/); "Battle of Kolombangara, 13 July 1943" (http://www.historyofwar.org/articles/battles_kolombangara.html: both accessed 11 February 2018).

CHAPTER 14 NOTES:

[1] Commander Task Group 36.2 (Commander Cruiser Division Twelve – Rear Admiral A. S. Merrill), Action Report – Minelaying and Bombardment, conducted in the Shortland-Faisi-Kolombangara Areas, night of 29-30 June, 1943, 16 July 1943.

[2] David D. Bruhn, *MacArthur and Halsey's "Pacific Island Hoppers": The Forgotten Fleet of World War II* (Berwyn Heights, Md: Heritage Books, 2016), 49.

[3] Ibid, 49-50.

[4] Ibid, 50-51

[5] CTG 36.2, Action Report – Minelaying and Bombardment, conducted in the Shortland-Faisi-Kolombangara Areas, night of 29-30 June, 1943, 16 July 1943.

[6] Commander South Pacific, Action Report - Minelaying and bombardment conducted in the Shortland-Faisi-Kolombangara Areas, night of 29-30 June, 20 August 1943.

[7] CTG 36.2, Action Report – Minelaying and Bombardment, conducted in the Shortland-Faisi-Kolombangara Areas, night of 29-30 June, 1943, 16 July 1943.

[8] Ibid.

[9] Ibid.

[10] Ibid.

[11] Ibid.

[12] Ibid.

[13] CTG 36.2, Action Report – Minelaying and Bombardment, conducted in the Shortland-Faisi-Kolombangara Areas, night of 29-30 June, 1943, 16 July 1943; Commander Task Unit 36.2.2 (Mining Detachment), Action Report, Mining Detachment Third Fleet Operation Plan 12-43, 1 July 1943.

[14] CTG 36.2, Action Report – Minelaying and Bombardment, conducted in the Shortland-Faisi-Kolombangara Areas, night of 29-30 June, 1943, 16 July 1943.

[15] Ibid.

[16] Ibid.

[17] "IJN Submarine *RO-100*: Tabular Record of Movement" (http://www.combinedfleet.com/RO-100.htm: accessed 12 February 2018).

[18] Commander in Chief, U.S. Pacific Fleet and Pacific Ocean Areas, Operations in Pacific Ocean Areas – August 1943, 20 November 1943; *Preble* War Diary, August 1943.

[19] Commander in Chief, U.S. Pacific Fleet and Pacific Ocean Areas, Operations in Pacific Ocean Areas – August 1943, 20 November 1943; ComDesRon 21, *Preble* War Diary, August 1943.

[20] Commander in Chief, U.S. Pacific Fleet and Pacific Ocean Areas, Operations in Pacific Ocean Areas – August 1943, 20 November 1943; ComDesRon 21, *Preble* War Diary, August 1943.

CHAPTER 15 NOTES:

[1] Bruhn, *Battle Stars for the "Cactus Navy": America's Fishing Vessels and Yachts in World War II*, 219-220.

[2] Ibid, 220.

[3] Ibid.

[4] Ibid, 220-222.

[5] *Sicard* War Diary, November 1943.

[6] *Breese, Sicard* War Diary, October 1943.

[7] *Sicard* War Diary, November 1943.

[8] *Sicard* War Diary, November 1943; Commanding Officer, USS *Pruitt* (DM-22), Mining Operations during Bougainville Campaign – Report on, 10 November 1943; *"Erskine M. Phelps"* (http://www.bruzelius.info/Nautica/Ships/Fourmast_ships/Erskine_M_Phelps(1898).html: accessed 12 February 2018).

[9] Bruhn, *Battle Stars for the "Cactus Navy"*, 220-222; Commander in Chief, U.S. Pacific Fleet and Pacific Ocean Areas, Operations in the Pacific Ocean Areas – November 1943, 28 February 1944.

[10] Bruhn, *Battle Stars for the "Cactus Navy"*, 220-222.

[11] Commanding Officer, USS *Pruitt* (DM-22), Mining Operations during Bougainville Campaign – Report on, 10 November 1943.

[12] *Tracy* War Diary, October-November 1943.

[13] *Tracy* War Diary, 1943; Commanding Officer, USS *Pruitt* (DM-22), Mining Operations during Bougainville Campaign – Report on, 10 November 1943.

[14] *Tracy* War Diary, 1943.

[15] *Tracy* War Diary, 1943; Commander Mine Division One, Offensive Mining Operation 7-8 November 1943 – Report of, 9 November 1943.

[16] Commander Mine Division One, Offensive Mining Operation 7-8 November 1943 – Report of, 9 November 1943.

[17] ComMinDiv 1 War Diary, 1-8 November 1943.

[18] Commander Mine Division One, Offensive Mining Operation 7-8 November 1943 – Report of, 9 November 1943; Commander Third Amphibious Force, Offensive Mining Operation, 7-8 November 1943 – Report of, 6 December 1943.

[19] Commander Third Amphibious Force, Offensive Mining Operation, 7-8 November 1943 – Report of, 6 December 1943.

[20] Bruhn, *Battle Stars for the "Cactus Navy"*, 222.

[21] Commander Mine Division One, Offensive Mining Operation 7-8 November 1943 – Report of, 9 November 1943.

[22] General Orders: Bureau of Naval Personnel Information Bulletin No. 327 (June 1944).

CHAPTER 16 NOTES:

[1] Bruhn, *Eyes of the Fleet*, 235.

[2] Ibid, 236.

[3] Ibid, 237.

[4] "USS *Terror* and her family" by Robert S. Egan, *Warship International*, vol. 48, issue 4, December 2011; *Terror, DANFS*.

[5] *Terror, DANFS*.

[6] Ibid.

[7] *Terror* War Diary, October-November 1943.

[8] Bruhn, *Eyes of the Fleet*, 235.

[9] Ibid, 239-240.

[10] Ibid, 240.

[11] Bruhn, *Eyes of the Fleet*, 240; *Terror* War Diary, November 1943.

[12] Commander Aircraft, Central Pacific Force War Diary, November 1943; Lott, *Most Dangerous Sea*, 136.

[13] Commander Aircraft, Central Pacific Force War Diary, November 1943.

[14] Bruhn, *Eyes of the Fleet*, 242.

[15] *Terror* War Diary, December 1943.

[16] Ibid.

[17] Ron Swart email, 18 February 2018.

[18] *Terror* War Diary, December 1943.

[19] Commander Aircraft, Central Pacific Force, Aircraft Mining in the Marshall Islands – Report of, 25 January 1944.

[20] Ibid.

[21] Ibid.

[22] Ibid.

CHAPTER 17 NOTES:

[1] "RAAF Command Headquarters - 7th Fleet reports on RAAF minelaying operations 1943 - 1945 National Archives of Australia A11093, 373/21A5 (https://recordsearch.naa.gov.au/SearchNRetrieve/Interface/ViewImage.asp x?B=3079547: accessed 17 February 2018).

[2] "No. 11 Squadron" (https://www.awm.gov.au/collection/U59376); "No. 20 Squadron" (https://www.awm.gov.au/collection/U59383: both accessed 19 February 2018).

[3] "No. 11 Squadron"; "No. 20 Squadron."

[4] Steve Larkins, "No. 43 Squadron (RAAF)" (https://rslvirtualwarmemorial.org.au/explore/units/1500); "No. 43 Squadron" (https://www.awm.gov.au/collection/U59405: both accessed 19 February 2018).

[5] Dr. John Charles Lane, flight lieutenant medical officer with 20 Squadron RAAF (Catalinas), interviewed on 31 October 1989 by Harry Martin for The Keith Murdoch Sound Archive of Australia in The War of 1939-45.

[6] Lane interview.

[7] Clifford Dent Hull, as a flying officer on Catalinas with 11 and 42 Squadrons RAAF, interviewed on 24 October 1989 by Harry Martin for The Keith Murdoch Sound Archive of Australia in The War of 1939-45.

[8] Hull interview.

[9] Lane and Hull interviews.

[10] Hull interview.

[11] Ibid.

[12] Ibid.

[13] Ibid.

[14] Ibid.

[15] Ibid.

[16] "No. 11 Squadron."

[17] "No. 20 Squadron."

[18] Larkins, "No. 42 Squadron (RAAF)"; "No. 42 Squadron" (https://www.awm.gov.au/collection/U59404: accessed 19 February 2018).
[19] Larkins, "No. 42 Squadron (RAAF)"; "No. 42 Squadron."
[20] Larkins, "No. 42 Squadron (RAAF)"; "No. 42 Squadron."
[21] Hull interview.
[22] Ibid.
[23] Ibid.
[24] Larkins, "No. 42 Squadron (RAAF)"; Hull interview.
[25] Hull interview.
[26] Larkins, "No. 42 Squadron (RAAF)"; Hull interview.
[27] Larkins, "No. 42 Squadron (RAAF).
[28] Ibid.
[29] Larkins, "No. 42 Squadron (RAAF); "No. 11 Squadron."
[30] "No. 20 Squadron"; Larkins, "No. 43 Squadron (RAAF)."
[31] Mike Turner and Hector Donohue, *Australian Minesweepers at War* (Canberra: Sea Power Centre - Australia, 2018), 70, 81, 289-290; Lott, *Most Dangerous Sea*, 215.
[32] "Rathmines – The Top Secret Catalina Operations" (https://www.catalinaflying.org.au/documents/Top%20secret%20Catalina%20ops.pdf: accessed 18 January 2018).
[33] Ibid.
[34] "No. 11 Squadron"; "No. 20 Squadron"; "No. 42 Squadron."

CHAPTER 18 NOTES:

[1] Bruhn, *Battle Stars for the "Cactus Navy"*, 245.
[2] Ibid.
[3] Bruhn, *Eyes of the Fleet*, 311.
[4] Bruhn, *Eyes of the Fleet*, 311-312.
[5] Bruhn, *Battle Stars for the "Cactus Navy"*, 252-253.
[6] Bruhn, *Eyes of the Fleet*, 312.
[7] Bruhn, *Battle Stars for the "Cactus Navy"*, 253.
[8] Morison, *The Two-Ocean War*, 347.
[9] Bruhn, *MacArthur and Halsey's "Pacific Island Hoppers"*, 160.
[10] *Montgomery*, *Preble* War Diary, September 1944.
[11] *Montgomery*, *Preble* War Diary, September 1944.
[12] *Preble* War Diary, September 1944.
[13] *Montgomery*, *Preble* War Diary, September 1944; Commander Task Group 32.9 (Commander Mine Squadron 2), Peleliu – Angaur and Kossol Passage Action Report – submission of, 11 October 1944.
[14] Commanding Officer, USS *Perry*, Action Report, 23 September 1944.
[15] *Preble* War Diary, September 1944; Commanding Officer, USS *Perry*, Action Report, 23 September 1944; Commander Task Group 32.9 (Commander Mine Squadron 2), Peleliu – Angaur and Kossol Passage Action Report – submission of, 11 October 1944.

[16] Commanding Officer, USS *Perry*, Action Report, 23 September 1944; Commander in Chief, US Pacific Fleet and Pacific Ocean Areas, Operations in the Pacific Ocean Areas – September 1944, 7 March 1945.

[17] Commander Task Group 32.9 (Commander Mine Squadron 2), Peleliu – Angaur and Kossol Passage Action Report – submission of, 11 October 1944.

[18] David D. Bruhn, *Wooden Ships and Iron Men: The U.S. Navy's Coastal and Motor Minesweepers, 1941-1953* (Westminster, Md: Heritage Books, 2009), 49.

[19] Ibid, 50.

[20] Lott, *Most Dangerous Sea*, 179.

[21] Commander Task Group 32.9 (Commander Mine Squadron 2), Peleliu – Angaur and Kossol Passage Action Report – submission of, 11 October 1944.

[22] Ibid.

[23] Commander Task Group 32.9 (Commander Mine Squadron 2), Peleliu – Angaur and Kossol Passage Action Report – submission of, 11 October 1944; Lott, *Most Dangerous Sea*, 428.

[24] Bruhn, *Battle Stars for the "Cactus Navy"*, 254.

CHAPTER 19 NOTES:

[1] *Planter* War Diary, April 1944.

[2] Ibid, April-May 1944.

[3] Ibid, May 1944.

[4] Lott, *Most Dangerous Sea*, 85.

[5] *Barbican* War Diary, April 1945.

[6] Ibid.

[7] Ibid, May-June 1945.

[8] *Chimo* War Diary, June-July 1945; Commander Mine Squadron 101 War Diary, August 1945.

[9] *Trapper* War Diary, March-August 1945.

[10] *Barricade*, *DANFS*.

[11] Commander LCS(L) Group One War Diary, December 1945.

[12] Daniel E. Barbey, *MacArthur's Amphibious Navy* (Annapolis, Md: U.S. Naval Institute, 1969), 343-344.

CHAPTER 20 NOTES:

[1] *British Mining Operations, 1939–1945, Vol. 1* (Ministry of Defence Director of Naval Warfare, 1973), 689.

[2] Ibid.

[3] Ibid, 689-670.

[4] Ibid, 724.

[5] *British Mining Operations, 1939–1945, Vol. 1*, 725, 757, Table L; "Chronological Sequence of Submarine Minelaying Operations 1939-45" (http://www.naval-history.net/xGM-Ops-Minelaying.htm; "HNMS *O 19* (N 54)" (https://www.uboat.net/allies/warships/ship/2890.html: both accessed 4 March 2018).

[6] *British Mining Operations, 1939–1945, Vol. 1*, 715-716.

[7] Ibid, 724.

[8] "United Kingdom/Britain Mines"
(http://www.navweaps.com/Weapons/WAMBR_Mines.php: accessed 3 March 2018).
[9] Paul Kemp, *The T-class submarine: the classic British design* (Annapolis, Md: Naval Institute Press, 1990, 18, 38; "T class"
(https://www.uboat.net/allies/warships/class/53.html: accessed 3 March 2018).
[10] *British Mining Operations, 1939–1945, Vol. 1*, 691-692;
[11] *British Mining Operations, 1939–1945, Vol. 1*, 692; "*Kasumi Maru* (+1945)"
(https://www.wrecksite.eu/wreck.aspx?135359: accessed 2 March 2018).
[12] "HIJMS Submarine *I-37*: Tabular Record of Movement"
(http://www.combinedfleet.com/I-37.htm: accessed 2 March 2018).
[13] "HMS *Rorqual* (N 74)"
(https://www.uboat.net/allies/warships/ship/3415.html: accessed 4 March 2018).
[14] "*Porpoise* class" (https://uboat.net/allies/warships/class/49.html); "Loss of the Submarine HMS *Narwhal*"
(https://doriccolumns.wordpress.com/ww2-1939-45/hmsub-narwhal/);
"*Porpoise* Class Prototype Minelaying Submarine
(http://britsub.x10.mx/html/boats/minelayers/porpoise_classml.html: all three accessed 2 March 2018).
[15] "HMS *Porpoise* (N 14)"
(https://uboat.net/allies/warships/ship/3412.html: accessed 2 March 2018); *British Mining Operations, 1939–1945, Vol. 1*, 706-707.
[16] *British Mining Operations, 1939–1945, Vol. 1*, 713-714.
[17] "HMS *Porpoise* (N 14)"
(https://uboat.net/allies/warships/ship/3412.html); "IJN Subchaser *CH-8*: Tabular Record of Movement" (http://www.combinedfleet.com/CH-8_t.htm: both accessed 2 March 2018).
[18] "HMS *Porpoise* (N 14)"
(https://uboat.net/allies/warships/ship/3412.html: accessed 4 March 2018).

CHAPTER 21 NOTES:

[1] Bruhn, Eyes of the Fleet: The U.S. Navy's Seaplane Tenders and Patrol Aircraft in World War II, 329.
[2] Ibid.
[3] Ibid, 330.
[4] Ibid.
[5] Lott, *Most Dangerous Sea*, 139-140; Commander Task Group 77.5, Minesweeping Operations in Surigao Strait and Leyte Gulf – report of, 29 October 1944.
[6] Lott, *Most Dangerous Sea*, 140; Commander Task Group 77.5, Minesweeping Operations in Surigao Strait and Leyte Gulf – report of, 29 October 1944.
[7] Commander Task Group 77.5, Minesweeping Operations in Surigao Strait and Leyte Gulf – report of, 29 October 1944.

[8] Ibid.
[9] Ibid.
[10] Ibid.
[11] Commanding Officer, USS *Breese* (DM-18), Action Report – USS *Breese*, 5 December 1944.
[12] Commanding Officer, USS *Preble* (DM-20), USS *Preble* (DM20) Action Report of the Central Philippines Landing, 20 October, 1944, 20 November 1944.
[13] Lott, *Most Dangerous Sea*, 141.
[14] Commanding Officer, USS *Requisite* (AM-109), Ship's History, - Submission of, 15 April 1946.
[15] Bruhn, *Eyes of the Fleet*, 341-342.
[16] Ibid, 342.
[17] Ibid, 343.
[18] Bruhn, *Eyes of the Fleet*, 343; Lott, *Most Dangerous Sea*, 144.
[19] Commander Task Group 77.6, Action Report of Minesweeping Operations in Lingayen Gulf, 14 January 1945.
[20] Lott, *Most Dangerous Sea*, 144.
[21] Commanding Officer, USS *Southard* (DMS-10), Action Report, Lingayen Gulf Operations 2-11 January 1945, 19 January 1945; *Brooks, Hovey, Long, Palmer, Southard, DANFS*.
[22] Bruhn, *Eyes of the Fleet*, 334.
[23] Lott, *Most Dangerous Sea*, 144.
[24] Commanding Officer, USS *Monadnock* (CM-9), Action Reports – Forwarding of, 20 January 1945.
[25] Lott, *Most Dangerous Sea*, 145; Commanding Officer, USS *Monadnock* (CM-9), Action Reports – Forwarding of, 20 January 1945.
[26] Lott, *Most Dangerous Sea*, 145.
[27] Ibid.
[28] Ibid, 145-146.
[29] Ibid, 147-148.
[30] *Breese, Preble* War Diary, January 1945.
[31] Lott, *Most Dangerous Sea*, 148.
[32] Ibid.
[33] Ibid.
[34] Ibid, 148-155.
[35] Ibid, 155.
[36] Hector Donohue, "Lieutenant (Sp) HL Billman DSC RANVR."
[37] Ibid.

CHAPTER 22 NOTES:

[1] Bruhn, *Eyes of the Fleet*, 353.
[2] Ibid, 353-354.
[3] CTG 52.3, Action Report – Minesweeping Operations at Iwo Jima, 6 March 1945.

[4] Commanding Officer, USS *Gamble* (DM-15), Action Report; Battle of Iwo Jima, 17-19 February 1945.
[5] Ibid.
[6] Ibid.
[7] Ibid.
[8] Ibid.
[9] Ibid.
[10] Commanding Officer, USS *Gamble* (DM-15), Action Report; Battle of Iwo Jima, 17-19 February 1945; *Gamble, DANFS*.
[11] Lott, *Most Dangerous Sea*, 181.
[12] Lott, *Most Dangerous Sea*, 182; *Thomas E. Fraser, DANFS*.
[13] Commanding Officer, USS *Thomas E. Fraser* (DM-24), Ship's History – submission of, 27 February 1946.
[14] Ibid.
[15] Lott, *Most Dangerous Sea*, 183.

CHAPTER 23 NOTES:

[1] Commander Task Group 32.2 (52.2), Report of Capture of Okinawa Gunto – Phases One and Two, 23 July 1945.
[2] Bruhn, *Eyes of the Fleet*, 361.
[3] Ibid, 361-362.
[4] Lott, *Most Dangerous Sea*, 229; Commanding Officer, USS *Dorsey* (DMS-1), Action Report, 12 April 1945.
[5] Commanding Officer, USS *Dorsey* (DMS-1), Action Report, 12 April 1945.
[6] Ibid.
[7] Ibid.
[8] Commanding Officer, USS *Dorsey* (DMS-1), Action Report, 12 April 1945; *Dorsey, DANFS*.
[9] Lott, *Most Dangerous Sea*, 230.
[10] Ibid, 230.
[11] Ibid.
[12] Ibid.
[13] Ibid.
[14] *Robert H. Smith* War Diary, March 1945.
[15] Ibid.
[16] Lott, *Most Dangerous Sea*, 231; "US Naval Technical Mission to Japan: Reports in the Navy Department Library"
(https://www.history.navy.mil/research/library/research-guides/us-naval-technical-mission-to-japan-reports-in-the-navy-department-library.html: accessed 17 March 2018).
[17] "Donald W. Panek – World War II Diary, Invasion of Okinawa"
(https://www.history.navy.mil/research/library/manuscripts/p-r/donald-w-panek-wwii-diary-invasion-of-okinawa.html: accessed 17 March 2018).
[18] Commanding Officer, USS *Skylark* (AM-63), Action Report, 2 April 1945.
[19] Ibid.

[20] Ibid.

[21] Commanding Officer, USS *Skylark* (AM-63), Action Report, 2 April 1945.

[22] Commander Mine Division Eight, Action Report, USS *Adams* (DM-27) for Period 19 March to 7 April 1945, 14 April 1945.

[23] Commander Minecraft, U.S. Pacific Fleet, Operation of 2200 Ton Destroyer Minelayers with Sweeper Groups, 18 March 1945.

[24] Commander Mine Division Eight, Action Report, USS *Adams* (DM-27) for Period 19 March to 7 April 1945, 14 April 1945.

[25] Commanding Officer, USS *Adams* (DM-27), Action Report, 9 March 1945-7 April 1945, 9 April 1945; Commander Mine Division Eight, Mount III, Casualty, USS *Adams* (DM-27) – Adminstrative Report of, 10 April 1945.

[26] Commanding Officer, USS *Adams* (DM-27), Action Report, 9 March 1945-7 April 1945, 9 April 1945.

[27] Ibid.

[28] Ibid.

[29] Ibid.

[30] Ibid.

[31] Ibid.

[32] Commander Amphibious Group One, USS *Adams* (DM-27), Action Report, Okinawa Operation, 19 March to 7 April 1945, 7 July 1945.

[33] Lott, *Most Dangerous Sea*, 232.

[34] Ibid.

[35] Ibid.

[36] Commanding Officer, USS *YMS-103*, Action Report – Okinawa Operation, 26 March to 8 April 1945.

[37] Commander Amphibious Group One, USS *YMS 103* Action Report, Okinawa, 26 March to 8 April 1945, 27 June 1945.

[38] Commanding Officer, *YMS-103*, Action Report – Okinawa Operation, 26 March to 8 April 1945, 25 April 1945.

[39] Ibid.

[40] Ibid.

[41] Ibid.

[42] Commanding Officer, *YMS-103*, Action Report – Okinawa Operation, 26 March to 8 April 1945, 25 April 1945; Commanding Officer, *PGM-18*, USS *PGM-18* action report of, 24 April 1945.

[43] Commanding Officer, *YMS-103*, Action Report – Okinawa Operation, 26 March to 8 April 1945, 25 April 1945; *Buoyant, DANFS*.

[44] Commanding Officer, *YMS-103*, Action Report – Okinawa Operation, 26 March to 8 April 1945, 25 April 1945.

[45] Bureau of Naval Personnel Information Bulletin No. 377 (July 1948)

[46] *Buoyant, DANFS*; Lott, *Most Dangerous Sea*, 236.

[47] Lieutenant John J. Griffin Jr., USNR, Action report and sinking of USS *Emmons* (DMS 22), 6 April 1945, 12 April 1945.

[48] Ibid.

[49] Ibid.

[50] Ibid.
[51] Ibid.
[52] Ibid.
[53] Ibid.
[54] Bureau of Naval Personnel Information Bulletin No. 387 (May 1949).
[55] Commanding Officer, USS *Swallow* (AM-65), Action Report, 29 April 1945.
[56] *Molala* War Diary, April 1945.
[57] *Gayety* War Diary, April 1945
[58] Commanding Officer, USS *Terror* (CM-5), Action Report, 24 May 1945.
[59] Ibid.
[60] Ibid.
[61] Commanding Officer, USS *Terror* (CM-5), Action Report, 24 May 1945; Terror, *DANFS*.
[62] Ibid.
[63] "Commander Dickie Reynolds" (https://www.telegraph.co.uk/news/obituaries/1346476/Commander-Dickie-Reynolds.html: accessed 17 March 2018).
[64] Hector Donohue, article "Australia's Corvettes in the British Pacific Fleet."
[65] Ibid.
[66] Ibid.
[67] Ibid.
[68] "Britain's Navy Fighting Ships - Operations – History" (http://www.britainsnavy.co.uk/Battle%20Honours/Okinawa%201945.htm: accessed 20 March 2018).
[69] "History of U.S. Navy Mine Disposal," presentation provided to the U.S. EOD Association on 20 October 1995, page 132.
[70] Ibid.
[71] Ibid.

CHAPTER 24 NOTES:

[1] Bruhn, *Eyes of the Fleet*, 375.
[2] Ibid.
[3] Commander Task Group 95.4 (CTF 39) War Diary, July 1945.
[4] Lott, *Most Dangerous Sea*, 266.
[5] Commander Task Group 95.4 (CTF 39), Commander Mine Division Fifteen, War Diaries, July 1945; United States Pacific Fleet Organization 1 May 1945.
[6] Hector Donohue, article "Australia's Corvettes in the British Pacific Fleet."
[7] *Robert H. Smith* War Diary, July 1945.
[8] Ibid.
[9] Ibid.
[10] CTG 95.4 War Diaries, July 1945.

CHAPTER 25 NOTES:

[1] Bruhn, *Eyes of the Fleet*, 391-392

[2] Ibid, 393.

[3] Ibid.

[4] "Allied Ships Present in Tokyo Bay During the Surrender Ceremony, 2 September 1945" (https://www.history.navy.mil/research/library/online-reading-room/title-list-alphabetically/a/allied-ships-present-in-tokyo-bay.html: accessed 14 March 2018).

[5] "Casualties: U.S. Navy and Coast Guard Vessels, Sunk or Damaged Beyond Repair during World War II, 7 December 1941-1 October 1945" (https://www.history.navy.mil/research/histories/ship-histories/casualties-navy-and-coast-guard-ships.html: accessed 14 March 2018).

[6] "Naval Ship Losses World War Two" (http://www.gunplot.net/casualties/ranww2shiplosses.html: accessed 14 March 2018).

POSTSCRIPT NOTES

[1] "Lieutenant Commander Leon Verdi Goldsworthy" (http://www.navy.gov.au/biography/lieutenant-commander-leon-verdi-goldsworthy: accessed 8 March 2018).

[2] David D. Bruhn and Rob Hoole, *Home Waters: Royal Navy, Royal Canadian Navy, and U.S. Navy Mine Forces Battling U-boats in World War II* (Berwyn Heights, Md: Heritage Books, 2018), 307-309.

[3] "Lieutenant Commander Leon Verdi Goldsworthy."

[4] Ibid.

[5] Ibid.

[6] "Lieutenant Commander Leon Verdi Goldsworthy."

[7] "Lieutenant Commander Leon Verdi Goldsworthy"; Article "Unpicking the Goldsworthy Myths" provided by its author, Hector Donohue.

[8] "Lieutenant Commander Leon Verdi Goldsworthy."

[9] "Cliff," AWM PR 89/184 Volume 1, Letter from Director of Torpedoes and Mining dated 11 September 1944.

[10] Bruhn, *Battle Stars for the "Cactus Navy,"* 270-272; Letter from Director of Torpedoes and Mining dated 11 September 1944.

[11] Robert Macklin, *One False Move: Bravest of the Brave The Australian Mine Defusers in World War II* (Hachette Australia, 2012), 317-319; "History of U.S. Mine Disposal" (https://www.scribd.com/doc/68400349/MD-History-Scanned-PDF: accessed 8 March 2018).

[12] Bruhn, *Battle Stars for the "Cactus Navy,"* 270-272.

[13] Ibid.

[14] Ibid.

[15] Ibid.

[16] Macklin, *One False Move: Bravest of the Brave The Australian Mine Defusers in World War II*.

[17] Ibid.

[18] Hector Donohue, article "Bomb and Mine Disposal (BMD) Operations in the Pacific."
[19] Ibid.
[20] Ibid.
[21] Hector Donohue, article "Lieutenant Commander Leon Verdi Goldsworthy, Unpicking the Goldsworthy Myths."
[22] Barry Davies, "Obituary: LT. CDR. Pakgrave Ebden Carr, D.F.C., RAN (Rtd.)," *Naval Historical Review*, September 1993.
[23] Ibid.
[24] Ibid.
[25] Ibid.
[26] Ibid.
[27] Ibid.
[28] Jess Bravin, "What War Captives Faced in Japanese Prison Camps, And How U.S. Responded" (http://www.wsj.com/public/resources/documents/SB11126507244539762 1.htm: accessed 6 March 2018); Davies, "Lt. Cdr. Palgrave Ebden Carr, DFC, RAN (Rtd), A Unique Australian Naval Officer."
[29] Davies, "Obituary: LT. CDR. Pakgrave Ebden Carr, D.F.C., RAN (Rtd.)," *Naval Historical Review*, September 1993.

APPENDIX A NOTES:

[1] "RAAF Command Headquarters - 7th Fleet reports on RAAF minelaying operations 1943 - 1945 National Archives of Australia A11093, 373/21A5; "Japanese Naval and Merchant Shipping Losses During World War II by All Causes," Prepared by The Joint Army-Navy Assessment Committee NAVEXOS P 468, February 1947.
[2] "Japanese Naval and Merchant Shipping Losses During World War II by All Causes," Prepared by The Joint Army-Navy Assessment Committee NAVEXOS P 468, February 1947.
[3] Ibid.
[4] Ibid.

APPENDIX B NOTE:

[1] "No. 11 Squadron" (https://www.awm.gov.au/collection/U59376); "No. 20 Squadron" (https://www.awm.gov.au/collection/U59383); "No. 42 Squadron" (https://www.awm.gov.au/collection/U59404); "No. 43 Squadron" (https://www.awm.gov.au/collection/U59405: accessed 19 February 2018).

Index

ABDA (American-British-Dutch-Australian) Fleet, xxx, 49, 72-74
Agnew, D. M., 87
Ainsworth, Walden Lee, 154-156
Akiyama, Teruo, 156
Albertson, James Paul, 214
Anderson, Gavin (Lt., RANVR), 240
Anderson Jr., Alexander, 250
Andrews Jr., John, 2, 24, 135, 148
Annen, Donald F., 274
Armstrong Jr., Henry Jacques, 22, 26, 257, 260, 275-276, 282
Armstrong, Warren Wilson, 24-25, 156, 164, 172
Arnold, Sydney Allan (Lt., RANVR), 300
Atkinson, Anthony Noel Lee, 194, 311
Australia/Australian
  Adelaide, 45, 47
  Bowen, 184
  Broken Hill, New South Wales, 292
  Brisbane, xviii, 38, 45, 189, 296, 303
  Broome, xxxi-xxxii, 80
  Cairns, xviii, 184, 190
  Darwin, Doctors Gully, 184, 190-193, 239
  Exmouth Gulf, 75
  Fremantle, xxxii, 45, 69, 80
  Great Barrier Reef, Torres Strait, xxviii
  Hobart, xii, 45
  Kurumba, 184
  Melbourne, 45, 302,
  Melville Island, 190, 193
  Queensland, xviii
  Royal Air Force
    No. 11, 20, 42, 43 Squadrons, xxviii, 48, 183-194, 313, 314
    No. 76 Wing, 192
    No. 100 Torpedo Bomber Squadron, 303
    PBY-5A Catalina flying boats, 21, 44, 48, 183-194, 303
    Pearce RAAF Base in Bullsbrook, 303
    RAAF Base Rathmines, 184-185, 193, 239
  Royal Australian Navy
    Flinders Naval Depot (Render Mines Safe training), 239, 299
    Minesweeping Flotilla
      21st, x, 272, 289
      22nd, 272
      25th, 132

Naval College at Jervis Bay, 302

Render Mine Safe, 238-239, 292, 296, 299

Sydney, xii, 46-47, 130, 167

Townsville, xii, xviii

Watsons Bay, New South Wales, 272

Bagley, David W., 106

Baker, Walter E., 296

Baldridge, Edward Francis, 25-26, 202, 208, 240

Baldwin, Charles E. (CPO, DSM, RN), 293

Barbey, Daniel E., 233

Barr, Noel Charles Edward, 194

Batterham, Maurice Samuel (Comdr., OBE, RANVR), 300-301

Battle of

Express Augusta Bay, 168

Guadalcanal, 1, 5, 151

Kula Gulf and Battle of Kolombangara, 156

Leyte Gulf, 225

Midway, 109, 114-177

Okinawa, 249, 262

Philippine Sea, 198, 225

Saipan, 198

Java Sea, xxx, 49, 74

Santa Cruz Islands, 150-151

Sunda Strait and Second Battle of Java Sea, xxx, 74, 80

Surigao Strait, 231

Tarawa, 174

Bayly, G. E. W. W., 70

Beard, F. W., 87

Bell, W. J., 206

Berkey, Russell S., 233

Bier, Charles D., 214

Billman, Harold Leon (Lt., DSC, MID, RANVR), 239, 300-301

Bird, John Arthur, 97, 99

Bjarnason, Paul Henrik, 22-28, 216, 275-276, 282, 284,

Blakeslee, Horace William, 25-26, 208, 248, 269, 276

Bloch, Claude C., 85

Boyington, Gregory, 167

Brander, J. M. (Lt., SSRNVR), 70

Beecher Jr., William G., 86, 253

Bremer, James John Gordon (Rear Adm. Sir, KCB, KCH, RN), 59

Brennan, John Edward, 235, 282

Britain/British

Army

1st Battalion, Middlesex Regiment, 2nd Battalion, Royal Scots, 2/14th Battalion, Punjab Regiment, 5/7th Battalion, Rajput Regiment, Royal Artillery, 64, 67

Mauritius Island (presently the Republic of Mauritius), 46
Royal Air Force
  Squadrons
    159 and 355, xxvii
  WAAF (Women's Auxiliary Air Force), 101-102
Royal and Dominion Air Forces
  Beaufort I, Halifax III, Hampden I, Lancaster III, Liberator I/II,
    Manchester I, Mosquito XIB, Stirling I, Wellington IC/X,
    Bristol Blenheim light bomber, 44
Royal Navy
  4th Flotilla, 217, 222
  8th Flotilla, xxxii, 217, 219
  British Pacific Fleet (Task Force 57), 265-277
  China Station, 53-54
  East Indies Station, xxxi, 46, 53
  Eastern Fleet, 53, 69, 217
  Enemy Mining Section, 294, 296
  Fleet Air Arm, Albacore, Avenger II, Baracuda II, Swordfish, 44
  HMS Vernon, 37-40, 292-296
  Mine Disposal Section (later Land Incident Section), 294
  mines
    A Mk 1, 44, 46, 69
    M Mk 2, 218-223
    Mk 14, xxvi, 55, 61
    Mk 16, 41, 218, 221, 223
    Mk 17, 56
  Naval Air Stations Arbroath and Crail, 303
  Port Clearance Parties (P Parties), 15, 294-295, 299
  Seafox floatplanes, 302-303
  Submarine Flotilla
    2nd, 4th, 8th, 217-224
  Technical Mission to Japan, 299
  Shoeburyness, 293
  Southampton, 39, 294
Brown, Albert, 124
Brown, Colin Stuart, 194, 313
Brown, Harold Gordon MacMurray, 194, 313
Bruce, John R., 112-113
Bryant, A. H., 113
Buckner Jr., Simon B., 117, 262, 284
Burgis, R. W., 180
Burke, Arleigh, 168-169
Burma (present day Myanmar), xxiii, xxvii, xxix, 42, 129, 218
Burnett, Norman R., 214
Burns (Lt.), 228
Burrage, Reginald Bruce, 194, 313

Butcher, H. C. (Lt., SSRNVR), 70
Byhre, E. B., 215
Calder, Norman (Comdr., OBE, RAN), 47
Calhoun, William L., 87, 177
Callin, John Elijah, 113
Canada/Canadian, 34, 64, 66, 116, 127, 272
   Royal Rifles, Winnipeg Grenadiers, 53
Caplan, Stanley, 235
Carlon, Charles B., 85
Carpenter, Richard Edward, 26-27, 248, 276
Carr, Palgrave Ebden, 189, 292, 302
Cassady, John H., 219
Chambers, Thomas Edward, 26-27, 248, 276
Chapman, Francis Blomfield, 194, 313
Cheney Jr., Wilbur Haines, 22, 27, 123-124, 216, 275-276, 282, 284
Chillingworth Jr., Charles F., 86
China/Chinese
   Amoy Harbor, Swatow, 215
   Kowloon, 58
   Wenchow, xxix, 193
Churchill, Winston, xv
Claggett, John H., 7
Clark, Robert Trevor, 194, 314
Clark, Robley Westfield, 179-181
Clay, Donald Noble, 26-27, 243, 248
Cleland, Arthur John, 194, 313
Cliff, Geoffrey John "Jack" (Lt. Comdr., OBE, MBE, GM, BEM, RANVR), 240, 296
Cochran, J. B., 87
Cohen, David Barney, 25, 228, 240
Cole, Benjamin Nooe, 235
Cole, John E., 26, 235, 240
Collis, John Leon, 2-10, 24, 106, 148
Conolly, Richard L., 233
Costello, Robert W., 85, 281
Coxe Jr., Alexander Bacon, 24-25, 91, 156, 164, 172
Crandell, Donald Allen, 24, 86, 93-94
Crichton, Charles Helmick, 24, 126
Croft, Charles George (Lt., MBE, RANVR), 300-301
Cronin, J. C., 233
Crowe Jr., John F., 115-124
Daly (Lt. (jg)), 228
Davidson, Howard C., 101
Dekker, J. P. A., 75
de Greeuw, C., 76
de Horsey, Algernon Frederick Rous (Adm. Sir, KCB, RN), xxi

de Jong, H., 76
Drain, Dan Thomas, 22, 25, 208
Duigan, Terence Lawless, 194, 313
Durgin, C. T., 346
Elphick, Thomas Roy, 192
Emslie, Alexander Ronald, 194, 313
Ensoy, L. B., 98
Estep, George M., 255
Etienne, Armand Andre, 192, 194, 314
Evans, Arthur Reginald (DSC, RANVR), 154
Farragut, David Glasgow, xxi
Farrow, Henry, 22, 26-27, 248, 275-276
Fatkin, Earl David, 214
Fields, Thomas M., 241
Fitch, Aubrey W., 140
Fitch, Howard Wesley, 25, 177, 182
Fletcher, Frank J., 114
Florance, J. E., 87
Foley, J. L., 87
Forster, Paton, 272
Fort, George H., 206
Foster, Edward Lee, 26-27, 248, 276
Fowler, T. F., 87
France/French
  Crozet Islands, Kerguelen Islands, xxviii, xxxiii, 46
  Le Havre, 13, 15, 19, 21
  Normandy, 15, 19, 23, 211, 225, 295
  North Africa
    Algeria, Oran, 15
    Morocco, Casa Blanca, 23
Fraser, Bruce Austin (1st Baron of North Cape, GCB, KBE, RN), 271-272
Fraser, Jean, 101-102
Freeze, Raymond Edsel, 236
Furlong, William R., 12, 86-90, 100
Gadrow, Robert Emmett, 26, 248
Garing, William Henry (Air Commodore, CBE, DFS, RAAF), 302
George, Albert Frederick Arthur (King George VI), 293
Ghormley, Robert L., 149
Glassford, William A., 74
Glenny, John Edward Maxwell (Capt., DSO, DSC, RN), 293
Glover, R. O., 233
Goldsworthy, Leon Verdi (Lt. Comdr., GC, DSC, GM, RANVR), 240, 285, 291-299
Granum, A. M., 233
Gray, Robert H., 118
Gray, Robin Henry, 194, 314

Green, Redmond Forest Michael, 194, 313
Greenlee Jr., D. G., 87
Griffin Jr., John J., 266-267
Gross, John Winston, 23, 27, 211, 214, 216
Guiot, Louis, 51
Guthrie, Richard Allen, 24, 86, 135
Gygax, Felix, 210
Habbell, M. H., 85
Haber, Charles Percy, 23, 211, 214
Halsey Jr., William F., 1, 9, 95-96, 114, 149-160, 174, 277, 287
Harrelson, George D., 214
Harwood, Robert F., 214
Hashimoto, Shintaro, 6-7
Hashizume, Hisao, 103-104
Hayes, John M., 252, 282
Hazzard, George Alonzo, 113
Helfrich, Conrad Emil Lambert (Lt. Adm., GNL, KCB, RNLN), 49, 74
Hemingway, W. E., 109
Heringa, A. D. H., 76
Herron, Edwin Warren, 24, 116
Hickok, Warren Paul, 97-98
Higgins, Brian Hartley, 194, 313
Hirohito, Michinomiya, 5, 287
Hirst, Robert Maxwell, 194, 313
Hisao, Okane, 163
Hodgkinson, Victor Allan, 194, 313
Hokanson, Leslie Harold, 194, 313
Holden, Andrew D., 131
Holm (Lt., RNR), 70
Holms, Robert Henderson, 25
Hoover, John H., 178-179
Huie, B. S., 233
Hull, Clifford Dent, 185-192
Hull, G. D., 100
Hunt, Robert Bagster Atlee (Comdr., OBE, RAN), 228
Hunter, John (Lt., RANVR), 240
Hurt, Lindley Maxwell, 194, 314
Hussey Jr., George F., 87
Hyakutake, Harukichi, 9
Ikin, Stuart Austin, 194, 313
Ilsemann, F. J., 87
Ingram II, William Thomas, 27, 216, 276, 282, 284
Ionides, Hugo Meynell Cyril (Capt., RN), 224
Jackson, Alfred Frank, 58
Japan/Japanese
   1st Mobile Fleet, 225

6th Fleet, 105
8th Fleet, 158
17th Army, 5
25th Army, 62
38th Infantry Division, 64, 66
331st Naval Air Group, 224
"Camote" aircraft laid mine, 240
Hong Kong prison camps, 66
Honshu, Hiroshima, 277, 285
Kurile Islands (Tankan Bay), 81
Kyushu, Nagasaki, 287
Military aircraft
    Aichi D3A Navy type 99 "Val" carrier bomber, 235-236, 253, 258, 266,
        349
    Kawanishi H8K1 "Emily" flying boat, 102-104
    Kawasaki Ki-45 Toryu Army type 2 "Nick" two-seat fighter, 244
    Kawasaki Ki-61 Hien Army type 3 "Tony" fighter, 258-259, 266
    Mitsubishi A6M Type Zero fighter, 82, 237, 258-259
    Mitsubishi G3M "Nell" medium bomber, 63
    Mitsubishi G4M Navy type 1 "Betty" land-based attack aircraft, 229, 252,
        284
    Mitsubishi Ki-21 type 97 "Sally" heavy bomber, 179, 235
    Mitsubishi Ki-46 Army type 100 "Dinah" reconnaissance aircraft, 284
    Nakajima B6N2 Tenzan torpedo-bomber, 224
    Yokosuka E14Y1 "Glen" floatplane, 221
Ofuna Interrogation Center, 304
Ryukyu Islands
    Amami, Osumi, and Tokara Islands, 249
    Okinawa Islands,
        Kerama Retto, 11, 18, 251, 253-254, 260, 264, 268-271
        Nakagusuku Wan ("Buckner Bay"), 253, 262-265,
    Sakishima Islands, 249, 271-272
Yano Battalion, 5
Jedeloo, Henri, 78
Jellema, Tobias, 52
Johnson, Clifford Arthur, 27, 276
Joos, H. W., 93
Kai-shek, Chiang, 215
Keith, George Richard, 97, 99
Kelly, James J., 7
Ketner, F. W., 237
Kimball, Whitefield F., 268
Kimberly, Lewis A., 131
Kimmel, Husband E., 82-83, 90
Kinkaid, Thomas C., 149, 232-233
King, Ernest J., 277, 288

Knapp, Richard Albert, 23, 211, 214
Knoops, Frederik Johan Adolf, 50
Koenraad, Pieter, 75, 78
Krueger, Walter, 232-233
Kuiper, L., 78
Küller, Gijsbertus Petrus, 50
Kurita, Takeo, 225
Lane, John Charles, 184
Latta, W. A., 228, 280, 282
Lawrence, Denis Russell, 194, 313
Lawrence, Sidney J., 181
Layton, Edwin T., 109
Layton, Geoffrey (Vice Adm. Sir, KCB, DSO, RN), 53, 72
Leatham, Ralph (Adm. Sir, KCB, RN), 46
Lebeau, J. R. L., 75
LeHardy, L. M., 87
Lewis, Richard E., 214
Lewis, Roger Curzon (Capt., DSO, OBE, RN), 293
Li-ming, Tu, 215
Lindsay Jr., William N., 202
Long, Elmer Cecil, 27, 216
Longmuir, John Kenneth, 194, 314
Longworth, Harold Richard, 194, 314
Loud, Wayne R., 199, 204, 206, 226, 228, 233-235, 279-283
Lundrum, Eugene M., 122
MacArthur, Douglas, 157-158, 161, 174-175, 208, 232-233, 287
MacKinnon, R. M., 87
Mahaffrey Jr., John K., 206
Maltby, Christopher, 53
Marr, Reginald Joseph, 194, 314
Marshall, George C., 215
Marsham, Hubert Anthony Lucius (Comdr, OBE, RN), 222
Masatake, Okumiya, 1
Masselink (Ens., RNLN), 78
McCain Sr., John S., 139-140, 286
McEathron, E. D., 87
McGuirk Jr., William Edward, 235
McKay, Baxter Morrison, 126
McKellar Jr., Clinton, 122
McKnight, George W., 26-27, 240, 248, 276, 282, 284
McLean, John B., 233
McMahon, Patrick John, 194, 314
McNally, Terry, 191
Merrill, Aaron Stanton, 157-168
Middleton, Gervase Boswell (Adm., CB, CBE, RN), 296
Miller, C. L., 181

Miller, L. H., 87
Miller, Ronald Nevill Damian, 194, 314
Momsen, Charles B., 106
Moody, Dwight Lyman, 25, 164
Morrison, C. H., 85
Mould, John Stuart (Comdr., GC, GM, KCBC, RANVR), 294
Mullins Jr., Henry, 126
Murray, George D., 149
Mutch, James Lionel, 194, 313
Nagumo, Chuichi, 81
Nankivell, Francis (Lt., MBE, RANVR), 301
Netherlands/Dutch
  Royal Netherlands Navy
    2nd Minesweeper Division, 75
    4th Minesweeper Division, xxxi, 78
    PBY-5 aircraft, 49, 52
    Submarine Division Four, 49
    Vickers T3 mine, xxxii, 41
New Zealand, xxviii, 38, 45, 47, 54, 129-130, 272
Nimitz, Chester W., 10, 109, 111, 116, 118, 149, 173-177, 196, 287-288
Noble, Percy Lockhart Harnam (Adm. Sir, GBE, KCB, CVO), 60
Noel Jr., John Vavasour, 172
Oakley, John Philip Holroyde (Comdr., DSC, RN), 221
O'Brien W. J., 87
Oldendorf, Jessie B., 233
Oliver, James Francis Burgess, 192
Operation
  CARTWHEEL (Allied grand strategy in the South Pacific), 157, 159
  CHERRY BLOSSOM (Allied assault landing at Cape Torokina, Bougainville), 167
  FORAGER (Assault landings in the Mariana Islands), 196
  K (Second aircraft attack on Pearl Harbor), 102
  KE (Ke-gō Sakusen, evacuation of Japanese forces on Guadalcanal), 5, 9
  KETSU-GO (plan to use Civilian Corps Voluneers to defend Japan), 285
  LANDCRAB (Allied invasion of Attu), 125
  MUSKETEER, 232
  STARVATION, xxiii
  TOENAILS (Allied invasion of New Georgia), 160
  TORCH (Allied invasion of North Africa), 13, 177
Ortlepp, Brian, 194, 314
Ouvry, John Garnault Delahaize (Comdr., DSO, RN), 293
Ozawa, Jisaburo, 30

Pacific Islands/Atolls
  Aleutian Islands, 24, 109, 115-128, 206
  Biak Island, Schouten Islands, 208
  Bismarck Archipelago

Admiralty Islands, Manus, 190, 226, 229
Emirau Island, 195
New Britain, 143, 157, 184, 300
Gasmatta, Lakunai, Rapopo, Vunakanau,184
Rabaul, xii, 5, 143-144, 156-159, 161, 165-165, 184
New Ireland, Kavieng, xxix, 190, 193
Bora Bora (Tearanui Harbor), 130
Caroline Islands
Ngulu, 17
Palau Islands, 21, 25, 195-208, 226, 290
Ponape, Puluwat, Satawan, Woleai, Yap, 196
Truk, 149, 195-196
Ulithi, 48, 252, 290
Ellice Islands, Funafuti, 177-179
Fiji, Suva, 130-134, 139
Gilbert Islands
Apamama, Makin, Tarawa, 174-182
Hawaiian Islands/Area
French Frigate Shoals (Tern Island), 101-114
Gardner Pinnacles, Laysan, Lisianski, 111
Kauai, Maui, 101
Kure, 108, 111, 277
Midway, 108-118
Necker, 111, 113
Oahu, 81-86, 98, 101-103, 112, 115, 177, 198, 274
Pearl and Hermes Reef, 108, 111
Indonesia (formerly Netherlands East Indies)
Greater Sundas
Sumatra, xxxi-xxxii, 42, 49, 51, 70, 192, 217-220
Java,
Babi Island, xxxii
Jakarta (formerly Batavia), Tandjong Priok, xxix-xxxii, 41, 51, 69
Madoera (Madura) Island, xxx-xxxi, 49, 51, 76-77, 80
Surabaya (formerly Soerabaja), xxix-xxxi, 49, 51-52, 73-80, 190, 192
Borneo, 49, 193, 227, 290, 296-299
Balikpapan, 227, 290, 297-299
Brunei Bay, 190, 192, 290, 297-298
Tarakan Island, xxx-xxxi, 49, 51, 190, 290, 297-298
Celebes (now Sulawesi), 190-193, 208, 304
Salajar Island, 191
Lesser Sundas
Bali, 77, 218
Flores, 192
Gili Islands, Kangean Islands, Sumbawa (Soembawa), 76-77, 80
Paternoster (Balabalagan) Islands, 80, 190
Timor, 190, 290

Maluku Islands (Moluccas), xxx, 73
  Aru Islands, Ambon Island, xxx, 184
  Halmahera Islands, Morotai Island, 190
Johnston Atoll, 86
Malaysia
  Kota Bharu, 50, 62-63
  Penang, 42-43, 69, 217-223
Mariana Islands, 25, 175-176, 195-197, 208, 225, 270, 290
  Farallon de Pajaros, 196
  Guam, 16, 20, 25, 129, 195-199, 208, 246, 260, 290
  Pagan, Rota, 196
  Saipan, 21, 195-198, 209, 225, 243, 246, 270, 290
  Tinian, 195-198
Marshall Islands
  Eniwetok, 20, 196, 270
  Jaluit, 104, 181
  Kwajalein, Majuro, 25, 174, 196
  Maloelap, Mille, 181
  Wotje, 102, 104-105, 181
New Caledonia (Noumea), 1, 8, 130, 160, 184
New Guinea (formerly Papua and New Guinea, today Papua-New Guinea)
  Aitape, 195
  Babo, Jefman, Kaimana, Langgoer, 184
  Hollandia, 195-196
  Milne Bay, 303
  Papua,
    Kaimania, Manokwari, Sorong, 190
    Port Moresby, xii, xxviii, 3, 239
  Sepik River, 185, 188
New Hebrides (today Vanuatu), 130
  Efate, 134, 136, 161
  Espiritu Santo, 3, 8, 139, 144-155, 168-170, 181
  Malo Island, 141
Nicobar Islands, 223
Philippine Islands/Sea
  Corregidor Island, 129, 239, 290
  Leyte Island, 25, 190, 208, 225-241, 277, 300
  Lingayen Gulf, 26, 230-240, 290
  Luzon Island, 26, 230-233, 240-241, 290
  Mindanao, 73-74, 175, 235
Samoa, 130
  Apia (British Samoa), 131
  Tutuila (Pago Pago Harbor), 86, 129-131
Solomon Islands
  Bougainville, Buin, xii, 1, 5, 7, 9, 29, 139, 156-172
  Guadalcanal area

Florida Island, 2, 142, 168
Gavutu Island, 4, 142
Guadalcanal Island, 1-10, 24, 112, 142-152, 160-161, 166, 168, 172, 176, 182, 209, 225, 283, 304
Malaita Island, Maramasike Island, 142-144
New Georgia Sound ("the Slot"), xxxiii, 3-8, 142, 156, 161, 166-168
Savo Island, 7-9, 144, 303
Tanambogo, 4, 142
Tulagi Island, 2-7, 24, 112, 142, 148-163
Munda, 1, 155-167
New Georgia
Faisi Island, 157-161, 168
Kolombangara Island, 152-164
Masamasa Island, 168, 170
Rendova Island, 1, 125, 159-164
Viru Harbor, 158
Nura Island, 147
Palmyra Island, 138
Russell Islands Group (Banika and Pavuvu), 160-170
San Cristobal, 6
Santa Cruz Group, 150-151
Shortlands, 7, 152, 157, 160-161
Upolu Island, Wallis Island, 130
Vangunu, 7, 25, 158, 164
Vella Lavella, 1, 167
Tonga Islands, Tongatapu, 130, 134-136
Wake Island, 129

Palliser, Arthur, 75
Panek, Donald D., 255
Parker, Ralph C., 117-118
Pears, Arthur Luard, 65
Peel, Reginald Frank (Able Seaman, BEM, RANR), 301
Pefley, Alfred R., 106
Pence, John Wallace, 97, 99
Pendleton, W. B., 87
Peskin, Michael, 106-109
Peters, Arthur Malcolm (Rear Adm. Sir, KCB, DSC), 60
Phelan, George R., 24, 86, 99, 106
Pierce Jr., H. R., 228, 281
Pikkaart, David James, 27, 216, 282, 284
Rasmussen Jr., S. C., 235
Rawlings, Henry Bernard Hughes (Adm. Sir, GBE KCB, RN), 277
Read, Gordon Neville, 194
Reeves Jr., John W., 118
Renfro, E. C., 181
Reynolds, C. D., 87

Reynolds, William, 110
Rhee, C. A. E., xxxi
Richards, Ralph L., 7
Richter, William Julius, 25, 116, 169-172
Riner, John D., 213
Roberts, D. G., 87
Roberts, Donald Arthur, 309
Roberts, W. L., 93
Robertson, Derek Fanshawe, 192
Robertson, Norman Valentine, 194, 309
Robson, Ian McCallister, 192
Rochefort, Joseph J., 101, 109
Roozen, P. A. H., 78
Rotgans, Johannes Pieter, 78
Rowe, Austin Edward, 23
Sakai, Takashi, 66
Sanders, E. C., 181
Sanders Jr., William Henry, 22, 26, 275-276,
Sasao, Shosuke, 103
Sax, Herman Nicolaas, 52
Schiano, C. B., 87
Schminke (Lt.), 228
Schmidt, Harry, 241
Schultz, William Christian, 24, 86
Scott, Henry Corbett, 194, 309
Scudder Jr., Theodore Thomas, 23, 211, 214
Scull, Herbert M., 178
Selby, Henry Charles Sylvester Collingwood, 66
Seymour, Robert Michael, 194, 309
Shands, C. (Capt.), 1
Sharp Jr., Alexander, 211, 242-253, 270
Shaw, B., (T/Lt., RNVR), 70
Shearon, Warren B., 214
Shigemitsu, Mamoru, 195

Ships and Craft
  Australian
    Ararat, xii
    Armidale, Patricia Cam, Warrnambool, 290
    Australia, xxviii, xxxiii, 45-46, 303
    Ballarat, 273, 289
    Bendigo, Burnie, Cairns, Kalgoorlie, Launceston, Lismore, Napier, Nepal, Nizam, Norman, Pirie, Quiberon, Quickmatch, Whyalla, 273
    Bungaree, xviii, 45, 47, 52,
    Canberra, 144
    Cessnock, Ipswich, 289
    Gascoyne, 228, 233

HDML-1074, 228
Medeau, Mercedes, 70
Manoora, Westralia, 303
Perth, 302
Sydney, 46
Vampire, 35
Warrego, 233
British
Abdiel, 29-30, 32-33
Adamant, 217
Adventure, xxvii, 29-31, 54
Agamemnon, 30-31
Alsey, 38
Apollo, Ariadne, 32-33
Ark Royal, Courageous, Furious, 302
Atreus, 38
Barlight, Tern, 66
Buffalo, Dauntless, 62
Cachalot, Grampus, 41
Changteh, Circle, Jerak, Medusa, 70
Corncrake, Redshank, 36
Darvel, Durban, Kedah, Kinta, Empire Star, Gorgon, Jupiter, Kedah, 69
Dittisham, Flintham, 40
Esmeralda, 295
Esk, 33-34
Express, xxvii, 33-34, 69
Formidable, 299
Hampton, 29, 31
Hermes, 303
Icarus, Impulsive, Intrepid, Ivanhoe, 33-34
Indefatigable, 271
Kelvin, 34
Kung Wo, 30-31, 35, 61, 63, 70
Latona, Manxman, Welshman, 32-33
Linnet, 29-30, 36-37
Li Wo, 70-71
M-1 (later *Miner I*), *M-2* (later *Miner II*), *M-3* (later *Miner III*), *M-4* (later *Miner IV*), *M-5* (later *Miner V*), *M-6* (later *Miner VI*), *Miner VII*, *Miner VIII*, 37-38
Maidstone, 217
Mao Lee, 60-61
Mao Yeung, 30, 32, 35, 60-61
Menestheus, 30, 32
ML-1063, Rahman, 70
Narwhal, 41
Neptune, 46

Nightingale, 30, 38, 40
Obdurate, Obedient, Opportune, Orwell, 34
Orion, 302-303
Plover, 29-30, 36
Porpoise, 41, 52, 217-224
Port Napier, Port Quebec, 30, 32
Princess Victoria, 29, 32
Redstart, 29, 37, 58, 60, 66-67
Ringdove, 29, 37
Robin, 60
Rorqual, 41, 217-221
Salmon, 222
Scout, Stronghold, 30, 35, 55-58, 61-69
*Sea Rover, Stoic, Tally-Ho, Tantalus, Tantivy, Taurus, Templar, Thorough, Thule,
   Tradewind, Trenchant, Trespasser, Truculent, Tudor*, 42-43, 52, 217-221
Seal, 41, 54
Shepperton, 29, 32
Sin Aik Lee, 70
Sirdhana, 57
Skylark (later *Vernon*, and *Vesuvius*), 38-39
Southern Prince, 30, 32
Stygian, 224
Sulphur, 59
Surf, 44
Tactician, 42, 217-218
Taitam, Waglan, 66
Tamar, 59
Tapah, 70
Tenedos, 30, 35, 55, 57-58
Teviot Bank, 29, 32, 35, 62-63
Thanet, Thracian, 30, 35, 55, 60-67
Wolfe, 217
Canadian
  Gatineau, 34
Dutch
  Abraham Crijnssen, viii, xx, xxxi-xxxii, 52, 75-80
  Bangkalan, 49, 52
  Eland Dubois, 76
  Gouden Leeuw, xxxi, 49, 51, 76
  Janssens, Petronella, 80
  K-8, K-9, K-11, K12, K-14, K15, 80
  Krakatau, xxxi, 50-15
  Merbaboe, Rindjani, Smeroe, xxxi, 78-80
  O-19, vii, xxxii, 49-52, 80, 217-218
  O-20, 49-50
  Pieter de Bitter, 75

Prins Van Oranje, Pro Patria, xxxi, 49-51
Rigel, xxxi, 50-51
Serdang, xxxi,
Sigli, Zaandam, 69
Soemenep, 50, 52
Tromp, Zuiderkruis, 80
Urania 2, 76
Willem van der Zaan, xxx-xxxi, 49-50, 63, 80
French, *Requin*, 136
German
  Atlantis, Komet, Kormoran, Pinguin, 46
  S-54, S-61, 33
  U-360, U-617, U-652, 32-34
Japanese
  Akagi, Hiryu, Kaga, Soryu, Zuikaku, 81
  Akigumo, Hamakaze, Isokaze, Kazegumo, Maikaze, Oshio, Satsuki, Tokitsukaze, Urakaze, Yukikaze, 6
  Arashi, 6, 69, 77
  Arashio, 6, 77
  Asagiri, Fubuki, 35
  Ayanami, Uranami, 50
  Chikuma, 151
  Fumizuki, Kawakaze, 6-7
  Hijo, 225
  Hyaski Maru, Shinko Maru No. 1, Yuno Maru, xxxii,
  I-9, I-15, I-19, I-23, I-26, 105
  I-37, 221
  I-123, 144-147
  Kagero, Oyashio, 154, 306
  Kasumi Maru, 221, 308
  Kurashio, Michishio, 154
  Lisbon Maru, 67-68
  Makigumo, Yugumo, 6, 9
  Makinami, 7
  Maya, Nowaki, 69
  Nagatsuki, Suzukaze, Tanikaze, 6, 156
  Niizuki, 156
  patrol boat *38, Yamakaze*, xxxi, 51
  patrol vessel *101*, 35
  RO-100, 163, 306
  Sendai, 112, 142
  Shirayuki, 6-7, 35
  Shokaku, 81-82, 150-151, 198, 225
  Taiho, 198, 225
  Yugiri, 35, 50
  Zuiho, 150

New Zealand
  Achilles, 136
  Leander, 156
  Matai, 132
Norwegian, *Höegh Transporter*, 57
United States
  Coast & Geodetic Survey, *Explorer*,
  Merchant Marine
    Benjamin Franklin, Paul Revere, 136
    Chester Sun, 170
    D. G. Schofield, President Garfield (later *Thomas Jefferson*), 130-131
    James B. McPherson, 152, 154
    Lewis L. Dyche, 236
    Nira Luckenbach, President Coolidge, 140-142
  Navy
    amphibious
      Brooks, 235, 237
      Crescent City, 267
      Grant, Heywood, 125
      Hinsdale, Pitt, 256
      LCI(G)-70, 237
      LCS(L)-30, LCS(L)-50, 215
      LCT-278, 182
      Little, 146-147
      LSM-126, 246
      Netrona, 268
      Panamint, 14
      Sands, 227-228
      Wayne, 256, 267
      William Ward Burrows, 144, 146
    auxiliaries
      Antares, 84
      Apache, 137
      APc-108, APc-109, 180-181
      Bridge, Dobbin, 130
      Cascade, Phaon, Vestal, 178
      Castor, 132
      Chickasaw, 228
      Clamp, 245
      Cowanesque, 235
      Cree, 263
      Cuyama, Kanawha, Whitney, 135
      Endymion, 260
      Erskine M. Phelps, 168
      Kaloli, Vireo, 111
      Kopara, 144, 146

Medusa, 92, 94
Molala, 268
Ontario, 96
Pawnee, 163
Tekesta, 260
Virgo, 274
combatants
  aircraft carriers
    Enterprise, 95-96, 149-150
    Essex, Sangamon, 259
    Gambier Bay, Princeton, 230
    Hornet, 149-150
    Saratoga, 143, 147, 219
  battleships
    California, 99
    Missouri, 287-288
    Nevada, 90, 245
    New Mexico, 250
    Pennsylvania, 97-99
    South Dakota, 149
    Utah (ex-battleship), 92-94
  cruisers
    Astoria, Quincy, Vincennes, 144
    Boise, 233
    Cleveland, Denver, 160
    Columbia, 160, 234
    Helena, 88-90, 156
    Honolulu, 155-156
    Indianapolis, 86, 118
    Minneapolis, 86
    Montpelier, 160-161
    New Orleans, 97-98
    St. Louis, 156
  destroyers
    Anderson, Wilson, 7
    Baine, 170
    Calhoun, Gregory, 146-147
    Chevalier, Nicholas, O'Bannon, Taylor, 164
    Clarence K. Bronson, 180
    Clark, 111
    Cooper, 231
    Cummings, 97-99
    Eaton, 169-170
    Halligan, 261
    Isherwood, Paul D. Bates, 268
    Macdonaugh, 128

Monaghan, 84, 93-94
Mullany, 258
Philip, Saufley, 160
Pringle, 157-163
Radford, 152, 154
Reid, 117
Renshaw, 160-161, 167-168
Tucker, 140-142
Waller, 160-161
motor torpedo boats
  PT-20, PT-21, PT-22, PT-24, PT-25, PT-26, PT-27, PT-28, PT-29, PT-30, 111
  PT-23, PT-42, 113
  PT-37, PT-111, PT-123, 7-8
  PT-109, 154
patrol craft, *Amherst* (PCER-853), 256
patrol gunboat,
  Charleston, 125
  PGM-10, 279
  PGM-16, PGM-20, PGM-21, PGM-23, PGM-25, PGM-29, PGM-30, PGM-31, PGM-32, 279-281
  PGM-11, 267, 282
  PGM-18, 262-263
  PGM-19, 239
patrol yacht, *Crystal*, 111
screw sloop-of-war, *Lackawanna*, 110
seaplane tenders
  Ballard, 111
  Curtis, 92-94, 178-179
  Mackinac, 142-144, 178, 182
  Orca, 237
  Thornton, 111
submarine chaser
  PC-800, 245-246
  PC-1179, 253, 256
  PC-1598, 253
yard patrol craft
  YP-237, YP-239, YP-277, YP-284, YP-290, YP-345, YP-348, YP-350, 111-113
  YP-346, 112, 140-142
  YP-421, 296
mine warfare
  minelayers
    Aroostook, Baltimore, Canandaigua, Canonicus, Housatonic, Quinnebaug, Roanoke, Saranac, Shawmut, 11-13
    Keokuk, 12-23

Miantonomah (later an auxiliary minelayer) 12-23, 211
Monadnock (later an auxiliary minelayer) 12-26, 233-240, 276
Oglala, 10-15, 24, 86-90, 100
Salem, Weehawken, 13-26, 276
San Francisco, 11-20
Terror, 12-15, 25-26, 176-182, 208, 244-248, 265-270, 276
auxiliary minelayers
Barbican, Trapper, 19, 213-214
Barricade, Planter, 19, 23, 210-214
Bastion, 19, 214
Buttress, 19-20
Chimo, 19, 211-216
Obstructor, 19, 28, 209, 213-216
Picket, 19-20, 214, 289
coastal minelayer
Wassuc, 18-19
light minelayers (ex-destroyers)
Aaron Ward, 17-18, 22, 26, 253, 276-276
Adams, 17, 22, 26-28, 253, 256-260, 270, 275-276
Breese, 17, 24-27, 86, 91-95, 104, 137-145, 152-172, 216, 227-229,
  237, 240, 248, 276, 282-284
Gamble, 16, 21-26, 86, 91-96, 130-139, 142-148, 153-156, 160-168,
  170-172, 243-248, 290
Gwin, 17, 22, 26-27, 156, 216, 246-250, 275-276, 282, 284, 289
Harry F. Bauer, 11, 17, 22, 26-27, 216, 248, 275-276
Henry A. Wiley, 17, 22, 26-28, 216, 248, 253, 275, 282, 284
J. William Ditter, 17, 22, 27, 275-276
Lindsey, 17, 26-27, 246, 248, 253, 263, 270-271, 276
Montgomery, xix, xxxiii, xxxv, 1-10, 17, 21-25, 86-96, 122, 132-136,
  148, 163-164, 198-208, 283, 290
Ramsay, 16, 24-25, 86-94, 122-136
Preble, xix, xxxiii, xxxv, 1-10, 17, 21-26, 86, 96-99, 106-121, 148-
  164, 198-208, 227-229, 237, 240, 283
Pruitt, 17, 24-25, 86, 96-99, 104, 106, 115-126, 169-172
Robert H. Smith, 17, 22-27, 216, 248, 253, 270-276, 282-284
Shannon, 17, 26-27, 216, 246, 248, 276, 282, 284
Shea, 17, 22, 27, 253, 275-276
Sicard, 17, 24-25, 86, 96-98, 104, 106-107, 115-128, 167-172
Thomas E. Fraser, 17, 26-27, 101-102, 246-248, 271-276, 282-289
Tolman, 17, 27, 256, 276
Tracy, viii, xix, xxxiii, xxxv, 1-10, 17, 24-27, 86, 96-99, 104-106,
  115, 137-148, 169-172, 248, 276, 283
coastal minesweepers
Cockatoo, Condor, Crossbill, Reedbird, 83-85, 99
high-speed minesweepers (ex-destroyers)
Boggs, 86-87

Butler, Harding, 253
Chandler, 86-87, 126, 228, 237, 246
Dorsey, 86-87, 237, 245, 251-253, 282
Elliot, 87, 126
Ellyson, 253, 267, 282, 289
Emmons, 253, 262, 264-267, 290
Gherardi, Hambleton, 253, 282, 289
Forrest, 250, 252
Hamilton, 199, 201, 205, 228, 237, 245
Hobson, 250
Hogan, 237
Hopkins, 86-87, 237, 289
Hovey, Long, 86-87, 199, 201-205, 228, 235, 237, 290
Howard, 228, 237
Jeffers, 253, 282, 289
Lamberton, 86-87, 125-126
Macomb, 250, 252, 265, 289
Palmer, 228, 235-237, 290
Perry, 21, 87, 100, 202-205, 290
Rodman, 253, 265
Southard, 7, 86-87, 199, 203-205, 228, 235, 237, 282
Trever, Wasmuth, Zane, 87, 100
minesweepers
Ardent, Champion, Defense, Heed, 252-253
Auk, 205
Bobolink, Grebe, Rail, 87, 100
Buoyant, 263
Competent, 205, 281
Density, Design, Device, Dextrous, Diploma, Dour, Dunlin, Execute, Hazard, Inflict, Nimble, Notable, Opponent, Phantom, Pioneer, Pivot, Prevail, Ransom, Rebel, Recruit, Reform, Scurry, Serene, Shelter, Signet, Skirmish, Specter, Staunch, Strategy, Strive, Steady, Success, Surfbird, Sustain, Toucan, 279-282
Devastator, Gladiator, Impeccable, 253, 281-282
Gayety, 268, 279
Kingfisher, 86-87, 132
Pirate, Pledge, 280
Pochard, 281, 289
Pursuit, 228, 255
Requisite, 228-231, 237, 255, 281
Revenge, 228, 255, 282, 289
Robin, 86-87
Sage, 228, 255, 282
Salute, 290, 298
Scout, 228
Sentry, 228, 237

Sheldrake, Spear, Vigilance, 253, 282

Skylark, 253-265, 290

Starling, 253

Swallow, 253, 265-268, 290

Tern, Turkey, Vireo, 87, 100

Token, Tumult, 205, 228, 281, 289

Triumph, 205, 253, 281

Velocity, Zeal, 205, 228, 281

yard minesweepers

  YMS-1, 205, 228, 262

  YMS-2, YMS-275, 205-206

  YMS-4, YMS-336, YMS-363, YMS-392, 215

  YMS-6, YMS-52, YMS-81, YMS-219, YMS-286, YMS-316,
    YMS-340, YMS-342, YMS-393, 228

  YMS-10, YMS-47, YMS-51, YMS-329, YMS-339, YMS-364,
    YMS-368, 298

  YMS-12, 207

  YMS-19, 206-207, 290

  YMS-33, YMS-180, 205

  YMS-39, 227-228, 290, 298

  YMS-48, YMS-385, 290

  YMS-49, YMS-314, YMS-334, YMS-335, 228, 298

  YMS-50, YMS-84, MS-365, YMS-481, 290, 297-298

  YMS-53, 235, 237

  YMS-70, 226, 228

  YMS-71, 228, 290

  YMS-73, 227-228

  YMS-89, YMS-93, YMS-94, YMS-147, YMS-283, YMS-311,
    YMS-323, YMS-327, YMS-372, YMS-376, YMS-403, YMS-418,
    YMS-426, YMS-443, YMS-449, YMS-468, YMS-471, YMS-475,
    YMS-478, 282

  YMS-92, 262-264

  YMS-103, 261-265, 290

  YMS-140, 205-206, 228, 279

  YMS-176, YMS-271, 279, 282

  YMS-177, 205, 289

  YMS-183, YMS-360, YMS-410, 279

  YMS-238, 205, 228

  YMS-243, 228, 282

  YMS-268, YMS-276, YMS-343, YMS-426, YMS-441, 289

  YMS-293, YMS-319, 206, 228

  YMS-299, 262, 282

  YMS-324, YMS-325, 205, 280-281

  YMS-327, 237, 281

  YMS-331, 280, 282

  YMS-341, 227, 281-282

       YMS-362, *YMS-371*, 281-282, 289

       YMS-388, *YMS-401*, 281-282

       YMS-390, *YMS-461*, *YMS-467*, *YMS-415*, 282, 289

       YMS-398, 228, 280, 282

       YMS-430, *YMS-434*, 281

   Submarines,

     Cod, xxxii

     Grouper, 67

Shunji, Izaki, 156

Sides, John H., 257

Sims, Gelzer Loyall, 24, 86

Singapore, 30, 35, 53-73, 129, 196, 221, 289

Skinner, Bob, 273

Slackett, Orson L., 226

Smith, Sammie, 28, 214, 216

Smith, W. W., 118

Snippe, Pieter Gerardus Johan, 50

Somerville, James (Adm. Sir, GCB, GBE, DSO, DL, RN), 217

South Africa, Durban, 46, 55

Spears, James Farrell, 194, 313

Spencer, Victor, 72

Spendlove, Albert, 72

Spooner, Ernest John (Vice Adm., DSO, RN), 69-70

Spruance, Raymond A., 114, 173-178, 250

Sri Lanka (formerly Ceylon), xxvii-xxxii, 35, 49, 55, 271

   Columbo, 49

   Trincomalee, 217-223

Stahli (Lt.), 228

Stanton, Ronald George Gladstone (Sub Lt., DSO, RN), 71-72

Steinke, Frederick Samuel, 2, 22-27, 148, 156, 164, 248, 275-276, 282, 284

Stokes, Thomas Vincent, 194, 313

Stout, Harold Franklin, 24, 86, 92

Struble, Arthur D., 211-212

Sullivan, John L., 11

Sweeney, John Joseph, 192

Tackney, Stephen Noel, 24, 145-148

Takahashi, Ibo, 30

Takeo, Ito, 66

Thailand (formerly Siam),

   Pattani, Songkhla, 63

   Phuket Island, 41, 218

Theobald, Robert A., 115-121

Theodore, Joaquin, 141

Thompson, Arthur William, 72

Thompson, Charles Frederick, 194, 314

Thompson, John William, 194, 314

Thompson Jr., Harry LeRoy, 126
Thornton, Leslie M., 261-264
Thurstun, Gilbert Robert, 194, 313
Tiemroth, H. H., 87
Tingman, Robert Richard, 192
Titshall, Benjamin Alfred, 194, 314
Townsend, Vernon Eric, 194, 313
Toyoda, Soemu, 195, 225
Truman, Harry S., 285
Turner, Hugh B., 222-223
Tumeth, Raymond Victor, 192
Turner, Hugh Bentley (A/Lt. Comdr., DSC, RN), 222-223
Turner, Richmond Kelly, 144
Turner, Victor William (Able Seaman, BEM, RAN), 301
Trist, George, 53
United States
  Army/Army Air Force
    Sixth Army, 225, 232-233, 239, 300
    Tenth Army, 249-250
    7th Fighter Command (formerly 14th Pursuit Wing), 101
    7th Infantry Division, 124
    27th Infantry Division, 180, 197
    37th Infantry Division, 167, 171
    77th Infantry Division, 198, 251, 254
    81st Infantry Division, 199, 207
    96th Infantry Division, 249
    164th Infantry Division, 4
    Airfield
      Bellows (Oahu), Hickam (Pearl Harbor), 81
      Wheeler (Pearl Harbor), 81, 103
    Coast Artillery Corps, 19, 210
    Fort Shafter (Oahu, Hawaii), 101-102
    WARD (Women's Air Raid Defense), 101-102
  Marine Corps
    1st Marine Division, 4-5, 142, 199, 206, 249
    2nd Marine Division, 180, 196, 198
    3rd Marine Division, 167, 171, 198, 242
    4th Marine Division, 196, 198
    5th Marine Division, 241
    6th Marine Division, 249
    4th Raider Battalion, 165
    Air Field/Station
      Ewa (Oahu), 81
      Henderson (Guadalcanal), 4-8, 149-151
    Squadron
      VMF-214, 167

VMF-222, 108
Military aircraft
  B17 Flying Fortress, 82
  B24 Liberator, xxix, 178-179, 192
  B25 Mitchell, 44
  B29 Superfortress, xxiii, 44, 196, 225, 241, 287
  C47 Skytrain, 179
  F4F Wildcat, P38 Lightning, P39 Airacobra, P40 Warhawk,
    SBD Dauntless, 7
  F4U Corsair, 283
  PBY-5 Catalina, 44, 48, 82-84, 92, 111, 117, 143, 181-182
  PB2Y-3 Coronado, 21, 44, 287
  PB4Y-1/PB4Y-2 Privateer, 21, 44, 181-182
  PV-1 Ventura, 21, 44, 181-182
  TBF Avenger, 7, 21, 44, 147
Navy/Naval
  Ammunition Depot
    Indian Head (Washington), 115, 118
    Lualualei (Oahu), 97-98
    West Loch (Oahu), 99, 106, 137, 274
  Aviation
    Aircraft, Battle Force, 96
    Air Field/Air Station
      Kanoehe (Oahu), 81
      Kodiak, 118
      Midway, 110
      Palmyra, 138
    CenCats Seaplane Transportation Service, 182
    Patrol Wing 4, 117
    Squadron
      VB-108, VB-109, VB-137, 181-182
      VP-14, 84
      VP-72, 181
  Cruiser Division Twelve, 160
  Destroyer Division
    Eight, Ten, Sixteen, Seventeen, 6
    Eleven, 116
    Forty-one, 163
  Destroyer Squadron Twenty-three, 168
  Mine Assembly Depot
    Darwin, Australia, xxviii-xxix, 193
    Espiritu Santo, New Hebrides, 169-170
    Noumea, New Caledonia, xxviii, 1
    Yorktown, Virginia, 176-177
  Mine Detail 19 (at Tarawa), 180-181
  Mine Division

One, 5, 9, 86, 96, 100, 104, 106, 115, 121, 123, 137, 169, 171-172
Two, 86, 90-96, 129-132, 137, 207, 228
Four, Ten, 87, 100
Five, Six, Eleven, 87
Eight, 257
Fifteen, 255-256
Fifty, 16
Mine Squadron
  One, 86, 100, 179,
  Two, 86-87, 199, 234
  Four, 86-87
  Five, 281
  Six, 100
  Eleven, 280
  Fifteen, 279
  Twenty, 282-283
  One Hundred One, One Hundred Two, 213-214
  One Hundred Four, One Hundred Five, One Hundred Six, One
    Hundred Seven, 214
Minecraft,
  Battle Force, 12, 86-90, 100
  U.S. Pacific Fleet, 14-15, 100, 211, 214, 213, 242-249, 273, 278
Mines
  Mk 6, ix, xxxiii, 1, 44, 106, 119, 121, 131-138, 152, 168, 170, 176-179,
    199, 201, 223, 257
  Mk 12, 43, 122-123, 181
  Mk 13, 181
Mobile Explosive Investigation Unit
  MEIU No. 1, 239-240, 296-297
  MEIU No. 4, 264, 273-274
Motor Torpedo Boat
  Flotilla One, 7
  Squadron One, 111-113
Naval District
  Fourteenth/Hawaiian Sea Frontier, 85, 106
  Philippine Sea Frontier, 239
Service Squadron Four, 178, 210
Service Squadron Six, 16, 20, 100, 211
Train Squadron Six, 100
Underwater Demolition Team 11, 297
van Versendaal, Anthonie Catharinus, 51
van der Horst, Eugenius Cornelis Johannes, 51
van Haga, Johan, 51
van Hooff, J. F. Drijfhout, 41
van Miert, Anthonie, 76-78
Vearncombe, Archibald L. (AB, DSM, RN), 293

Veeder, W. S., 87
Vernon, David, 194, 313
Vian, Philip Louis (Adm. Sir, GCB, KBE, DSO & Two Bars, RN), 272
Volckmann, Russel W., 238
Warton, Albert Leslie, 192
Wearne, Athol Galway Hope, 194, 313
Welles, Gideon, xxi
White, William George Searle, 194, 313
Whitemarsh, Ross P., 86, 130, 137, 139, 146
Whittemore, Robert B., 274
Wilkinson, Thomas (Lt., VC, RNR), 70-72
Wilkinson Jr., Theodore S., 233
Williams Jr., Richard Claggett, 22-27, 126, 172, 216, 248, 275-276
Willfong, J. L., 87
Wilson Jr., Ralph C., 214
Woodhouse, E. W., 228, 281
Woodman, Ronald Joseph, 26-27, 248, 276, 282, 284
Yamamoto Isoroku, 114, 116
Yamashita, Tomoyuki, 62-63
Yano, Keiji, 5
Yasuyo, Yamazaki, 124, 126
Young, Mark Aitchison (Sir, GCMG), 63, 65
Zacek, Laddie John, 97, 99
Zinn, F. K., 213

About the Author

Rob Hoole joined the Royal Navy as a 'seaman officer' (later termed 'warfare officer') in 1971. He qualified as a Ships' Diving Officer in 1975 and as a Minewarfare & Clearance Diving Officer in 1976. In 1991, he co-founded the Royal Naval Minewarfare & Clearance Diving Officers' Association (MCDOA) of which he is Vice Chairman. Rob acquired an MBA from Oxford Brooks University in 2000 and left the Royal Navy in late 2002 after serving 32 years in all manner of operational and training roles at sea and ashore including Command of diving teams and mine countermeasures vessels. Rob lives in Waterlooville near Portsmouth with his wife Linda. They have three grown children and two young grandsons.

Aside from his family, his passions include country pubs & real ale, sailing, reading and maritime history, particularly the development of naval & military diving, minewarfare and explosive ordnance disposal (EOD) otherwise known as bomb & mine disposal. He has contributed to many books, journals and other publications covering different aspects of his favourite subjects and was the editor of the book *Last of the Wooden Walls – An Illustrated History of the TON Class Minesweepers & Minehunters*. Last but not least, he is a fierce supporter of Project Vernon, the campaign to erect a monument at Gunwharf Quays in Portsmouth, Hampshire, to commemorate the minewarfare and diving heritage of HMS Vernon which previously occupied the site.

About the Author

Commander David D. Bruhn, U.S. Navy (Retired) served twenty-two years on active duty and two in the Naval Reserve, as both an enlisted man and as an officer, between 1977 and 2001.

Following completion of basic training, he served as a sonar technician aboard USS *Miller* (FF 1091) and USS *Leftwich* (DD 984). He was commissioned in 1983 following graduation from California State University at Chico. His initial assignment was to USS *Excel* (MSO 439), serving as supply officer, damage control assistant, and chief engineer. He then served in USS *Thach* (FFG 43) as chief engineer and Destroyer Squadron Thirteen as material officer.

After graduation from the Naval Postgraduate School, Commander Bruhn was assigned to Secretary of the Navy and Chief of Naval Operation staffs as a budget analyst and resources planner before attending the Naval War College in 1996, following which he commanded the mine countermeasures ships USS *Gladiator* (MCM 11) and USS *Dextrous* (MCM 13) in the Persian Gulf.

Commander Bruhn's final assignment was executive assistant to a senior (SES 4) government service executive at the Ballistic Missile Defense Organization in Washington, D.C.

Following military service, he was a high school teacher and track coach for ten years, and is now a USA Track & Field official. He lives in northern California with his wife Nancy and has two sons, David and Michael.

www.ingramcontent.com/pod-product-compliance
Lightning Source LLC
Chambersburg PA
CBHW060959220326
41599CB00023B/3768